Micro Electro Mechanical System Design

MECHANICAL ENGINEERING
A Series of Textbooks and Reference Books

Founding Editor

L. L. Faulkner

*Columbus Division, Battelle Memorial Institute
and Department of Mechanical Engineering
The Ohio State University
Columbus, Ohio*

1. *Spring Designer's Handbook*, Harold Carlson
2. *Computer-Aided Graphics and Design*, Daniel L. Ryan
3. *Lubrication Fundamentals*, J. George Wills
4. *Solar Engineering for Domestic Buildings*, William A. Himmelman
5. *Applied Engineering Mechanics: Statics and Dynamics*, G. Boothroyd and C. Poli
6. *Centrifugal Pump Clinic*, Igor J. Karassik
7. *Computer-Aided Kinetics for Machine Design*, Daniel L. Ryan
8. *Plastics Products Design Handbook, Part A: Materials and Components; Part B: Processes and Design for Processes*, edited by Edward Miller
9. *Turbomachinery: Basic Theory and Applications*, Earl Logan, Jr.
10. *Vibrations of Shells and Plates*, Werner Soedel
11. *Flat and Corrugated Diaphragm Design Handbook*, Mario Di Giovanni
12. *Practical Stress Analysis in Engineering Design*, Alexander Blake
13. *An Introduction to the Design and Behavior of Bolted Joints*, John H. Bickford
14. *Optimal Engineering Design: Principles and Applications*, James N. Siddall
15. *Spring Manufacturing Handbook*, Harold Carlson
16. *Industrial Noise Control: Fundamentals and Applications*, edited by Lewis H. Bell
17. *Gears and Their Vibration: A Basic Approach to Understanding Gear Noise*, J. Derek Smith
18. *Chains for Power Transmission and Material Handling: Design and Applications Handbook*, American Chain Association
19. *Corrosion and Corrosion Protection Handbook*, edited by Philip A. Schweitzer
20. *Gear Drive Systems: Design and Application*, Peter Lynwander
21. *Controlling In-Plant Airborne Contaminants: Systems Design and Calculations*, John D. Constance
22. *CAD/CAM Systems Planning and Implementation*, Charles S. Knox
23. *Probabilistic Engineering Design: Principles and Applications*, James N. Siddall
24. *Traction Drives: Selection and Application*, Frederick W. Heilich III and Eugene E. Shube

25. *Finite Element Methods: An Introduction*, Ronald L. Huston and Chris E. Passerello
26. *Mechanical Fastening of Plastics: An Engineering Handbook*, Brayton Lincoln, Kenneth J. Gomes, and James F. Braden
27. *Lubrication in Practice: Second Edition*, edited by W. S. Robertson
28. *Principles of Automated Drafting*, Daniel L. Ryan
29. *Practical Seal Design*, edited by Leonard J. Martini
30. *Engineering Documentation for CAD/CAM Applications*, Charles S. Knox
31. *Design Dimensioning with Computer Graphics Applications*, Jerome C. Lange
32. *Mechanism Analysis: Simplified Graphical and Analytical Techniques*, Lyndon O. Barton
33. *CAD/CAM Systems: Justification, Implementation, Productivity Measurement*, Edward J. Preston, George W. Crawford, and Mark E. Coticchia
34. *Steam Plant Calculations Manual*, V. Ganapathy
35. *Design Assurance for Engineers and Managers*, John A. Burgess
36. *Heat Transfer Fluids and Systems for Process and Energy Applications*, Jasbir Singh
37. *Potential Flows: Computer Graphic Solutions*, Robert H. Kirchhoff
38. *Computer-Aided Graphics and Design: Second Edition*, Daniel L. Ryan
39. *Electronically Controlled Proportional Valves: Selection and Application*, Michael J. Tonyan, edited by Tobi Goldoftas
40. *Pressure Gauge Handbook*, AMETEK, U.S. Gauge Division, edited by Philip W. Harland
41. *Fabric Filtration for Combustion Sources: Fundamentals and Basic Technology*, R. P. Donovan
42. *Design of Mechanical Joints*, Alexander Blake
43. *CAD/CAM Dictionary*, Edward J. Preston, George W. Crawford, and Mark E. Coticchia
44. *Machinery Adhesives for Locking, Retaining, and Sealing*, Girard S. Haviland
45. *Couplings and Joints: Design, Selection, and Application*, Jon R. Mancuso
46. *Shaft Alignment Handbook*, John Piotrowski
47. *BASIC Programs for Steam Plant Engineers: Boilers, Combustion, Fluid Flow, and Heat Transfer*, V. Ganapathy
48. *Solving Mechanical Design Problems with Computer Graphics*, Jerome C. Lange
49. *Plastics Gearing: Selection and Application*, Clifford E. Adams
50. *Clutches and Brakes: Design and Selection*, William C. Orthwein
51. *Transducers in Mechanical and Electronic Design*, Harry L. Trietley
52. *Metallurgical Applications of Shock-Wave and High-Strain-Rate Phenomena*, edited by Lawrence E. Murr, Karl P. Staudhammer, and Marc A. Meyers
53. *Magnesium Products Design*, Robert S. Busk
54. *How to Integrate CAD/CAM Systems: Management and Technology*, William D. Engelke
55. *Cam Design and Manufacture: Second Edition; with cam design software for the IBM PC and compatibles*, disk included, Preben W. Jensen
56. *Solid-State AC Motor Controls: Selection and Application*, Sylvester Campbell

57. *Fundamentals of Robotics*, David D. Ardayfio
58. *Belt Selection and Application for Engineers*, edited by Wallace D. Erickson
59. *Developing Three-Dimensional CAD Software with the IBM PC*, C. Stan Wei
60. *Organizing Data for CIM Applications*, Charles S. Knox, with contributions by Thomas C. Boos, Ross S. Culverhouse, and Paul F. Muchnicki
61. *Computer-Aided Simulation in Railway Dynamics*, by Rao V. Dukkipati and Joseph R. Amyot
62. *Fiber-Reinforced Composites: Materials, Manufacturing, and Design*, P. K. Mallick
63. *Photoelectric Sensors and Controls: Selection and Application*, Scott M. Juds
64. *Finite Element Analysis with Personal Computers*, Edward R. Champion, Jr. and J. Michael Ensminger
65. *Ultrasonics: Fundamentals, Technology, Applications: Second Edition, Revised and Expanded*, Dale Ensminger
66. *Applied Finite Element Modeling: Practical Problem Solving for Engineers*, Jeffrey M. Steele
67. *Measurement and Instrumentation in Engineering: Principles and Basic Laboratory Experiments*, Francis S. Tse and Ivan E. Morse
68. *Centrifugal Pump Clinic: Second Edition, Revised and Expanded*, Igor J. Karassik
69. *Practical Stress Analysis in Engineering Design: Second Edition, Revised and Expanded*, Alexander Blake
70. *An Introduction to the Design and Behavior of Bolted Joints: Second Edition, Revised and Expanded*, John H. Bickford
71. *High Vacuum Technology: A Practical Guide*, Marsbed H. Hablanian
72. *Pressure Sensors: Selection and Application*, Duane Tandeske
73. *Zinc Handbook: Properties, Processing, and Use in Design*, Frank Porter
74. *Thermal Fatigue of Metals*, Andrzej Weronski and Tadeusz Hejwowski
75. *Classical and Modern Mechanisms for Engineers and Inventors*, Preben W. Jensen
76. *Handbook of Electronic Package Design*, edited by Michael Pecht
77. *Shock-Wave and High-Strain-Rate Phenomena in Materials*, edited by Marc A. Meyers, Lawrence E. Murr, and Karl P. Staudhammer
78. *Industrial Refrigeration: Principles, Design and Applications*, P. C. Koelet
79. *Applied Combustion*, Eugene L. Keating
80. *Engine Oils and Automotive Lubrication*, edited by Wilfried J. Bartz
81. *Mechanism Analysis: Simplified and Graphical Techniques, Second Edition, Revised and Expanded*, Lyndon O. Barton
82. *Fundamental Fluid Mechanics for the Practicing Engineer*, James W. Murdock
83. *Fiber-Reinforced Composites: Materials, Manufacturing, and Design, Second Edition, Revised and Expanded*, P. K. Mallick
84. *Numerical Methods for Engineering Applications*, Edward R. Champion, Jr.
85. *Turbomachinery: Basic Theory and Applications, Second Edition, Revised and Expanded*, Earl Logan, Jr.
86. *Vibrations of Shells and Plates: Second Edition, Revised and Expanded*, Werner Soedel
87. *Steam Plant Calculations Manual: Second Edition, Revised and Expanded*, V. Ganapathy

88. *Industrial Noise Control: Fundamentals and Applications, Second Edition, Revised and Expanded*, Lewis H. Bell and Douglas H. Bell
89. *Finite Elements: Their Design and Performance*, Richard H. MacNeal
90. *Mechanical Properties of Polymers and Composites: Second Edition, Revised and Expanded*, Lawrence E. Nielsen and Robert F. Landel
91. *Mechanical Wear Prediction and Prevention*, Raymond G. Bayer
92. *Mechanical Power Transmission Components*, edited by David W. South and Jon R. Mancuso
93. *Handbook of Turbomachinery*, edited by Earl Logan, Jr.
94. *Engineering Documentation Control Practices and Procedures*, Ray E. Monahan
95. *Refractory Linings Thermomechanical Design and Applications*, Charles A. Schacht
96. *Geometric Dimensioning and Tolerancing: Applications and Techniques for Use in Design, Manufacturing, and Inspection*, James D. Meadows
97. *An Introduction to the Design and Behavior of Bolted Joints: Third Edition, Revised and Expanded*, John H. Bickford
98. *Shaft Alignment Handbook: Second Edition, Revised and Expanded*, John Piotrowski
99. *Computer-Aided Design of Polymer-Matrix Composite Structures*, edited by Suong Van Hoa
100. *Friction Science and Technology*, Peter J. Blau
101. *Introduction to Plastics and Composites: Mechanical Properties and Engineering Applications*, Edward Miller
102. *Practical Fracture Mechanics in Design*, Alexander Blake
103. *Pump Characteristics and Applications*, Michael W. Volk
104. *Optical Principles and Technology for Engineers*, James E. Stewart
105. *Optimizing the Shape of Mechanical Elements and Structures*, A. A. Seireg and Jorge Rodriguez
106. *Kinematics and Dynamics of Machinery*, Vladimír Stejskal and Michael Valásek
107. *Shaft Seals for Dynamic Applications*, Les Horve
108. *Reliability-Based Mechanical Design*, edited by Thomas A. Cruse
109. *Mechanical Fastening, Joining, and Assembly*, James A. Speck
110. *Turbomachinery Fluid Dynamics and Heat Transfer*, edited by Chunill Hah
111. *High-Vacuum Technology: A Practical Guide, Second Edition, Revised and Expanded*, Marsbed H. Hablanian
112. *Geometric Dimensioning and Tolerancing: Workbook and Answerbook*, James D. Meadows
113. *Handbook of Materials Selection for Engineering Applications*, edited by G. T. Murray
114. *Handbook of Thermoplastic Piping System Design*, Thomas Sixsmith and Reinhard Hanselka
115. *Practical Guide to Finite Elements: A Solid Mechanics Approach*, Steven M. Lepi
116. *Applied Computational Fluid Dynamics*, edited by Vijay K. Garg
117. *Fluid Sealing Technology*, Heinz K. Muller and Bernard S. Nau
118. *Friction and Lubrication in Mechanical Design*, A. A. Seireg
119. *Influence Functions and Matrices*, Yuri A. Melnikov

120. *Mechanical Analysis of Electronic Packaging Systems*, Stephen A. McKeown
121. *Couplings and Joints: Design, Selection, and Application, Second Edition, Revised and Expanded*, Jon R. Mancuso
122. *Thermodynamics: Processes and Applications*, Earl Logan, Jr.
123. *Gear Noise and Vibration*, J. Derek Smith
124. *Practical Fluid Mechanics for Engineering Applications*, John J. Bloomer
125. *Handbook of Hydraulic Fluid Technology*, edited by George E. Totten
126. *Heat Exchanger Design Handbook*, T. Kuppan
127. *Designing for Product Sound Quality*, Richard H. Lyon
128. *Probability Applications in Mechanical Design*, Franklin E. Fisher and Joy R. Fisher
129. *Nickel Alloys*, edited by Ulrich Heubner
130. *Rotating Machinery Vibration: Problem Analysis and Troubleshooting*, Maurice L. Adams, Jr.
131. *Formulas for Dynamic Analysis*, Ronald L. Huston and C. Q. Liu
132. *Handbook of Machinery Dynamics*, Lynn L. Faulkner and Earl Logan, Jr.
133. *Rapid Prototyping Technology: Selection and Application*, Kenneth G. Cooper
134. *Reciprocating Machinery Dynamics: Design and Analysis*, Abdulla S. Rangwala
135. *Maintenance Excellence: Optimizing Equipment Life-Cycle Decisions*, edited by John D. Campbell and Andrew K. S. Jardine
136. *Practical Guide to Industrial Boiler Systems*, Ralph L. Vandagriff
137. *Lubrication Fundamentals: Second Edition, Revised and Expanded*, D. M. Pirro and A. A. Wessol
138. *Mechanical Life Cycle Handbook: Good Environmental Design and Manufacturing*, edited by Mahendra S. Hundal
139. *Micromachining of Engineering Materials*, edited by Joseph McGeough
140. *Control Strategies for Dynamic Systems: Design and Implementation*, John H. Lumkes, Jr.
141. *Practical Guide to Pressure Vessel Manufacturing*, Sunil Pullarcot
142. *Nondestructive Evaluation: Theory, Techniques, and Applications*, edited by Peter J. Shull
143. *Diesel Engine Engineering: Thermodynamics, Dynamics, Design, and Control*, Andrei Makartchouk
144. *Handbook of Machine Tool Analysis*, Ioan D. Marinescu, Constantin Ispas, and Dan Boboc
145. *Implementing Concurrent Engineering in Small Companies*, Susan Carlson Skalak
146. *Practical Guide to the Packaging of Electronics: Thermal and Mechanical Design and Analysis*, Ali Jamnia
147. *Bearing Design in Machinery: Engineering Tribology and Lubrication*, Avraham Harnoy
148. *Mechanical Reliability Improvement: Probability and Statistics for Experimental Testing*, R. E. Little
149. *Industrial Boilers and Heat Recovery Steam Generators: Design, Applications, and Calculations*, V. Ganapathy
150. *The CAD Guidebook: A Basic Manual for Understanding and Improving Computer-Aided Design*, Stephen J. Schoonmaker

151. *Industrial Noise Control and Acoustics*, Randall F. Barron
152. *Mechanical Properties of Engineered Materials*, Wolé Soboyejo
153. *Reliability Verification, Testing, and Analysis in Engineering Design*, Gary S. Wasserman
154. *Fundamental Mechanics of Fluids: Third Edition*, I. G. Currie
155. *Intermediate Heat Transfer*, Kau-Fui Vincent Wong
156. *HVAC Water Chillers and Cooling Towers: Fundamentals, Application, and Operation*, Herbert W. Stanford III
157. *Gear Noise and Vibration: Second Edition, Revised and Expanded*, J. Derek Smith
158. *Handbook of Turbomachinery: Second Edition, Revised and Expanded*, edited by Earl Logan, Jr. and Ramendra Roy
159. *Piping and Pipeline Engineering: Design, Construction, Maintenance, Integrity, and Repair*, George A. Antaki
160. *Turbomachinery: Design and Theory*, Rama S. R. Gorla and Aijaz Ahmed Khan
161. *Target Costing: Market-Driven Product Design*, M. Bradford Clifton, Henry M. B. Bird, Robert E. Albano, and Wesley P. Townsend
162. *Fluidized Bed Combustion*, Simeon N. Oka
163. *Theory of Dimensioning: An Introduction to Parameterizing Geometric Models*, Vijay Srinivasan
164. *Handbook of Mechanical Alloy Design*, edited by George E. Totten, Lin Xie, and Kiyoshi Funatani
165. *Structural Analysis of Polymeric Composite Materials*, Mark E. Tuttle
166. *Modeling and Simulation for Material Selection and Mechanical Design*, edited by George E. Totten, Lin Xie, and Kiyoshi Funatani
167. *Handbook of Pneumatic Conveying Engineering*, David Mills, Mark G. Jones, and Vijay K. Agarwal
168. *Clutches and Brakes: Design and Selection, Second Edition*, William C. Orthwein
169. *Fundamentals of Fluid Film Lubrication: Second Edition*, Bernard J. Hamrock, Steven R. Schmid, and Bo O. Jacobson
170. *Handbook of Lead-Free Solder Technology for Microelectronic Assemblies*, edited by Karl J. Puttlitz and Kathleen A. Stalter
171. *Vehicle Stability*, Dean Karnopp
172. *Mechanical Wear Fundamentals and Testing: Second Edition, Revised and Expanded*, Raymond G. Bayer
173. *Liquid Pipeline Hydraulics*, E. Shashi Menon
174. *Solid Fuels Combustion and Gasification*, Marcio L. de Souza-Santos
175. *Mechanical Tolerance Stackup and Analysis*, Bryan R. Fischer
176. *Engineering Design for Wear,* Raymond G. Bayer
177. *Vibrations of Shells and Plates: Third Edition, Revised and Expanded*, Werner Soedel
178. *Refractories Handbook*, edited by Charles A. Schacht
179. *Practical Engineering Failure Analysis*, Hani M. Tawancy, Anwar Ul-Hamid, and Nureddin M. Abbas
180. *Mechanical Alloying and Milling*, C. Suryanarayana
181. *Mechanical Vibration: Analysis, Uncertainties, and Control, Second Edition, Revised and Expanded*, Haym Benaroya

182. *Design of Automatic Machinery*, Stephen J. Derby
183. *Practical Fracture Mechanics in Design: Second Edition, Revised and Expanded*, Arun Shukla
184. *Practical Guide to Designed Experiments*, Paul D. Funkenbusch
185. *Gigacycle Fatigue in Mechanical Practive*, Claude Bathias and Paul C. Paris
186. *Selection of Engineering Materials and Adhesives*, Lawrence W. Fisher
187. *Boundary Methods: Elements, Contours, and Nodes*, Subrata Mukherjee and Yu Xie Mukherjee
188. *Rotordynamics*, Agnieszka (Agnes) Muszńyska
189. *Pump Characteristics and Applications: Second Edition*, Michael W. Volk
190. *Reliability Engineering: Probability Models and Maintenance Methods*, Joel A. Nachlas
191. *Industrial Heating: Principles, Techniques, Materials, Applications, and Design*, Yeshvant V. Deshmukh
192. *Micro Electro Mechanical System Design*, James J. Allen
193. *Probability Models in Engineering and Science*, Haym Benaroya and Seon Han
194. *Damage Mechanics*, George Z. Voyiadjis and Peter I. Kattan

Micro Electro Mechanical System Design

James J. Allen

CRC Press is an imprint of the
Taylor & Francis Group, an **informa** business

A TAYLOR & FRANCIS BOOK

CRC Press
Taylor & Francis Group
6000 Broken Sound Parkway NW, Suite 300
Boca Raton, FL 33487-2742

CRC Press is an imprint of Taylor & Francis Group, an Informa business
© 2005 by Taylor & Francis Group, LLC

No claim to original U.S. Government works

ISBN 13: 978-0-8247-5824-0 (hbk)

Library of Congress Card Number 2005041771

This book contains information obtained from authentic and highly regarded sources. Reprinted material is quoted with permission, and sources are indicated. A wide variety of references are listed. Reasonable efforts have been made to publish reliable data and information, but the author and the publisher cannot assume responsibility for the validity of all materials or for the consequences of their use.

No part of this book may be reprinted, reproduced, transmitted, or utilized in any form by any electronic, mechanical, or other means, now known or hereafter invented, including photocopying, microfilming, and recording, or in any information storage or retrieval system, without written permission from the publishers.

For permission to photocopy or use material electronically from this work, please access www.copyright.com (http://www.copyright.com/) or contact the Copyright Clearance Center, Inc. (CCC) 222 Rosewood Drive, Danvers, MA 01923, 978-750-8400. CCC is a not-for-profit organization that provides licenses and registration for a variety of users. For organizations that have been granted a photocopy license by the CCC, a separate system of payment has been arranged.

Trademark Notice: Product or corporate names may be trademarks or registered trademarks, and are used only for identification and explanation without intent to infringe.

Visit the Taylor & Francis Web site at
http://www.taylorandfrancis.com

and the CRC Press Web site at
http://www.crcpress.com

Library of Congress Cataloging-in-Publication Data

Allen, James J.
   Micro electro mechanical system design / James J. Allen.
       p. cm. -- (Mechanical engineering ; 192)
   Includes bibliographical references and index.
   ISBN 0-8247-5824-2 (alk. paper)
   1. Microelectromechanical systems--Design and construction. 2. Engineering design. I. Title. II. Mechanical engineering (Taylor & Francis) ; 192.

TK153.A47 2005
621--dc22
                                             2005041771

Dedication

To Susan and Nathan

Preface

This book attempts to provide an overview of the process of microelectromechanical system (MEMS) design. In order to design a MEMS device successfully, an appreciation for the full spectrum of issues involved must be considered. The designer must understand

- Fabrication technologies
- Relevant physics for a device at the micron scale
- Computer-aided design issues in the implementation of the design
- Engineering of the MEMS device
- Evaluation testing of the device
- Reliability and packaging issues necessary to produce a quality MEMS product

These diverse issues are interrelated and must be considered at the initial stages of a design project in order to be completely successful and timely in product development. This book has ten chapters and eight appendices:

Chapter 1. Introduction
Chapter 2. Fabrication Processes
Chapter 3. MEMS Technologies
Chapter 4. Scaling Issues for MEMS
Chapter 5. Design Realization Tools for MEMS
Chapter 6. Electromechanics
Chapter 7. Modeling and Design
Chapter 8. MEMS Sensors and Actuators
Chapter 9. Packaging
Chapter 10. Reliability
Appendices

The MEMS field is very exciting to many people for a variety of reasons. MEMS is a multiphysics technology that provides many new, innovative ways of implementing devices with functionality previously undreamed of. One of the challenges facing the people entering this field is the breadth of knowledge required to develop a MEMS product; many of them are from a variety of technical fields that may be tangential to the spectrum of MEMS design issues enumerated here. This book is written for the new entrant into the field of MEMS design. This person may be a senior or first-year graduate student in engineering or science, as well as a practicing engineer or scientist exploring a new field to develop a new device or product.

The organization of the book is meant to be a logical sequence of topics that a new MEMS designer would need to learn. At the end of each chapter, questions and problems provide a review and promote thought into the subject matter. The Appendices provide succinct information necessary in the various stages of a MEMS design project. The chapter on modeling, actuation, and sensing focuses primarily on the mechanical and electrical aspects of MEMS design. However, MEMS design projects frequently involve many other realms of science and engineering, such as optics, fluid mechanics, radio frequency (RF) devices, and electromagnetic fields. These topics are mentioned when appropriate, but this book focuses on an overview of the breadth of the MEMS designs technical area and the specific topics required to develop a MEMS device or product.

Acknowledgments

I am privileged to be a part of the Microsystems Science, Technology and Components Center at Sandia National Laboratories, Albuquerque, New Mexico, whose management and staff provide a collegial atmosphere of research and development of MEMS devices for the national interest. Many references and examples cited in this book come from their published research. I apologize in advance if I have overlooked any one particular contribution.

I am very indebted to Dr. David R. Sandison, manager of the Microdevices Technology Department, who encouraged the pursuit of this project and gave much of his time to reviewing the entire manuscript. I also am grateful to Victor Yarberry, Dr. Robert Huber, and Dr. Andrew Oliver, who reviewed sections of the manuscript.

About the Author

James J. Allen attended the University of Arkansas in Fayetteville, Arkansas, and received a B.S. degree in mechanical engineering in 1971. He spent 6 years in the U.S. Navy nuclear propulsion program and served aboard the fast attack submarines, USS *Nautilus* (SSN-571), USS *Haddock* (SSN-621), and USS *Barb* (SSN-596). After completion of his naval service, he returned to graduate school and received an M.S. in mechanical engineering from the University of Arkansas (1977) and a Ph.D. in mechanical engineering from Purdue University (1981). Dr. Allen taught mechanical engineering at Oklahoma State University for 3 years prior to joining Sandia National Laboratories, where he has worked for 20 years. He is also a registered professional engineer in New Mexico.

Dr. Allen is currently in the MEMS Device Technology Department at Sandia National Laboratories, where he holds eight issued patents in MEMS devices and has several patents pending. He has been active in the American Society of Mechanical Engineers (ASME), where he is a fellow of ASME and he has been the MEMS track manager for the International Mechanical Engineering Congress for 3 years. He is also the vice chair of the ASME MEMS division.

Contents

Chapter 1 Introduction 1

1.1 Historical Perspective 1
1.2 The Development of MEMS Technology 3
1.3 MEMS: Present and Future 6
1.4 MEMS Challenges 12
1.5 The Aim of This Book 13
Questions 14
References 14

Chapter 2 Fabrication Processes 17

2.1 Materials 17
    2.1.1 Interatomic Bonds 17
    2.1.2 Material Structure 18
    2.1.3 Crystal Lattices 19
    2.1.4 Miller Indices 21
    2.1.5 Crystal Imperfections 23
2.2 Starting Material — Substrates 25
    2.2.1 Single-Crystal Substrate 25
        2.2.1.1 Czochralski Growth Process 25
        2.2.1.2 Float Zone Process 27
        2.2.1.3 Post-Crystal Growth Processing 27
    2.2.2 Silicon on Insulator (SOI) Substrate 28
2.3 Physical Vapor Deposition (PVD) 30
    2.3.1 Evaporation 32
    2.3.2 Sputtering 34
2.4 Chemical Vapor Deposition (CVD) 35
2.5 Etching Processes 38
    2.5.1 Wet Chemical Etching 38
    2.5.2 Plasma Etching 39
    2.5.3 Ion Milling 43
2.6 Patterning 43
    2.6.1 Lithography 43
    2.6.2 Lift-Off Process 48
    2.6.3 Damascene Process 50
2.7 Wafer Bonding 50
    2.7.1 Silicon Fusion Bonding 51
    2.7.2 Anodic Bonding 51
2.8 Annealing 51

2.9   Chemical Mechanical Polishing (CMP) 53
2.10 Material Doping 54
     2.10.1 Diffusion 56
     2.10.2 Implant 60
2.11 Summary 61
Questions..62
References...63

Chapter 3   MEMS Technologies..65

3.1   Bulk Micromachining 68
     3.1.1 Wet Etching 70
     3.1.2 Plasma Etching 72
     3.1.3 Examples of Bulk Micromachining Processes 74
           3.1.3.1   SCREAM 75
           3.1.3.2   PennSOIL...76
3.2   LIGA 79
     3.2.1 A LIGA Electromagnetic Microdrive 80
3.3   Sacrificial Surface Micromachining 83
     3.3.1 SUMMiT™...88
3.4   Integration of Electronics and MEMS Technology (IMEMS)....................94
3.5   Technology Characterization 95
     3.5.1 Residual Stress...98
     3.5.2 Young's Modulus..101
     3.5.3 Material Strength 102
     3.5.4 Electrical Resistance..103
     3.5.5 Mechanical Property Measurement for Process Control 105
3.6   Alternative MEMS Materials...106
     3.6.1 Silicon Carbide 106
     3.6.2 Silicon Germanium..108
     3.6.3 Diamond...108
     3.6.4 SU-8 109
3.7   Summary 109
Questions..110
References...110

Chapter 4   Scaling Issues for MEMS...115

4.1   Scaling of Physical Systems..115
     4.1.1 Geometric Scaling 115
     4.1.2 Mechanical System Scaling...117
     4.1.3 Thermal System Scaling..121
     4.1.4 Fluidic System Scaling..124
     4.1.5 Electrical System Scaling..129
     4.1.6 Optical System Scaling 134
     4.1.7 Chemical and Biological System Concentration 135

4.2	Computational Issues of Scale	137
4.3	Fabrication Issues of Scale	139
4.4	Material Issues	141
4.5	Newly Relevant Physical Phenomena	144
4.6	Summary	145
Questions		149
References		152

Chapter 5   Design Realization Tools for MEMS 155

5.1	Layout		155
5.2	SUMMiT Technology Layout		158
	5.2.1	Anchoring Layers	159
	5.2.2	Rotational Hubs	164
	5.2.3	Poly1 Beam with Substrate Connection	170
	5.2.4	Discrete Hinges	170
5.3	Design Rules		176
	5.3.1	Manufacturing Issues	176
		5.3.1.1 Patterning Limits	176
		5.3.1.2 Etch Pattern Uniformity	178
		5.3.1.3 Registration Errors	178
		5.3.1.4 Etch Compatibility	179
		5.3.1.5 Stringers	179
		5.3.1.6 Floaters	180
		5.3.1.7 Litho Depth of Focus	180
		5.3.1.8 Stiction (Dimples)	181
		5.3.1.9 Etch Release Holes	181
		5.3.1.10 Improper Anchor (Area of Anchor)	182
	5.3.2	Design Rule Checking	182
5.4	Standard Components		183
5.5	MEMS Visualization		184
5.6	MEMS Analysis		186
5.7	Summary		188
Questions			189
References			190

Chapter 6   Electromechanics 193

6.1	Structural Mechanics		194
	6.1.1	Material Models	194
	6.1.2	Thermal Strains	200
	6.1.3	Axial Rod	201
	6.1.4	Torsion Rod	203
	6.1.5	Beam Bending	205
	6.1.6	Flat Plate Bending	208
	6.1.7	Columns	211

| | 6.1.8 | Stiffness Coefficients | 213 |

6.1.8 Stiffness Coefficients 213
6.2 Damping 216
    6.2.1 Oscillatory Mechanical Systems and Damping 217
    6.2.2 Damping Mechanisms 220
    6.2.3 Viscous Damping 222
    6.2.4 Damping Models 224
        6.2.4.1 Squeeze Film Damping Model 224
        6.2.4.2 Slide Film Damping Model 226
6.3 Electrical System Dynamics 228
    6.3.1 Electric and Magnetic Fields 229
    6.3.2 Electrical Circuits — Passive Elements 234
        6.3.2.1 Capacitor 234
        6.3.2.2 Inductor 235
        6.3.2.3 Resistor 236
        6.3.2.4 Energy Sources 238
        6.3.2.5 Circuit Interconnection 238
Questions 240
References 241

Chapter 7 Modeling and Design 243

7.1 Design Synthesis Modeling 243
7.2 Lagrange's Equations 244
    7.2.1 Lagrange's Equations with Nonpotential Forces 246
    7.2.2 Lagrange's Equations with Equations of Constraint 247
    7.2.3 Use of Lagrange's Equations to Obtain Lumped Parameter
        Governing Equations of Systems 248
    7.2.4 Analytical Mechanics Methods for Continuous Systems 257
7.3 Numerical Modeling 262
7.4 Design Uncertainty 267
Questions 270
References 271

Chapter 8 MEMS Sensors and Actuators 273

8.1 MEMS Actuators 273
    8.1.1 Electrostatic Actuation 273
        8.1.1.1 Parallel Plate Capacitor 273
        8.1.1.2 Interdigitated Comb Capacitor 278
        8.1.1.3 Electrostatic Actuators 278
    8.1.2 Thermal Actuation 285
    8.1.3 Lorentz Force Actuation 288
8.2 MEMS Sensing 290
    8.2.1 Capacitative Sensing 290
    8.2.2 Piezoresistive Sensing 298
        8.2.2.1 Piezoresistivity 298

| | | 8.2.2.2 | Piezoresistance in Single-Crystal Silicon | 299 |

8.2.2.2 Piezoresistance in Single-Crystal Silicon 299
8.2.2.3 Piezoresistivity of Polycrystalline and Amorphous Silicon 304
8.2.2.4 Signal Detection 304
8.2.3 Electron Tunneling 306
8.2.4 Sensor Noise 308
     8.2.4.1 Noise Sources 311
8.2.5 MEMS Physical Sensors 314
     8.2.5.1 Accelerometer 314
     8.2.5.2 Gyroscope 319
     8.2.5.3 Pressure Sensors 324
8.2.6 Chemical Sensors 328
     8.2.6.1 Taguchi Gas Sensor 330
     8.2.6.2 Combustible Gas Sensor 331
Questions 332
References 333

Chapter 9 Packaging 339

9.1 Packaging Process Steps 339
  9.1.1 Postfabrication Processing 340
      9.1.1.1 Release Process 341
      9.1.1.2 Drying Process 341
      9.1.1.3 Coating Processes 342
      9.1.1.4 Assembly 345
      9.1.1.5 Encapsulation 348
  9.1.2 Package Selection/Design 350
  9.1.3 Die Attach 352
  9.1.4 Wire Bond and Sealing 353
9.2 Packaging Case Studies 353
  9.2.1 R&D Prototype Packaging 355
  9.2.2 DMD Packaging 357
  9.2.3 Electrical-Fluidic Packaging 359
9.3 Summary 361
Questions 362
References 363

Chapter 10 Reliability 367

10.1 Reliability Theory and Terminology 367
10.2 Essential Aspects of Probability and Statistics for Reliability 370
10.3 Reliability Models 380
  10.3.1 Weibull Model 380
  10.3.2 Lognormal Model 383
  10.3.3 Exponential Model 386
10.4 MEMS Failure Mechanisms 386

10.4.1　Operational Failure Mechanisms..388
　　　　10.4.1.1　Wear 388
　　　　10.4.1.2　Fracture 390
　　　　10.4.1.3　Fatigue...391
　　　　10.4.1.4　Charging 391
　　　　10.4.1.5　Creep 391
　　　　10.4.1.6　Stiction and Adhesion 391
10.4.2　Degradation Mechanisms 392
10.4.3　Environmental Failure Mechanisms 392
　　　　10.4.3.1　Shock and Vibration...392
　　　　10.4.3.2　Thermal Cycling 393
　　　　10.4.3.3　Humidity 393
　　　　10.4.3.4　Radiation 393
　　　　10.4.3.5　Electrostatic Discharge (ESD)...................................393

10.5　Measurement Techniques for MEMS Operational, Reliability, and
　　　Failure Analysis Testing...394
　　　10.5.1　Optical Microscopy 394
　　　10.5.2　Scanning Electron Microscopy..396
　　　10.5.3　Focused Ion Beam 396
　　　10.5.4　Atomic Force Microscope 397
　　　10.5.5　Lift-Off..397
　　　10.5.6　Stroboscopy...397
　　　10.5.7　Blur Envelope 398
　　　10.5.8　Video Imaging 399
　　　10.5.9　Interferometry 399
　　　10.5.10　Laser Doppler Velocimeter (LDV)..400
10.6　MEMS Reliability and Design 400
10.7　MEMS Reliability Case Studies 403
　　　10.7.1　DMD Reliability 403
　　　10.7.2　Sandia Microengine 407
10.8　Summary 412
Questions..412
References...413

Appendix A — Glossary 417

Appendix B — Prefixes 419

Appendix C — Micro–MKS Conversions...421

Appendix D — Physical Constants..423

Appendix E — Material Properties..425

Appendix F — Stiffness Coefficients of Frequently Used MEMS Flexures 427

Appendix G — Common MEMS Cross-Section Properties 433

Appendix H 437

Index 453

1 Introduction

1.1 HISTORICAL PERSPECTIVE

Making devices small has long had engineering, scientific, and aesthetic motivations. For example, John Harrison's quest [1] to make a small (e.g., hand-sized) *chronometer* in the 1700s for nautical navigation was motivated by the desire to have an accurate time-keeping instrument that was insensitive to temperature, humidity, and motion. A small chronometer could meet these objectives and allow for multiple instruments on a ship for redundancy and error averaging. A number of technological firsts came from this work, such as the development of the roller bearing. Driven by the need for portability, the miniaturization of many mechanical devices has advanced over the years.

The 20th century saw the rise of electrical and electronic devices that had an impact on daily life. Until the advent of the *point contact transistor* in 1947 by Bardeen and Brattain [2] and, later, the *junction transistor* by Shockley [3], electronic devices were based upon the *vacuum tube* invented in 1906 by Lee de Forest. The transistor was a great leap forward in reducing size, power requirements, and portability of electronic devices.

By the mid 20th century, electronic devices were produced by connecting individual components (i.e., vacuum tubes, switches, resistors and capacitors). This resulted in large devices that consumed significant power and were costly to produce. The reliability of these devices was also poor due to the need to assemble the multitude of components. The state of the art was epitomized by the world's first digital computer [4], ENIAC (electronic numerical integrator and computer), which was developed at the University of Pennsylvania [5] for the Army Ordnance Department to carry out ballistics calculations. The need for ENIAC illustrates the need for computers to assist in the development of engineering devices that was emerging at the time. However, ENIAC consisted of thousands of electronic components, which needed to be replaced at frequent intervals, consumed significant power, and wasted heat.

Several key events occurred in the late 1950s that would motivate development of electronics at an increased pace beyond the discrete transistor. The development of the planar silicon transistor [6,7] and the planar fabrication process [8,9] set the stage for development of fabrication processes and equipment to achieve electronic devices monolithically integrated on a single substrate with small feature sizes. The development of this technology for integrated circuits started the *microelectronics revolution*, which led to the production of microelectronic devices with smaller and smaller features and continues to the present day.

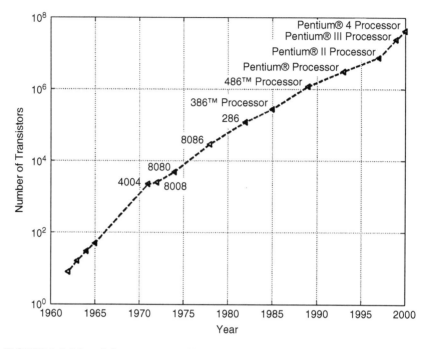

FIGURE 1.1 Moore's law as expressed by the number of transistors in integrated circuits vs. time. (These data are a compilation of data taken from several sources.)

Microelectronic technology developed rapidly, as can be seen by the paper presented by Gordon Moore [10] in 1965 in which he predicted the rapid growth of microelectronics. At this point, microelectronics was producing integrated circuits with 50 transistors on 1-in. wafers, which could be spaced 50 µm apart. Silicon had emerged as the microelectronic material of choice due to the ability to produce a high-quality, stable silicon dioxide layer, which is essential to the fabrication of transistors. In his paper, Moore stated,

> The complexity of minimum component costs has increased at a rate of roughly a factor of two per year. Certainly over the short term this rate can be expected to continue, if not increase. Over the longer term, the rate of increase is a bit more uncertain, although there is no reason to believe it will not remain nearly constant for at least 10 years.

The pace of microelectronic development has been maintained over the years, as can be seen in Figure 1.1.

Dr. Richard Feynman presented a seminal talk, "There's Plenty of Room at the Bottom" on December 29, 1959, at the annual meeting of the American Physical Society at the California Institute of Technology (Caltech); the text was first published in the 1960 issue of Caltech's engineering and science magazine [11] and has since been reprinted several times [12,13]. In the talk, Dr. Feynman

Introduction

conceptually presented, motivated, and challenged people with the desire and advantages of exploring engineered devices at the small scale. This talk is frequently sited as the conceptual beginnings of the fields of *microelectromechanical systems* (MEMS) and *nanotechnology*. Dr. Feynman provided some very insightful comments on the scaling of physical phenomena as size is reduced as well as some prophetic uses of the small-scale devices upon which he was speculating.

- Scaling of physical phenomena
  - "The effective viscosity of oil would be higher and higher in proportion as we went down" in size.
  - "Let the bearings run dry; they won't run hot because the heat escapes away from such a small device very, very rapidly."
- Miniaturizing the computer
  - "...the possibilities of computers are very interesting — if they could be made to be more complicated by several orders of magnitude. If they had millions of times as many elements, they could make judgments."
  - "For instance, the wires should be made 10 or 100 atoms in diameter, and the circuits should be a few thousand angstroms across."
- Use of small machines
  - "...it would be interesting in surgery if you could swallow the surgeon. You put the mechanical surgeon inside the blood vessel and it goes into the heart and looks around."

During this presentation, Dr. Feynman offered two $1000 prizes for the following achievements:

- Build a working electric motor no larger than a 1/64-in. (400-μm) cube
- Print text at a scale (1/25,000) that the Encyclopedia Britannica could fit on the head of a pin

In less than a year, a Caltech engineer, William McLellan, constructed a 250-μg, 2000-rpm electric motor using 13 separate parts to collect his prize [14]. This illustrated that technology was constantly moving toward miniaturization and that aspects of the technology already existed. However, the second prize was not rewarded until 1985, when T. Newman and R.F.W. Pease used e-beam lithography to print the first page of *A Tale of Two Cities* within a 5.9-μm square [14]. The achievement of the second prize was enabled by the developments of the microelectronics industry in the ensuing 25 years. Images of these achievements are available in references 16 and 17.

1.2 THE DEVELOPMENT OF MEMS TECHNOLOGY

Microelectromechanical system (MEMS) technology (also known as microsystems technology [MST] in Europe) has been inspired by the development of the microelectronic revolution and the vision of Dr. Feynman. MEMS and MST were built upon the technological and commercial needs of the latter part of the 20th century, as well as the drive toward miniaturization that had been a driving force for a number of reasons over a much longer period of time. The development of MEMS technology synergistically used to a large extent the materials and fabrication methods developed for microelectronics. Table 1.1 is a historical time line of some of the key events in the development of MEMS technology.

MEMS technology is a result of a long history of technology development starting with machine and machining development through the advent of microelectronics. In fact, in a continuum of devices and fabrication process MEMS occupies the size range from 1 mm to 1 μm. In this book, size scales are referred to as *macro*, *meso*, *micro*, and *nano* scale. Table 1.2 attempts to provide a more definitive definition of these terms.

The development of the discrete transistor and its use began to replace the vacuum tube in electronic applications in the 1950s. In the early days of the development of the transistor, the piezoresistive properties of the semiconductor materials used to develop the transistor, silicon and germanium, were researched [18]. This advance provided a link between the electronic materials and mechanical sensing. This link was exploited early in the time line of MEMS development to produce strain gages and pressure sensors.

The key technical advances that precipitated the microelectronic revolution were the development of the planar silicon transistor [6,7] and fabrication process [8,9]. The planar silicon fabrication process provided a path that enabled the integration of large numbers of transistors to create many different electronic devices and, through continuous technical advancement of the fabrication tools (lithography, etching, diffusion, and implantation), a continual reduction in size of the transistor. This ability to increasingly miniaturize the electronic circuitry over a long period of time was predicted by Moore in 1965 in what was to become known as *Moore's law*. The effects of this law continue today and at least for the next 20 years [19]. This development of fabrication tools for increasingly smaller dimensions is a key enabler for MEMS technology.

In 1967, Nathanson et al. developed the resonant gate transistor [20], which showed the possibilities of an integrated mechanical–electrical device and silicon micromachining. In the early days of microelectronics and through the 1970s, *bulk micromachining*, which utilizes deep etching techniques, was developed and used to produce pressure sensors and accelerometers. In 1982, Petersen [21] wrote a seminal paper, "Silicon as a Mechanical Material." Thus, silicon was considered and utilized to an even greater extent to produce sensors that needed a mechanical element (inertial mass, pressure diaphragm) and a transduction mechanism (mechanical–electrical) to produce a sensor. Bulk micromachining was also utilized to make ink nozzles, which were becoming a large commercial market due to the computer revolution's need for low-cost printers.

In 1983, Howe and Muller [22] developed the basic scheme for surface micromachining; this utilizes two types of material (structural, sacrificial) and the tools developed for microelectronics to create a fabrication technology capable

Introduction

TABLE 1.1
A Time Line of Key MEMS Developments and Other Contemporary Technological Developments

Time	Event	Company	Ref.
1947	ENIAC (electronic numerical integrator and computer)	University of Pennsylvania	
1947	Invention of the bipolar transistor		2
1954	Piezoresistive effect in germanium and silicon		18
1958	First commercial bare silicon strain gages	Kulite Semiconductor	
1959	"There's plenty of room at the bottom"		11,12
1959	Planar Silicon Transistor		6,7
1959	Planar fabrication process for microelectronics		8,9
1960	Feynman prize awarded for electric motor no larger than a 1/64-in. cube		14,16
1961	Silicon pressure sensor demonstrated	Kulite Semiconductor	
1965	Moore's law		10
1967	Resonant gate transistor		19
1974	First high-volume pressure sensor	National Semiconductor	
1977–1979	Micromachined ink-jet nozzle	International Business Machines, Hewlett-Packard	
1982	Silicon as a mechanical material		20
1982	Disposable blood pressure transducer	Foxboro/ICT, Honeywell	
1985	Feynman prize awarded for producing text at a 1/25,000 scale		15,17
1983	Surface micromachining process		21
1987	Digital micromirror device (DMD) invented	Hornbeck	
1988	Micromechanical elements		22
1986	LIGA process		25
1989	Lateral comb drive		23
1991	Polysilicon hinge		24
1993	ADXL50 accelerometer commercially sold	Analog Devices Inc.	
1996	Digital light processor (DLP™) containing DMD commercially sold	Texas Instruments	
2002	Analog Devices ADXRS gyroscope introduced	Analog Devices Inc.	

of producing complex mechanical elements without the need for postfabrication assembly. Many of the essential actuation and mechanical elements were demonstrated in the ensuing years [23–25].

Also in the 1980s, the LIGA (Lithographie Galvanoformung Abformung) process [26] was developed in Germany. The material set that LIGA uses is significantly different from bulk and surface micromachining, which tend to use

6   Micro Electro Mechanical System Design

TABLE 1.2
A Definition of Size Scale Terminology

Size scale	Fabrication technology	Devices	Measurement methods
Macroscale (>10 mm)	Conventional machining	Conventional devices and machines	Attachable sensors (strain gauges, accelerometers); visual and optical measurements
Mesoscale (10 mm ↔ 1 mm)	Precision machining	Miniature parts, devices, and motors	Combination of macroscale, and microscale measurement methods
Microscale (1 mm ↔ 1 μm)	LIGA; bulk micromachining; sacrificial surface micromachining	MEMS devices	Optical microscopy; SEM
Nanoscale (1 μm ↔ 1 nm)	Biochemical engineering	Molecular scale devices	AFM, SEM; Scanning probe microscopy

the microelectronic fabrication tools and materials. LIGA can be used to make parts or molds from electroplateable materials or use the molds to make injection molded plastics.

The 1990s saw the development of commercial products that require the integration of MEMS mechanical and electrical fabrication (IMEMS) technologies due to the need for high-resolution sensing of mechanical elements or the addressing and actuation of large arrays of mechanical elements. Analog Devices, Inc. developed an IMEMS technology [27] to facilitate the development of inertial sensors (accelerometer, gyroscope) for automotive applications. Texas Instruments developed an IMEMS technology [28] to produce a large array (~10^6) of mirrors used in projectors, cinema, and televisions. The development of IMEMS technologies is discussed in detail in Chapter 3.

1.3 MEMS: PRESENT AND FUTURE

The 1980s to the mid 1990s saw the development of three categories of fabrication technologies for MEMS. *Bulk micromachining, sacrificial surface micromachining,* and *LIGA* have unique capabilities based on the fabrication materials utilized, ability to integrate with electronics, assembly, and thickness of materials. These technologies enable many different types of applications and will be discussed in detail in Chapter 3. The information available on MEMS technology has grown as it has matured. Sample lists of journals, periodicals, and Web sites is provided in Table 1.3 through Table 1.5; these offer a wealth of information and a starting point for further research into the world of MEMS.

Introduction

TABLE 1.3
MEMS Journals

Journal	Publisher
Journal of Microelectromechanical Systems	IEEE/ASME
Journal of Micromechanics and Microengineering	Institute of Physics
Sensors and Actuators	Elsevier Science Ltd
Microsystem Technologies	Springer-Verlag

TABLE 1.4
MEMS Magazines and Newsletters

Magazine/newsletter	Frequency	Publisher
smalltimes	bimonthly	Small Time Media LLC http://www.smalltimes.com/
Micro/Nano	monthly	Reed Business Information
mstnews: International Newsletter on Microsystems and MEMS	bimonthly	VDI/VDE-IT GmbH

The mid 1990s to the present day has seen a shift in the emphasis of MEMS technology research from fabrication process development and the demonstration of prototype sensors and actuators to the commercialization of MEMS products. The impact of MEMS technology is very broad as can be seen by the brief list of MEMS applications in Table 1.6. These MEMS products range from physical sensors (e.g., pressure, inertial), biological, optical, and robotics to radio frequency (RF) devices. MEMS applications span the range of physics. As a result, the MEMS field affects a wide swath of engineers, physicists, chemists, and biologists.

Today's automobile is one area in which the world of MEMS [29] has a direct impact on daily life. A number of locations within the automobile contain MEMS technology, for example:

- *Accelerometers* are used for multiple functions, such as air bag deployment, vehicle security, and seat belt tension triggers.
- *Gyroscopes* are used — possibly in conjunction with accelerometers — in car stability control systems to correct the yaw of a car before this becomes a problem for the driver.
- *Pressure sensors*: the manifold absolute pressure sensor is used to control the fuel–air mixture in the engine. Tire pressure monitoring has also been recently mandated for use in automobiles.
- The *wheel speed sensor* is a component of the ABS braking system that can also be used as an indirect measure of tire pressure.
- The *oil condition sensor* detects oil temperature, contamination, and level.

TABLE 1.5
A Sample of MEMS Web Sites

Organization/name	Topic
Research and information	
MEMS and Nanotechnology Clearinghouse http://www.memsnet.org/	MEMS information, material database, universities and companies
Berkeley Sensor and Actuator Center http://www-bsac.eecs.berkeley.edu	University research
Fabrication	
Sandia National Laboratories http://mems.sandia.gov/scripts/index.asp	Government research foundry (SUMMiT™) process
MEMS Exchange http://www.mems-exchange.org/	Foundry processing
Fairchild Semiconductor http://www.fairchildsemi.com/	Foundry (SUMMiT) processing
Products	
Analog Devices Incorporated http://www.analog.com/	MEMS inertial sensors
Texas Instruments http://www.ti.com/	MEMS display technology
Kulite Semiconductor www.kulite.com	MEMS pressure sensors
Software	
MEMSCap http://www.memscap.com/	Software, design, foundry processing
Coventor http://www.coventor.com/	Software
ANSYS http://www.ansys.com/industry/mems/	Software
Intellisense http://www.intellisensesoftware.com/	Software, design, consulting
Marketing and trade associations	
MEMS Industry Group http://www.memsindustrygroup.org/	North American MEMS trade association
NEXUS http://www.nexus-mems.com/	European microsystems network
Yole Development http://www.yole.fr/	MEMS and high-tech marketing

Introduction 9

TABLE 1.6
MEMS Applications

Device	Use
Pressure sensors	Automotive, medical, industrial
Accelerometer	Automotive and industrial motion sensing
Gyroscope	Automotive and industrial motion sensing
Optical displays	Cinema and business projectors, home theater, television
RF devices	Switches, variable capacitors, filters
Robotics	Sensing, actuation
Biology and medicine	Chemical analysis, DNA sequencing, drug delivery, implantable prosthetics

The automotive market is a mass market in which MEMS is playing an ever increasing role. For example, 90 million air bag accelerometers and 30 million manifold absolute pressure sensors were supplied to the automotive market in 2002 [30].

Another mass market in which MEMS has an increasing impact is the biological medical market. MEMS technology enables the production of a device of the same scale as biological material. Figure 1.2 shows a comparison of a MEMS device and biological material. An example of MEMS' impact on the medical market is the DNA sequencing chip, GeneChip®, developed by Affymetrix Inc. [31], which allows medical testing in a fraction of the time and cost previously available. In addition, MEMS facilitates direct interaction at the cellular level [32]. Figure 1.3 shows cells in solution flowing through the cellular manipulator, which could disrupt the cell membrane to allow easier insertion of genetic and chemical materials. Also shown in Figure 1.3 are chemical entry and extraction ports that allow the injection of genetic material, proteins, etc. for processing in

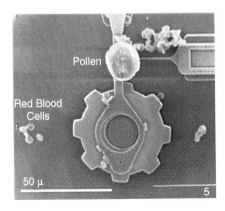

FIGURE 1.2 MEMS device and biological material comparison. (Courtesy of Sandia National Laboratories.)

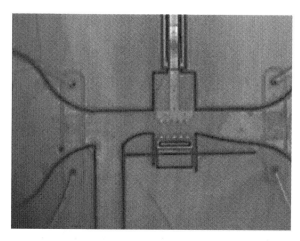

FIGURE 1.3 Red blood cells flowing through a cellular manipulator with chemical entry/extraction ports. (Courtesy of Sandia National Laboratories.)

a continuous fluid flow system. An additional illustration of the impact of MEMS that would have been thought to be science fiction a few years ago is the retinal prosthesis [33] under development that will enable the blind to see.

MEMS also has a significant impact on space applications. The miniaturization of sensors is an obvious application of MEMS. The use of MEMS for thermal control of microsatellites is somewhat unanticipated. MEMS louvers [34] are micromachined devices similar in function and design to conventional mechanical louvers used in satellites; here, a mechanical vane or window is opened and closed to vary the radiant heat transfer to space. MEMS is applicable in this context because it is small and consumes little power, but produces the physical effect of variable thermal emittance, which controls the temperature of the satellite. The MEMS louver consists of an electrostatic actuator that moves a louver to control the amount of gold surface exposed (i.e., variable emittance). Figure 1.4 shows the MEM louvers that will be demonstrated on an upcoming NASA satellite mission.

The integration of MEMS devices into automobiles or satellites enables attributes such as smaller size, smaller weight, and multiple sensors. The use of MEMS in systems can also allow totally different functionality. For example, a miniature robot with a sensor, control circuitry, locomotion, and self-power can be used for chemical or thermal plume detection and localization [35]. In this case, MEMS technology enables the group behavior of a large number of small robots capable of simple functions. The group interaction ("swarming") of these simple expendable robots is used to search an area to locate something that the sensor can detect, such as a chemical or temperature.

One vision of the future direction of MEMS is expressed in Picraux and McWhorter [36], who propose that MEMS applications will enable systems to *think, sense, act, communicate,* and *self-power.* Many of the applications discussed in this section indeed integrate some of these attributes. For example, the

Introduction

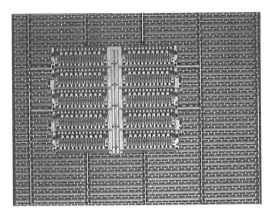

FIGURE 1.4 MEMS variable emittance lover for microsatellite thermal control. The device was developed under a joint project with NASA, Goddard Spaceflight Center, The Johns Hopkins Applied Physics Laboratory, and Sandia National Laboratories.

FIGURE 1.5 A small robot with a sensor, locomotion, control circuitry, and self power. (Courtesy of Sandia National Laboratories.)

small robot shown in Figure 1.5 has a sensor, can move, and has a self-contained power source. To integrate all of these functions on one chip may not be practical due to financial or engineering constraints; however, integration of these functions via packaging may be a more viable path.

MEMS is a new technology that has formally been in existence since the 1980s when the acronym MEMS was coined. This technology has been focusing on commercial applications since the mid 1990s with significant success [37]. The MEMS commercial businesses are generally organized around three main models: MEMS manufacturers; MEMS design; and system integrators. In 2003, 368 MEMS fabrication facilities existed worldwide, with strong centers in North America, Japan, and Europe. There are 130 different MEMS applications in production consisting of a few large-volume applications in the automotive (iner-

TABLE 1.7
Comparison of MEMS and Microelectronics

Criteria	Microelectronics	MEMS
Feature size	Submicron	1–3 µm
Device size	Submicron	~50 µm–1 mm
Materials	Silicon based	Varied (silicon, metals, plastics)
Fundamental devices	Limited set: transistor, capacitor, resistor	Widely varied: fluid, mechanical, optical, electrical elements (sensors, actuators, switches, mirrors, etc.)
Fabrication process	Standardized: planar silicon process	Varied: three main categories of MEMS fabrication processes plus variants: Bulk micromachining Surface micromachining LIGA

tial, pressure); ink-jet nozzles; and medical fields (e.g., Affymetrix GeneChip). The MEMS commercial market is growing at a 25% annual rate [37].

1.4 MEMS CHALLENGES

MEMS is a growing field applicable to many lines of products that has been synergistically using technology and tools from the microelectronics industry. However, MEMS and microelectronics differ in some very fundamental ways. Table 1.7 compares the devices and technologies of MEMS and microelectronics, and Figure 1.6 compares the levels of device integration of MEMS and microelectronics. The most striking observation is that microelectronics is an enormous industry based on a few fundamental devices with a standardized fabrication process. The microelectronics industry derives its commercial applicability from the ability to connect a multitude of a few fundamental types of electronic devices (e.g., transistors, capacitors, resistors) reliably on a chip to create a plethora of new microelectronic applications (e.g., logic circuits, amplifiers, computer processors, etc.). The exponential growth predicted by Moore's law comes from improving the fabrication tools to make increasingly smaller circuit elements, which in turn enable faster and more complex microelectronic applications.

The MEMS industry derives its commercial applicability from the ability to address a wide variety of applications (accelerometers, pressure sensors, mirrors, fluidic channel); however, no one fundamental *unit cell* [38,39] and *standard fabrication process* to build the devices exists. In fact, the drive toward smaller devices for microelectronics, which increased speed and complexity, does not necessarily have the same impact on MEMS devices [40] due to scaling issues (Chapter 4). MEMS is a new rapidly growing [37] technology area in which contributions are to be made in fabrication, design, and business.

Introduction

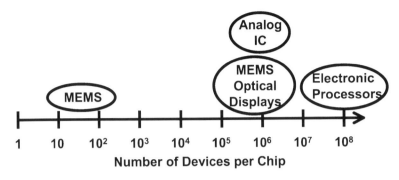

FIGURE 1.6 Levels of device integration of MEMS vs. microelectronics.

1.5 THE AIM OF THIS BOOK

This book is targeted at the practicing engineer or graduate student who wants an introduction to MEMS technology and the ability to design a device applicable to his or her area of interest. The book will provide an introduction to the basic concepts and information required to engage fellow professionals in the area and will aid in the design of a MEMS product that addresses an application area. MEMS is a very broad technical area difficult to address in detail within one book due to this breadth of material. It is the hope that this text coupled with an engineering or science educational background will enable the reader to become a *MEMS designer*. The chapters (topics) of this book are organized as follows. They can be taken in whole or as needed to fill the gaps in an individual's background.

- Chapter 2: Fabrication Processes — offers an overview of the individual fabrication process applicable to MEMS.
- Chapter 3: MEMS Technologies — is an overview of the combination of fabrication processes necessary to produce a technology suitable for the production of MEMS devices and products.
- Chapter 4: Scaling Issues for MEMS — covers the physics and device operation issues that arise due to the reduction in size of a device.
- Chapter 5: Design Realization Tools for MEMS — discusses the computer-aided design tools required to interface a design with the fabrication infrastructure encountered in MEMS.
- Chapter 6: Electromechanics — provides an overview of the physics of electromechanical systems encountered in MEMS design.
- Chapter 7: Modeling and Design — is an introduction to modeling for MEMS design with an emphasis on low-order models for design synthesis.
- Chapter 8: MEMS Sensors and Actuators — offers an overview of sensors and actuators utilized in MEMS devices.

14 Micro Electro Mechanical System Design

- Chapter 9: Packaging — is a review of the packaging processes and how the packaging processes and fabrication processes interact; three packaging case studies are presented.
- Chapter 10: Reliability — covers the basic concepts of reliability and the aspects of reliability unique to MEMS, such as failure mechanisms and failure analysis tools.

QUESTIONS

1. Use the Web as a tool to explore what is happening in the world of MEMS.
2. Pick an application and research how it is used. What type of fabrication process is used and how many companies have products in this area?
3. Look at a MEMS application that existed before MEMS technology existed. How did MEMS technology have an impact on this application in performance, cost, or volume production?

REFERENCES

1. D. Sobel, *Longitude, The True Story of a Lone Genius Who Solved the Greatest Scientific Problem of His Time*, Penguin Books, New York, 1995.
2. J. Bardeen, W. H. Brattain, The transistor, a semiconductor triode, *Phys. Rev.*, 74, 130–231, 1948.
3. W. Shockley, A unipolar field-effect transistor, *Proc. IRE*, 40, 1365, 1952.
4. ENIAC (electronic numerical integrator and computer) U.S. Patent No. 3,120,606, filed 26 June 1947.
5. ENIAC Museum: http://www.seas.upenn.edu/~museum/.
6. J.A. Hoerni, Planar silicon transistors and diodes, *IRE Transactions Electron Devices*, 8, 2, March 1961.
7. J.A. Hoerni, Method of manufacturing semiconductor devices, U.S. Patent 3,025,589, issued March 20, 1962.
8. J.S. Kilby, Miniaturized electronic circuits, U.S. Patent 3,138,743, filed February 6, 1959.
9. R.N. Noyce, Semiconductor device and lead structure, U.S. Patent 2,918,877, filed July 30, 1959.
10. G.E. Moore, Cramming more components onto integrated circuits, *Electronics*, 38(8), April 19, 1965.
11. R.P. Feynman, There's plenty of room at the bottom, *Eng. Sci.* (California Institute of Technology), February 1960, 22–36.
12. R.P. Feynman, There's plenty of room at the bottom, *JMEMS*, 1(1), 60–66, March 1992.
13. R.P. Feynman, There's plenty of room at the bottom, http://nano.xerox.com/nanotech/feynman.html.
14. E. Regis, *Nano: The Emerging Science of Nanotechnology*, Little, Brown and Company, New York, 1995.
15. N. Maluf, *An Introduction to Microelectromechanical Systems Engineering*, Artech House Inc., Boston, 2000.

Introduction

16. The Caltech Institute Archives: http://archives.caltech.edu/index.html.
17. Pease Group Homepage: http://chomsky.stanford.edu/docs/home.html.
18. C.S. Smith, Piezoresistive effect in germanium and silicon, *Phys. Rev.* 94(1), 42–49, April, 1954.
19. J.D. Meindel, Q. Chen, J.A. Davis, Limits on silicon nanoelectronics for terascale integration, *Science*, 293, 2044–2049, September 2001.
20. H.C. Nathanson, W.E. Newell, R.A. Wickstrom, J.R. Davis, The resonant gate transistor, *IEEE Trans. Electron Devices*, ED-14, 117–133, 1967.
21. K.E. Petersen, Silicon as a mechanical material, *Proc. IEEE*, 70(5), 420–457, May 1982.
22. R.T. Howe and R.S. Muller, Polycrystalline silicon micromechanical beams, *J. Electrochem. Soc.: Solid-State Sci. Technol.*, 130(6), 1420–1423, June 1983.
23. L-S. Fan, Y-C Tai, R.S. Muller, Integrated movable micromechanical structures for sensors and actuators, *IEEE Trans. Electron Devices*, 35(6), 724–730, 1988.
24. W.C. Tang, T.C.H. Nguyen, R.T. Howe, Laterally driven polysilicon resonant microstructures, *Sensors Actuators*, 20(1–2), 25–32, November 1989.
25. K.S.J. Pister, M.W. Judy, S.R. Burgett, R.S. Fearing, Microfabricated hinges, *Sensors Actuators A*, 33, 249–256, 1992.
26. E.W. Becker, W. Ehrfeld, P. Hagmann, A. Maner, and D. Muchmeyer, Fabrication of microstructures with high aspect ratios and great structural heights by synchrotron radiation lithography, galvanoforming, and plastic molding (LIGA process), *Microelectron. Eng.*, 4, 35, 1986.
27. Analog Devices IMEMS technology: http://www.analog.com/.
28. Texas Instrument DLP™ technology: http://www.ti.com/.
29. D. Forman, Automotive applications, *smalltimes*, 3(3), 42–43, May/June 2003.
30. R. Grace, Autos continue to supply MEMS "killer apps" as convenience and safety take a front seat, *smalltimes*, 3(3), 48, May/June 2003.
31. Affymetrix, Inc. http://www.affymetrix.com GeneChip®.
32. M. Okandan, P. Galambos, S. Mani, J. Jakubczak, Development of surface micromachining technologies for microfluidics and BioMEMS, *Proc. SPIE*, 4560, 133–139, 2001.
33. D. Sidawi, Emerging prostheses attempt vision restoration, *R&D Mag.*, 46(6), 30–32, June 2004.
34. R. Osiander, J. Champion, A. Darrin, D. Douglass, T. Swanson, J. Allen, E. Wyckoff, MEMS shutters for spacecraft thermal control, NanoTech 2002, Houston, TX. 9–12 September 2002.
35. R. H. Byrne, D. R. Adkins, S. E. Eskridge, H. H. Harrington, E. J. Heller, J. E. Hurtado, Miniature mobile robots for plume tracking and source localization research, *J. Micromechatronics*, 1(3), 253–261, 2002.
36. S.T. Picraux and P.J. McWhorter, The broad sweep of integrated microsystems, *IEEE Spectrum*, 35(12), 24–33, December 1998.
37. MEMS not so small after all, *Micro Nano*, 8(8), 6, Aug 2003
38. M.W. Scott and S.T. Walsh, Promise and problems of MEMS or nanosystem unit cell, *Micro/Nano Newslett.*, 8(2), 8, February 2003.
39. M. Scott, MEMS and MOEMS for national security applications, *Proc. SPIE*, 4979, 26–33, 2003.
40. S.D. Senturia, Microsensors vs. ICs: a study in contrasts, *IEEE Circuits Devices Mag.*, 20–27, November 1990.

2 Fabrication Processes

This chapter will present an overview of the various processes used in the fabrication of MEMS devices. The first section will present an introduction to materials and their structure. The processes that will be discussed in subsequent sections include deposition, patterning, and etching of materials as well as processes for annealing, polishing, and doping, which are used to achieve special mechanical, electrical, or optical properties. Many of the processes used for MEMS are adapted from the microelectronics industry; however, the conceptual roots for some of the fabrication processes (e.g., sputtering, damascene) significantly predate that industry.

2.1 MATERIALS

2.1.1 INTERATOMIC BONDS

The material structure type is greatly influenced by the interatomic bonds and their completeness. There are three types of interatomic attractions: *ionic* bonds, *covalent* bonds, and *metallic* bonds (Figure 2.1). The *ionic bonds* occur in materials where the interatomic attractions are due to electrostatic attraction between adjacent ions. For example, a sodium atom (Na) has one electron in its valence shell (i.e., outer electron shell of an atom), which can be easily released to produce a positively charge sodium ion (Na^+). A chlorine atom (Cl) can readily accept an electron to complete its valence shell, which will produce a negatively charged chlorine ion (Cl^-). The electrostatic attraction of an ionic bond will cause the negatively charged chlorine ion to surround itself with positively charged sodium ions.

The electronic structure of an atom is stable if the outer valence shells are complete. The outer valence shell can be completed by sharing electrons between adjacent atoms. The *covalent bond* is the sharing of valence electrons. This bond is a very strong interatomic force that can produce molecules such as hydrogen (H_2) or methane (CH_4), which have very low melting temperature and low attraction to adjacent molecules, or diamond, which is a covalent bonded carbon crystal with a very high melting point and great hardness. The difference between these two types of covalent bonded materials (i.e., CH_4 vs. diamond) is that the covalent bond structure of CH_4 completes the valence shell of the component atoms within one molecule, whereas the valence shell of the carbon atoms in diamond are

17

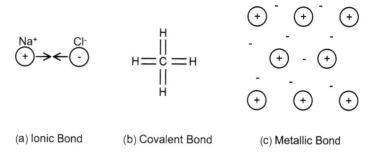

FIGURE 2.1 Simplified representation of interatomic attractions of the ionic bond, covalent bond, metallic bond.

completed via a repeating structure of a large number of carbon atoms (i.e., crystal/lattice structure).

A third type of interatomic bond is the *metallic bond*. This type of bond occurs in the case when only a few valence electrons in an atom may be easily removed to produce a positive ion (e.g., positively charged nucleus and the nonvalence electrons) and a free electron. Metals such as copper exhibit this type of interatomic bond. Materials with the metallic bond have a high electrical and thermal conductivity.

Another, weaker group of bonds is called *van der Waals* forces. The mechanisms for these forces come from a variety of mechanisms arising from the asymmetric electrostatic forces in molecules, such as molecular polarization due to electrical dipoles. These are very weak forces that frequently only become significant or observable when the ionic, covalent, or metallic bonding mechanisms cannot be effective. For example, ionic, covalent, and metallic bonding is not effective with atoms of the noble gases (e.g., helium, He), which have complete valence electron shells, and rearrangements of the valence electrons cannot be done.

2.1.2 Material Structure

The atomic structure of materials can be broadly classified as *crystalline*, *polycrystalline*, and *amorphous* (illustrated in Figure 2.2). A *crystalline* material has a large-scale, three-dimensional atomic structure in which the atoms occupy specific locations within a lattice structure. Epitaxial silicon and diamond are examples of materials that exhibit a crystalline structure. A *polycrystalline* material consists of a matrix of grains, which are small crystals of material with an interface material between adjacent grains called the grain boundary. Most metals, such as aluminum and gold, as well as polycrystalline silicon, are examples of this material structure.

The widely used metallurgical processes of cold working and annealing greatly affect the material grains and grain boundary and the resulting material properties of strength, hardness, ductility, and residual stress. Cold working uses

Fabrication Processes

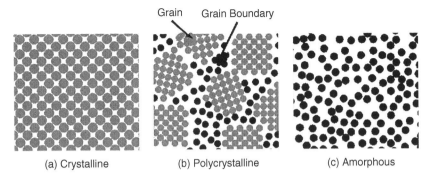

(a) Crystalline (b) Polycrystalline (c) Amorphous

FIGURE 2.2 Schematic representation of crystalline, polycrystalline, and amorphous material structures.

mechanical deformation to reduce the material grain size; this will increase strength and hardness, but reduce ductility. Annealing is a process that heats the material above the recrystallization temperature for a period of time, which will increase the grain size. Annealing will reduce residual stress and hardness and increase material ductility. A noncrystalline material that exhibits no large-scale structure is called *amorphous*. Silicon dioxide and other glasses are examples of this structural type.

2.1.3 CRYSTAL LATTICES

The structure of a crystal is described by the configuration of the basic repeating structural element, the unit cell. The unit cell is defined by the manner in which space within the crystal lattice is divided into equal volumes using intersecting plane surfaces. The crystal unit cell may be in one of seven crystal systems. These crystal systems are *cubic*; *tetragonal*; *orthorhombic*; *monoclinic*; *triclinic*; *hexagonal*; and *rhombohedral*. They include all the possible geometries into which a crystal lattice may be subdivided by the plane surfaces. The crystalline material structure is greatly influenced by factors such as the number of valance electrons and atomic radii of the atoms in the crystal (Table 2.1). The cubic crystal system is a very common and highly studied system that includes most of the common engineering metals (e.g., iron, nickel, copper, gold) as well as some materials used in semiconductors (e.g., silicon, phosphorus).

The cubic crystal system has three common variants: *simple cubic* (SC), *body-centered cubic* (BCC), and *face-centered cubic* (FCC), which are shown in Figure 2.3. The properties of crystalline material are influenced by the structural aspects of the crystal lattice, such as the number of atoms per unit cell; the number of atoms in various directions in the crystal; and the number of neighboring atoms within the crystal lattice, as shown in Table 2.2. The unit cells depicted are shown with the fraction of the atom that would be included in the unit cell (i.e., the simple cubic has one atom per unit cell; the body-centered cubic has two atoms per unit cell; face-centered cubic has four atoms per unit cell). As can be surmised,

TABLE 2.1
Atomic and Crystal Properties for Selected Elements

Element	Atomic number	Atomic mass (g/g-atom)	Crystal	Valence	Atomic radius (Å)
Boron (B)	5	10.81	Orthorhombic	3	0.46
Aluminum (Al)	13	26.98	FCC	3	1.431
Silicon (Si)	14	28.09	Diamond	4	1.176
Phosphorus (P)	15	30.97	Cubic	5	—
Iron (Fe)	26	55.85	BCC	2	1.241
Nickel (Ni)	28	58.71	FCC	2	1.245
Copper (Cu)	29	63.54	FCC		1.278
Gallium (Ga)	31	69.72	Ortho	3	1.218
Germanium (Ge)	32	72.59	Diamond	4	1.224
Arsenic (As)	33	74.92	Rhombic	5	1.25
Indium (In)	49	114.82	Tetra	3	1.625
Antimony (Sb)	51	121.75	Rhombic	5	1.452
Tungsten (W)	74	183.9	BCC	—	1.369
Gold (Au)	79	197.0	FCC	—	1.441

Notes: BCC — body-centered cubic; FCC — face-centered cubic.

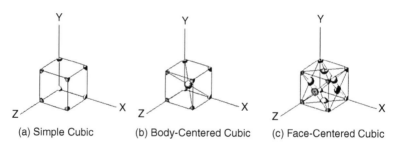

(a) Simple Cubic (b) Body-Centered Cubic (c) Face-Centered Cubic

FIGURE 2.3 Cubic crystal structures.

TABLE 2.2
Properties of Different Forms of the Cubic Lattice

Crystal structure	Number of nearest neighbors	Atoms/Cell	Packing factor[a] (atom vol/cell vol)
Cubic	6	1	0.52
Body-centered cubic	8	2	0.68
Face-centered cubic	12	4	0.74
Diamond cubic	4	8	—

[a] Assuming only one atom type in the lattice.

Fabrication Processes

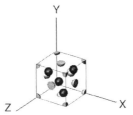

FIGURE 2.4 The diamond cubic lattice can be formed by adding four atoms (shaded dark) to the face-centered cubic lattice.

the crystal structure and the unit cell size (i.e., lattice constant) will greatly influence the density of the material. For example, dense materials such as metals crystallize in the body-centered cubic (e.g., iron, tungsten) or the face-centered cubic (e.g., aluminum, cooper, gold, nickel), which contain more atoms per unit cell instead of the simple cubic crystal, which contains only one atom per unit cell.

Silicon and germanium are Group IV elements on the periodic table; these have four valence electrons and need four more electrons to complete the outer electron shell. This can be accomplished by forming covalent bonds with four nearest neighbor atoms in the lattice. However, none of the basic cubic lattice forms have four nearest neighbors (Table 2.2). Elements such as silicon and germanium form a diamond structure, which can be conceptually thought of as two interlocking face-centered cubic lattices with a one-fourth lattice constant diagonal offset. This means that the diamond cubic lattice has four additional atoms within a face-centered cubic-like lattice structure (Figure 2.4). The gallium arsenide and indium phosphide compounds also use a version of the diamond cubic lattice, called the *zincblende*, which has a reduced level of symmetry due to the different atom sizes. Every atom in the diamond cubic lattice is tetrahedrally bonded to its four neighbors. For example, in the zincblende lattice, each gallium atom is tetrahedrally bonded to four arsenic atoms, and each arsenic atom is tetrahedrally bonded to four gallium atoms.

The properties of crystalline materials such as mechanical strength or chemical etch rates are affected by the lattice structure, and they may depend upon the directionality of the lattice structure. For example, a cubic lattice is uniform in all directions (i.e., the same number of atoms on any plane or in any direction). However, the diamond lattice has a different number of atoms in any plane or direction. The anisotropy of silicon material properties and etch rates can be somewhat attributed to its crystal structure.

2.1.4 MILLER INDICES

The *Miller indices* is nomenclature to express directions or planes in a crystal structure. Figure 2.5 shows the Miller index notation for direction in a orthorhombic lattice. An orthorhombic lattice is defined by orthogonal planes spaced differently in each direction. Miller index notation is based on the lattice unit cell intercepts within square brackets (e.g., [1 1 1]) vs. the Cartesian distances. For

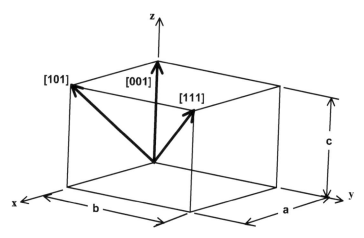

FIGURE 2.5 Crystal directions in an orthorhombic lattice.

example, the Miller index [1 1 1] denotes the direction from the origin of the *unit* cell through the opposite corner of the unit cell (i.e., not the Cartesian direction vector; Figure 2.5). Note that the [2 2 2] direction is identical to the [1 1 1] direction and the lowest combination of integers is used (e.g., [1 1 1]).

The planes within a lattice also need to be identified. The planes are denoted with labels within curved brackets — e.g., (1 0 0) — as illustrated in Figure 2.6. The (1 0 0) plane is orthogonal to the [1 0 0] direction. The numbers used in the Miller notation for planes are the reciprocals of the intercepts of the axes in unit cell distances from the origin. The Miller index notation includes not only the (1 0 0) plane shown in Figure 2.6, but also all equivalent planes. In a simple cubic lattice structure, the point of origin is arbitrarily chosen, and the (1 0 0) plane

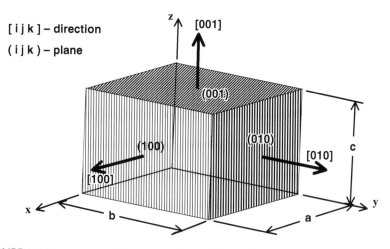

FIGURE 2.6 Crystal plane directions utilizing Miller indices.

Fabrication Processes   23

will have the same properties as the (0 1 0) and the (0 0 1) planes. The (1 0 0) refers to all three planes. Conversely, in an orthorhombic lattice, the planes (1 0 0), (0 1 0), and (0 0 1) are unique.

2.1.5 CRYSTAL IMPERFECTIONS

The symmetry of the crystal is broken at the surface of the material. The atoms at the surface are not bound to the other atoms in the same way as the bulk material. Therefore, the surface will behave differently than the bulk crystal. For example, the surface can chemically react and form an oxide or the surface can become electrically charged. Integrated circuit manufacturers frequently build the circuits upon a single-crystal silicon wafer with a (100) orientation (i.e., the [100] plane is the wafer surface) because this orientation minimizes surface charges.

In addition to the surface differences, imperfections in the crystal lattices can also be found. These can influence many characteristics of the material such as mechanical strength, electrical properties, and chemical reactivity. The lattice imperfections can be due to missing, displaced, or extra atoms in the lattice, which are called *point defects*. *Line defects* have an edge due to an extra plane of atoms.

Figure 2.7 illustrates several types of point defects, which include *substitutional*, *vacancy*, and *interstitial* types of defects. A *substitutional* defect is due to an impurity atom occupying a lattice site for the bulk material. In a *vacancy* defect, a lattice site is not occupied. An *interstitial* defect involves an atom of the bulk material or an impurity atom occupying space between the lattice sites. These defects can arise from the imperfect lattice formation during crystallization or due to impurities in the material during crystallization. The defects can also arise from thermal vibrations of the lattice atoms at elevated temperatures. Vacancies may be a single or they may condense into a larger vacancy. Conversely, defects within a single-crystal lattice structure may be intentionally created via

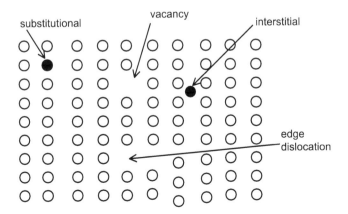

FIGURE 2.7 Schematic of lattice point and line defects.

24 Micro Electro Mechanical System Design

the processes of diffusion or implantation to produce effects in the electronic structure of the material for MEMS or microelectronics manufacturing.

The most common type of line defect is an *edge dislocation*, which is the edge of an extra plane of atoms within a crystal structure (Figure 2.7). This type of dislocation distorts the lattice, thus increasing the energy along the edge dislocation. There can also be surface defects, which are basically the transition region, *grain* boundaries, in a polycrystalline material. Each grain of a polycrystalline material is a crystal oriented differently, and the grain boundary is the transition between the grains (Figure 2.2b).

Atoms can move within a solid material as shown in Figure 2.8. However, energy is required to facilitate the movement. The energy required for the movement of the atoms is called the *activation energy* and depends on a number of factors, such as atom size and type of movement. A vacancy movement requires less energy than an interstitial movement. Atoms can move within a lattice without point or line defects using a method called *ring diffusion* (Figure 2.9). These various methods of atomic movement within a crystal are utilized in *diffusion* processes.

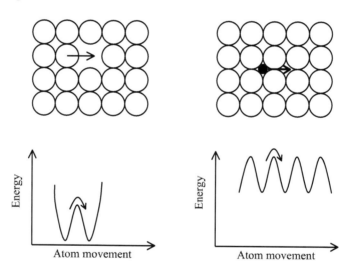

FIGURE 2.8 Atomic movements within a material.

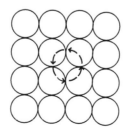

FIGURE 2.9 Ring diffusion of atoms.

2.2 STARTING MATERIAL — SUBSTRATES

A substrate is needed for mechanical support or a platform upon which to build the MEMS device. The substrate could be made of any material; however, consideration of subsequent processing and the applications of the device that are to build upon it require careful selection of the substrate material. MEMS devices are generally built using the fabrication processes developed in the microelectronics industry and the MEMS device may need to be integrated with electronics.

To a large extent, the microelectronics industry has been developed using silicon-based materials. Silicon dominates this industry because silicon forms a stable oxide essential in the formation of a MOS-FET (metal-oxide semiconductor field effect transistor). Another popular material for electronics is gallium arsenide (GaAs). GaAs has a higher electron mobility than silicon, but the hole mobility is lower and GaAs has a poor thermal oxide. GaAs-based microelectronics is generally limited to high-speed analog circuits; however, GaAs has found applications in optical devices [5] and MEMS in recent years.

Because the MEMS industry is heavily leveraging the materials and processes of the microelectronics industry, MEMS substrates generally come from the microelectronics infrastructure as well. Two substrates of particular interest for MEMS applications are single-crystal substrates and silicon-on-insulator (SOI) substrates.

2.2.1 SINGLE-CRYSTAL SUBSTRATE

2.2.1.1 Czochralski Growth Process

Czochralski growth is the method used to produce most of the single-crystal substrates used in microelectronics and MEMS. The process was developed by Czochralski in the early 1900s, and Teal [1] developed the process for use in the microelectronics industry. Czochralski growth (Figure 2.10) involves the solidification of a crystal from a molten bath.

High-grade polycrystalline silicon is loaded into a fused silica crucible that is purged with an inert gas. The crucible and its contents are heated to approximately 1500°C to form a molten bath. A *seed* crystal is then lowered into contact with the molten bath. This crystal is approximately 0.5 cm in diameter, and it

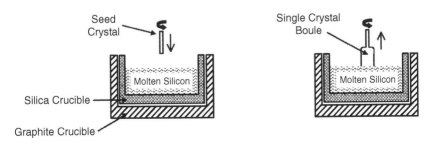

FIGURE 2.10 Schematic of Czochralski growth.

26 Micro Electro Mechanical System Design

has been carefully etched and oriented because it will serve as a template for crystal growth. The solidification or crystal growth is accomplished by the reduction in temperature as the seed crystal is gradually withdrawn from the molten bath. A simple heat transfer analysis of the liquid–solid interface can be performed, as depicted in Equation 2.1, which shows that the speed of withdrawal, which is proportional to dm/dt, is limited by the transfer of the latent heat of fusion across the interface:

$$KA\frac{dT}{dx} = L\frac{dm}{dt} \tag{2.1}$$

where
K = thermal conductivity
L = latent heat of fusion
T = temperature
A = area
m = mass
x = pull direction of boule
t = time

In reality, the pull rate is slower than the heat transfer limit and changes during the process. At the beginning of the process, the pull rate is rapid to form a *tang*, which is a narrow, highly perfect crystal that will trap crystal imperfections. The crucible and the seed crystal are then counter-rotated; the pull rate and temperature of the furnace are lowered to form a *boule* of the desired size. Boules of up to 300 mm in diameter can be produced.

Silicon in its pure or intrinsic state is a semiconductor with an electrical resistance between that of a conductor and an insulator. The resistance can be significantly varied by introducing a small amount of impurities into the silicon crystal lattice. These impurities or dopants are added to the molten bath to obtain wafers of a particular resistivitiy.

Silicon is in group IV of the periodic table and it has four valence electrons, which can form four covalent bonds with all four neighboring silicon atoms in single-crystal silicon. If silicon is doped with a small amount of a group V element, an excess of valance electrons will be present. Frequently used group V dopants are phosphorus (P), arsenic (As), or antimony (Sb). Silicon doped with these impurities is referred to as *n-type*, in which electrons are the majority carriers. If silicon is doped with a small amount of a group III element such as Boron (B), holes will be the majority carrier; this is referred to as *p-type*.

However, the dopant materials that are added to the charge of materials in the Czochralski growth process have different solubility in the liquid and solid phases. A segregation coefficient, K, is a metric defined as the ratio of the impurity concentration in the solid phase (C_s) and phase liquid (C_L) (see Equation 2.2). Table 2.3 lists the segregation coefficients of some commonly used impurities in

Fabrication Processes

TABLE 2.3
Segregation Coefficients of Impurities in Silicon

Impurities	P	As	Sb	O	B
K_{Si}	0.35	0.3	0.023	0.25	0.8

silicon. The segregation coefficients for impurities in silicon are less than one, which means that the dopants in the molten bath of the Czochralski growth process are increasing as the boule is drawn from the bath. As a result, the dopant concentration in the boule will also vary; however, refinements to the Czochralski process attempt to mitigate these effects.

$$k = \frac{C_S}{C_L} \tag{2.2}$$

The fused silica (SiO_2) crucible used in the Czochralski process releases a significant amount of oxygen into the molten silicon, which will be incorporated into the boule as shown by the segregation coefficient of oxygen in silicon. However, oxygen precipitates in silicon have several beneficial features:

- Oxygen helps localize crystal defects.
- Oxygen increases the mechanical strength of silicon.
- Oxygen traps mobile impurities.

2.2.1.2 Float Zone Process

The float zone technique is used when very high purity silicon is required. Figure 2.11 is a schematic of a float zone system, in which localized heating is done using a high-power RF coil. The RF heater is moved along the length of the silicon rod, where eddy current heating causes localized melting and crystallization of the silicon. A crucible is not required in this process and the crystal orientation is set by a seed crystal. The float zone method is used for producing high-purity, high-resistance silicon. It is difficult to introduce a uniform distribution of dopants with this process; it is generally limited to production of smaller diameter wafers and not generally used for GaAs.

2.2.1.3 Post-Crystal Growth Processing

Processing still remains to convert the boule of grown crystal into a polished wafer suitable for use in microelectronic or MEMS processing (Figure 2.12). The boule will have an undulating surface along its length due to the nature of the growth process. First, the boule will have crystallographic and resistivity inspections after which the seed crystals will be removed and the boule ground to the

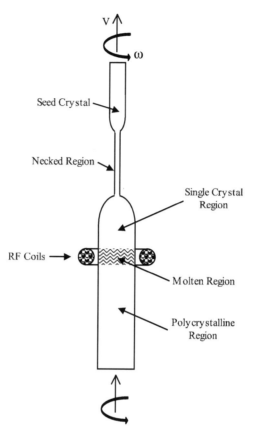

FIGURE 2.11 Schematic of a float zone system.

proper diameter. Silicon and gallium arsenide are brittle materials that can be sawed and ground using diamond-bonded wheels. Flats will be ground into the boules to identify crystallographic plane (Figure 2.13). For wafers greater than 150 mm, a notch will be ground into the edge. The boule will then be sawed into wafers that are typically 625 to 725 μm thick. The edges of the wafers are rounded by grinding to minimize chipping from subsequent mechanical handling. The wafers are then lapped and polished, followed by subsequent etching to remove any mechanical damage. Then, the wafers are laser marked for identification and quality-control purposes. Silicon wafers in use are typically 100 to 300 mm, with commercial IC manufacturing currently working toward the use of 300-mm wafers. GaAs wafers are typically 100 to 150 mm.

2.2.2 Silicon on Insulator (SOI) Substrate

Silicon on insulator (SOI) wafers have found increased application in recent years in the microelectronics industry. An SOI wafer consists of three layers: a base

Fabrication Processes

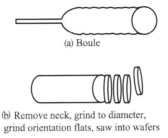

(a) Boule

(b) Remove neck, grind to diameter, grind orientation flats, saw into wafers

(c) Round edges, lap and polish

FIGURE 2.12 Post-crystal growth processing operations.

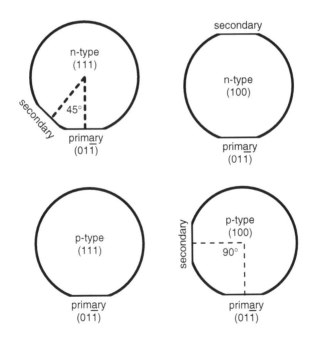

FIGURE 2.13 Standard flat orientations of silicon wafers.

single-crystal silicon layer or handling wafer; a buried silicon dioxide (BOX) layer; and the silicon on insulator layer, as illustrated in Figure 2.14. The thickness of the various layers can be specified when ordering SOI wafers.

Use of SOI wafers offers advantages for microelectronics and MEMS applications. In microelectronics, the active region (transistor junction) of the wafer consists of only the top couple of microns. The rest of the wafer thickness

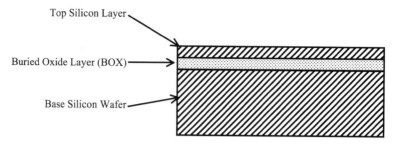

FIGURE 2.14 Silicon on insulator wafer layers.

(typically ~700 μm) is for mechanical rigidity during processing and handling. If the transistor could be fabricated on a very thin layer of single-crystal silicon with an insulator below, the capacitance of the transistor could be reduced, thus enabling higher speed switching cycles and lower power consumption. This approach also reduces the microelectronic sensitivities to radiation, which can cause data corruptions. This is a growing issue as operating voltages decrease.

If the SOI layer can be made thick (10 to 100 μm), MEMS devices that require very flat stiff surfaces can be enabled. Optical MEMS devices frequently require metalization or optical coatings to produce desired properties; however, these layers can induce stresses in the optical structure that frequently have flatness constraints. Use of a thick SOI layer for this application is very attractive [6]. Currently, two manufacturing processes are available for production of SOI wafers: SIMOX and Unibond®.

The SIMOX process, shown in Figure 2.15, produces SOI wafers by implantation of oxygen. High-energy oxygen atoms are implanted into a single-crystal silicon wafer. The depth of implantation of the oxygen atoms is controlled by their energy. The implantation of oxygen will damage the silicon crystalline structure. Then, the wafer is annealed, which will heal the damage induced by the oxygen implantation as well as oxidize the silicon to create the BOX layer of silicon dioxide.

An SOI wafer produced by the Unibond process involves the fusion bonding of two wafers (Figure 2.16). One silicon wafer has an implanted subsurface layer of hydrogen; the other has an outer layer of silicon dioxide. During the bonding process, the heat causes the implanted hydrogen layer to fracture, yielding a thin SOI layer.

2.3 PHYSICAL VAPOR DEPOSITION (PVD)

Physical deposition processes are a class of material deposition methods that do not require a chemical reaction for the deposition process to occur. Physical deposition methods have the capability to deposit thin films of conductors and insulators that are used in MEMS application for optical coatings or electrical conductors. The two physical deposition processes that will be discussed are evaporation and sputtering.

Fabrication Processes

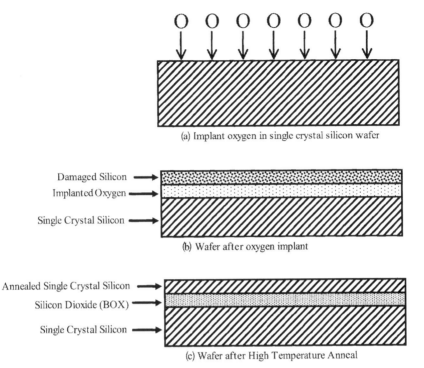

FIGURE 2.15 SIMOX process for SOI wafers.

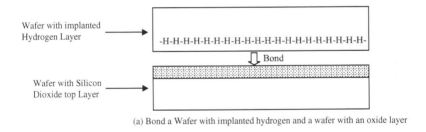

FIGURE 2.16 Unibond® process for SOI wafers.

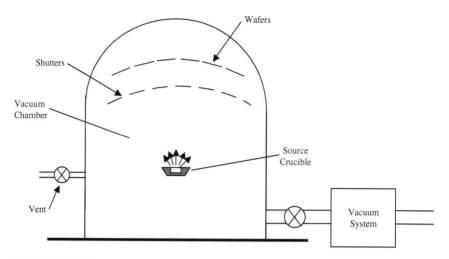

FIGURE 2.17 Evaporator schematic.

2.3.1 EVAPORATION

A schematic of an evaporation chamber is shown in Figure 2.17. The key features of an evaporator are:

- High-vacuum chamber with an associated pumping system
- Crucible containing the material to be deposited with an associated heating system
- Wafer support structure for holding the samples to be coated

The evaporator may also have a shutter system for control of the deposition time and multiple crucibles for depositing multiple layers or alloys.

The crucible is frequently made of boron nitride (BN) and contains the molten charge of material to be deposited. Several methods are available for heating the charge of material. The simplest is resistive heating; however, for extremely high-temperature evaporation, the resistive elements may also evaporate, leading to contamination. Inductive heating and electron beam heating are alternative methods for these applications.

Evaporation is a *"line of sight"* deposition phenomena from the molten material source to the wafer. Several wafers may be fixed around the crucible in various orientations to increase throughput or enhance deposition on particular features. Alloys or multilayer films can be deposited via evaporation using an evaporator equipped with multiple crucibles and a shutter system to control deposition times of the various materials.

At low pressures and elevated temperatures, materials exhibit a vapor pressure, P_v. The physical process for material loss from a molten sample due to the elevated vapor pressure is *evaporation*. The process for material loss from a solid due to an elevated vapor pressure is *sublimation*. Most practical processes

Fabrication Processes

TABLE 2.4
Melting Point and Temperatures Required to Achieve 10^{-3} torr Vapor Pressure for Selected Elements

Material	Al	Cr	Si	Au	Ti	Pt	Mo	Ta	W
Melting point (°C)	660	1900	1410	1063	1668	1774	2622	2996	3382
Temperature (°C) to produce a P_v = 10^{-3} torr	889	1090	1223	1316	1570	1904	2295	2820	3016

involve evaporation of material from molten samples. For materials of interest in MEMS fabrication, vapor pressures less than a millitorr (i.e., 1 torr = 1 mm Hg) are typical. Table 2.4 shows melting temperatures for various materials as well as the range of temperature necessary for these materials to exhibit a vapor pressure of 10^{-3} torr. These data show that the required temperature to achieve a vapor pressure of 10^{-3} torr ranges from 889°C for aluminum (Al) to 3016°C for tungsten (W). The higher temperature materials require specialized equipment for heating and minimization of contamination due to the elevated vapor pressure of other materials in the chamber at these temperatures.

The kinetic of theory gases (Equation 2.3) can relate the evaporator chamber pressure, P_v, and temperature, T, to the flux of atoms leaving the surface of the molten sample, J:

$$J = \sqrt{\frac{P_v^2}{2\pi kTM}} \tag{2.3}$$

where

P_v = vapor pressure
k = Boltzmann constant (1.38×10^{-23} J/°K)
T = temperature (°K)
M = atomic mass
J = atomic flux

The mass flux of deposition in an evaporation process can be calculated from the preceding equation and a geometric "view factor" from the molten sample to the deposition surface because evaporation is a line of sight deposition process. This information can be used to determine deposition times and material thickness.

The line of sight nature of the evaporation deposition process leads to the issue of *step coverage* of topographic features on a wafer. In any MEMS processing sequence, topography will be generated on the wafer due to the sequence of deposition, patterning, and etching that has preceded the evaporation process. This issue for MEMS is accentuated due to the thickness of the layers involved. Because evaporation is a line of sight phenomena, the rate of material deposition

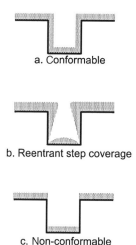

FIGURE 2.18 Step coverage of topographic features.

on the top and bottom of a topographic feature is greater than on the side walls of the feature (see Figure 2.18). This leads to thinner coverage of the side walls and possibly very thin coverage in the corners of the topographic features.

Methods such as rotating the wafer during deposition or heating the wafer to increase the surface mobility of the deposited atoms have been used to mitigate the step coverage issues encountered in evaporation. However, step coverage issues of a particular process can sometimes be used to advantage — for example, in the development of a "lift-off" process for patterning of deposited layers (see Section 2.6.2). Also, a *self-shadowing* design feature can be used in a MEMS device to allow a blanket evaporation of a conductive material such as gold and yet maintain electrical isolation of different portions of a design.

2.3.2 SPUTTERING

Sputtering is a process that has its roots as far back as 1852 [2]. The sputtering process utilizes a plasma formed by a large voltage in a low pressure gas (0.1 torr) across a closely spaced electrode pair. The target material (source material to be deposited) is on the cathode. The ions come from an inert gas within the chamber. Bombardment of the cathode by energetic ions gives rise to the sputtering process. When ions strike a material surface, several things can happen, depending on the energy of the ions:

- Bouncing off the surface
- Absorption by the surface to produce heat
- Penetration of the surface to deposit the energy within the material
- Ejection of surface atoms from the cathode (sputtered)

Fabrication Processes

Sputtered atoms have more energy than evaporated atoms, which increase the surface mobility of the sputtered atoms. Increased surface mobility produces better step coverage than is attainable with the evaporation process. Because ion collisions give rise to the sputtering of the target material, a gas with a high atomic weight is advantageous. Argon is a frequently used inert gas in a sputtering process.

Sputtered films can be deposited at ambient temperature. The sputter deposition does not depend on the substrate temperature; however, substrate may be heated to promote adhesion or prevent film cracking.

There are several variations to the sputtering process to achieve special effects.

- *Reactive sputtering*. A chemical combination between the sputtered material and the ambient gas reacts to form a compound (e.g., sputter silicon with a nitrogen ambient to form silicon nitride films).
- *Triode sputtering*. An additional filament in the chamber is used to increase the sputter rate by producing additional electrons.
- *Magnetron sputtering*. A magnetic field is used to increase density of electrons, which will increase the sputter rate.

2.4 CHEMICAL VAPOR DEPOSITION (CVD)

Chemical vapor deposition (CVD) processes involve a chemical reaction in the deposition of a material. The chemical reactions can occur in the gas phase or on the surface; however, the reaction that occurs on the surface is generally more useful. The reactions that occur in the gas phase tend to produce particles, which is not usually beneficial. The CVD chemical reactions can have the following variants:

- A compound decomposes at temperature. For example, silane gas (SiH_4) can decompose under heating at greater than 400°C to produce silicon and hydrogen.
- A combination of compounds react to produce a film. Silane gas (SiH_4) and oxygen (O_2) react to form silicon dioxide (SiO_2). Silane is very reactive and it can spontaneously ignite or explode.

A large number of CVD reactions can be useful for deposition (Table 2.5). CVD reactions can be used for deposition of silicon, silicon dioxide, and silicon nitride films, as well as tungsten, aluminum, titanium nitride, and copper films.

A CVD reactor is a complicated system in which pressure, energy, and flow of reactants must be carefully controlled to produce a good film (Figure 2.19). An evacuation system is also needed for the reactor to remove the reaction by-products and to remove contaminants. The reactor is carefully designed with the fluid and thermal transport issues in mind to produce a carefully controlled, uniform deposition process. The fluid dynamic issues in CVD reactor design include:

TABLE 2.5
CVD Reactions

SiO_2	$SiH_4 + O_2 \rightarrow SiO_2 + H_2O$
	$SiH_4 + N_2O \rightarrow SiO_2 + NH_3 + H_2O$
	$SiO(CH_3)_4 + O_2 \rightarrow SiO_2 + CH_3 + O_2$
Si_3N_4	$SiH_4 + NH_3 \rightarrow Si_3N_4 + H_2$
	$SiH_4Cl_2 + NH_3 \rightarrow Si_3N_4 + HCl$
$Si^{(poly)}$	$SiH_4 \rightarrow Si + H_2$
W	$WF_6 + SiH_4 \rightarrow W + SiF_4 + H_2 + F_2$
TiN	$TCl_4 + NH_3 \rightarrow TiN + Cl_2 + H_2$

- *Fluid boundary layer control.* This greatly influences the diffusion of reactants to the wafer surface.
- *Nonuniformity of gas flows.* Phenomena such as recirculation cells can have a great impact on uniformity of deposition across the wafer surface.

These considerations lead to design of wafer mounting and the flow paths within the reactor. The resultant CVD processes must be controlled kinetically or by mass transport. *Kinetically controlled* processes are limited by the reaction rate that can take place at the wafer surface. A *mass transport-controlled* process is limited by the flow of reactant gases in the CVD reactor.

Several versions of CVD processes are available:

- *Atmospheric pressure CVD (APCVD).* Due to its high deposition rates, APCVD is primarily used to deposit thick dielectrics such as silicon nitride. A drawback of APCVD is particulate contamination.
- *Low-pressure CVD (LPCVD).* LPCVD operates at 0.1 to 1 torr pressure and produces high-quality conformal films. Due to the low pressure, the diffusion effects in the process are minimized. LPCVD is used to deposit silicon dioxide, polycrystalline silicon, tungsten, and silicon nitride.
- *Plasma-enhanced CVD (PECVD).* The plasma decomposes the incoming reactant gases generating ions and radicals, which recombine to form a surface film. PECVD is a low-temperature process (i.e., <400°C), but the low temperature also reduces surface mobility of the reactants, which frequently leads to amorphous films. The PECVD process is frequently used for deposition of silicon nitride passivation layers.

The choice of process generally depends on the maximum processing temperature, film stress, and film quality (number of pinholes).

Fabrication Processes

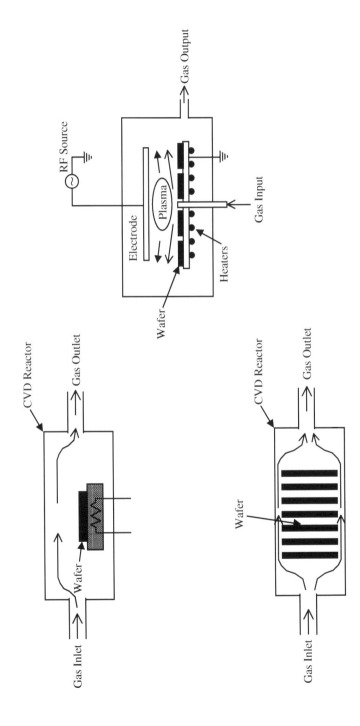

FIGURE 2.19 CVD reactor schematic.

38   Micro Electro Mechanical System Design

2.5 ETCHING PROCESSES

Etching processes are a fundamental process used in microelectronics and MEMS to impress a desired pattern into a material. There are several options, depending upon the various features that are important to be achieved. The considerations for selection of a particular etch process include:

- Etch rate. The speed of the etch needs to be fast enough to be viable for production but controllable.
- Uniformity. The etch is not location dependent.
- Selectivity. The etch rate ratio of the material desired to be etched vs. the material that is not desired to be etched is important in selecting the mask to be used in the etch process and the layer or material upon which the etch will stop.
- Directionality. The etch can be isotropic (omnidirectional), or anisotropic (directional).

Etching processes can be characterized by the method used to achieve the material etch. Etching can be done by physical or chemical attack of the material. Ion milling is an example of a physical attack of a material to achieve an etch. The material is physically bombarded by ions that will remove material similar to sputtering of a target. A chemical etch involves a chemical reaction that can be isotropic or anisotropic. Wet etching, plasma etching, and ion milling will be discussed further.

2.5.1 WET CHEMICAL ETCHING

Wet etching is purely a chemical process that is an isotropic etch in an amorphous material such as silicon dioxide and can be directional in crystalline materials such as silicon. Contaminants and particulate in this type of process are purely a function of the chemical purity and chemical system cleanliness. Figure 2.20 illustrates the phases involved in chemical etching:

- Movement of the chemical reactants to the surface
- The chemical reaction that performs the etching
- Removal of the chemical reaction by-products

The slowest step in this process is called the *rate limiting step*. Agitation of the wet chemical bath is frequently used to aid in the movement of reactants and by-products to and from the surface. Agitation will also aid the uniformity of the etch because the by-product may form solids or gases that must be moved. A modern wet chemical bench will usually have agitation, temperature, and time controls as well as filtration to remove particulate.

The etching of SiO_2 is a common wet etch process employed in surface micromachining (e.g., release etch, etch of isotropic features), which may be done

Fabrication Processes

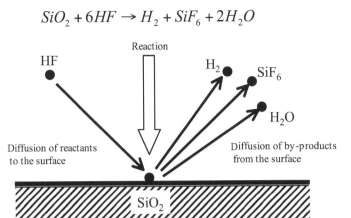

FIGURE 2.20 Wet chemical-etching process phases.

with a 6:1 by volume water to HF mixture. Equation 2.4 is the chemical reaction involved in this etch. Because hydrofluoric (HF) acid is consumed in this reaction, the concentration will decrease as the etch proceeds; this would require that more HF be added to maintain concentration. Alternatively, a buffering agent could be used to help maintain a concentration and pH in this reaction. Equation 2.5 shows the chemical reaction that would enable NH_4F to be used as a buffering agent in the HF etch.

$$SiO_2 + 6HF \rightarrow H_2 + SiF_6 + 2H_2O \quad (2.4)$$

$$NH_4F \leftrightarrow NH_3 + HF \quad (2.5)$$

Wet etching methods can be used on crystalline material to achieve anisotropic directional etches. For example, a common directional wet etchant for crystalline silicon is potassium hydroxide (KOH). KOH etches 100 times faster in the (100) direction than the (111) direction. Patterned silicon dioxide can be used as an etch mask for these types of etches. Very directional etches can be achieved with these techniques, as illustrated in Figure 2.21. Note the angular features (54.7°) that can be etched in silicon. Table 2.6 lists some of the common etchants for crystalline silicon and their selectivity.

Boron-doped silicon has a greatly reduced etch rate in KOH. Boron-doped diffused or implanted regions have been used to form features or as an etch stop (Figure 2.22).

2.5.2 PLASMA ETCHING

Plasma etching offers a number of advantages compared to wet etching, including:

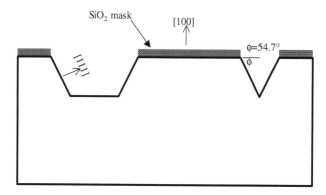

FIGURE 2.21 Directional etching of crystalline silicon.

TABLE 2.6
Common Crystalline Silicon Etchant Selectivity and Etch Rates

Etchant	Etch rate
18HF + 4HNO$_3$ + 3Si → 2H$_2$SiF$_6$ + 4NO + 8H$_2$O	Nonselective
Si + H$_2$O$_2$ + 2KOH → K$_2$SiO$_3$ + 2H$_2$	{100} 0.14 μm/min
	{111} 0.0035 μm/min
	SiO$_2$ 0.0014 μm/min
	SiN$_4$ not etched
Ethylene diamine pyrocatechol (EDP)	{100} 0.75 μm/min
	{111} 0.021 μm/min
	SiO$_2$ 0.0002 μm/min
	SiN$_4$ 0.0001 μm/min
Tetramethylammonium hydroxide (TMAH)	{100} 1.0 μm/min
	{111} 0.029 μm/min
	SiO$_2$ 0.0002 μm/min
	SiN$_4$ 0.0001 μm/min

- Easy to start and stop the etch process
- Repeatable etch process
- Anisotropic etches
- Few particulates

Plasma etching includes a large variety of etch processes and associated chemistries that involve varying amounts of physical and chemical attack. The plasma provides a flux of ions, radicals, electrons, and neutral particles to the surface to be etched. Ions produce physical and chemical attack of the surface,

Fabrication Processes

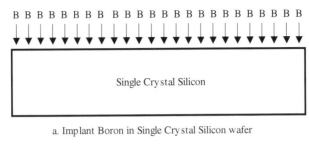

a. Implant Boron in Single Crystal Silicon wafer

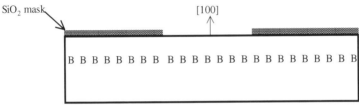

b. Deposit and Pattern Silicon Dioxide Etch Mask

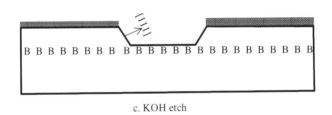

c. KOH etch

FIGURE 2.22 Boron-doped silicon used to form features or an etch stop.

and the radicals contribute to chemical attack. The sequence of events that occur in a plasma etch chamber (Figure 2.23) are listed below:

- Plasma breaks down the feed gases into chemically reactive species.
- Reactive species diffuse to the wafer surface and are adsorbed.
- Surface diffusion of reactive species takes place until they chemically react.
- Reaction product desorption occurs.
- Reaction products diffuse away from the surface.
- Reaction products are transported out of the chamber

The details and types of etch chemistries involved in plasma etching are varied and quite complex. This topic is too voluminous to be discussed in detail here, but a number of excellent references on this subject are available [3,4]. The proper choice of these chemistries produces various etch rates and selectivity of material etch rates, which is essential to the integration of processes to produce micro-

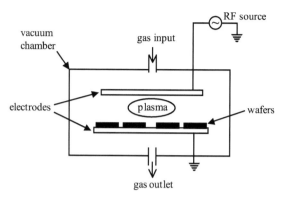

FIGURE 2.23 Schematic of a plasma etch chamber.

electronics or MEMS devices. Fluoride etch chemistries are one of the most widely studied for silicon etches. Equation 2.6 through Equation 2.8 illustrate some of the fluoride reactions involved in the etching of silicon, silicon dioxide, and silicon nitride, respectively. A number of feed gases can produce the free radicals involved in these reactions.

$$Si + 4F^* \rightarrow SiF_4 \tag{2.6}$$

$$3SiO_2 + 4CF_3^+ \rightarrow 2CO + 2CO_2 + 3SiF_4 \tag{2.7}$$

$$Si_3N_4 + 12F^* \rightarrow 3SiF_4 + 2N_2 \tag{2.8}$$

The anisotropy of the plasma etch can be increased by the formation of nonvolatile fluorocarbons that deposit on the sidewalls. This process is called *polymerization* and is controlled by the ratio of fluoride to carbon in the reactants. The side wall deposits produced by polymerization can only be removed by physical ion collisions. Etch products from the resist masking are also involved in the polymerization.

End point detection of an etch is important in controlling the etch depth or minimizing the damage to underlying films. This detection is accomplished by analysis of the etch effluents or spectral analysis of the plasma glow discharge.

Types of plasma etches include reactive ion etching (RIE) and high-density plasma etching (HDP). RIE etching utilizes a low-pressure plasma. Chlorine (Cl)-based plasmas are commonly used to etch silicon, GaAs, and Al. RIE etching may damage the material due to the impacts of the ions; this can be removed by annealing at high temperatures. HDP etches utilize magnetic and electric fields

Fabrication Processes

to increase dramatically the distance that free electrons can travel in the plasma. HDP etches have good selectivity of Si to SiO_2 and resist.

2.5.3 Ion Milling

Ion milling is a purely physical etching process; no chemical reactions are involved. Ion milling uses noble gases with significant mass such as argon in a process analogous to sputtering. This process is very isotropic because the ion impinges the surface nearly vertically. However, the process has an etch rate selectivity of the material to be etched to the mask material of nearly 1:1 because the process is purely physical. Ion milling is not widely used for production applications, and it is generally limited to the smaller wafer sizes (<200 mm). The etch rates can be increased by increasing the ion densities impacting the surface through the use of magnetic fields.

2.6 PATTERNING

The ability to pattern deposited layers is an essential capability required in microelectronics and MEMS processing. Three widely utilized methods of patterning will be discussed: lithography, the lift-off process, and the damascene process.

Lithography is the mainstream process utilized for patterning in MEMS processes. Lift-off and damascene are processes used for patterning materials in which a reliable etch process such as metallization layers or optical coating layers does not exist. These are frequently required in the postprocessing of MEMS devices.

The current research and development in patterning for very fine line widths (<0.35 μm) involve the development of sophisticated tools such as x-ray lithography [7] or direct-write E-beam lithography [8]. Microelectronics will need the capability to pattern features (line widths) of this size in the future in order to continue development of microelectronic devices of increasing speed and capability. However, mainstream MEMS technology does not currently require such fine features, so these methods will not be discussed here.

2.6.1 Lithography

Lithography is the most widely used method to pattern layers in microelectronic and MEMS processing. Figure 2.24 is a schematic of a basic lithography system. The basic components of a photolithographic system include:

- Illumination source
- Shutter
- Mask
- Wafer alignment/support system
- Photosensitive layer (photoresist or "resist") on a wafer

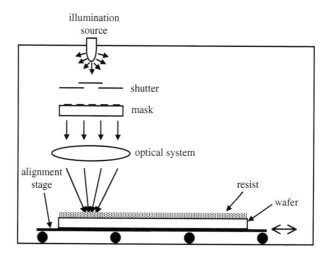

FIGURE 2.24 Lithography system schematic.

Lithography is the most critical process in microelectronics and MEMS processes, and the equipment is generally the most costly in a microelectronics or a MEMS fabrication facility. For example, a microelectronic process will require 20 or more lithography steps, and a surface micromachine process will require 10 or more lithography steps. A lithography process step will generally require application and prebaking of the photoresist to harden the resist; the exposure of the photoresist in the lithography tool; development of the photoresist; and a postbake of the photoresist to fully harden the resist to define the feature accurately. Thus, a lithography step requires several subprocesses that are repeatedly performed to fabricate a MEMS or microelectronic device. The other processing steps required in a MEMS or microelectonic fabrication may go through different tools for a specific deposition or a particular type of etch, but lithography is the common tool that will always be used. Therefore, lithography is the critical path in the fabrication facility and much attention is paid to the development of technology and enhancements that can speed this process step.

The performance metrics most important to lithographic processing are resolution, registration, and processing throughput. The resolution for optical lithography is very closely tied to the wavelength of the illumination source. The development of lithographic equipment has used ever decreasing wavelength illumination from the visible spectrum to ultraviolet (UV) and on to the research and development use of extreme ultraviolet (EUV).

The design of the optical system of the lithographic equipment is very complex, as can be illustrated by the discussion of a few key parameters of the optical system. The minimum line width, W_{min}, capability of the lithographic system can be expressed by Equation 2.9, which is very similar to the Raleigh criteria [9] for optical resolution:

Fabrication Processes

$$W_{\min} \approx K \frac{\lambda}{NA} \qquad (2.9)$$

where
$W_{\min}$ = minimum line width
K = a measure of the ability of the photoresist to distinguish changes in intensity
λ = illumination source wavelength
NA = numerical aperture

The numerical aperture defined by Equation 2.10 is a function of the refractive index, n, of the medium between the objective and wafer and the half angle of the image, α:

$$NA = n \sin \alpha \qquad (2.10)$$

Another optical parameter of interest in the design of the lithographic system is the depth of focus, σ (Equation 2.11). The depth of focus is an issue in MEMS fabrication due to the thickness of the films involved and the possible wafer warpage due to residual stress of the deposited films. Films involved in MEMS processes can be several microns thick. Patterning of the various layers will give rise to topographic features on the wafers. When photoresist is spread on a wafer containing these topographical features or the wafer is warped due to film residual stress, the lithographic process will attempt to expose the photoresist at various heights, thus making the depth of focus capability a critical issue.

$$\sigma = \frac{\lambda}{NA^2} \qquad (2.11)$$

As can be seen from this limited subset of the optical design parameters, the optical design is complex and the design parameters interrelated. For example, to make $W_{\min}$ smaller, utilizing a smaller wavelength source, λ, and a larger NA would be beneficial; however, the depth of focus, σ, will be reduced as a result.

Masks contain the patterns that need to be etched into the material to implement the MEMS design. The masks can be the same size (1:1) as the patterns to be transferred and etched into the MEMS material. Depending on the lithographic system, the masks may be larger than the patterns to be etched into the material. Masks are typically 1×, 5×, or 10× larger than the patterns to be imaged and etched. The mask is made of materials (e.g., fused silica) that are transparent at the illumination wavelength, with the patterns defined by an opaque material (e.g., chromium) at the illumination wavelength. The mask will need to be very flat and insensitive to changes in temperature (e.g., small coefficient of thermal

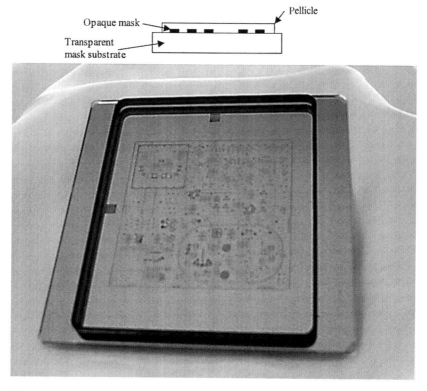

FIGURE 2.25 Photomask with pellicle.

expansion, α_T). Contamination control is also an issue in the lithography process. For example, masks may have a *pellicle* membrane (Figure 2.25) held above the patterned area to keep particles off the mask surface and out of the image plane of the mask to prevent degradation of the lithographic image.

Photoresist is a photosensitive organic compound applied to the wafer surface. The photoresist consists of three components:

- *Resin material* is organic material that forms the bulk material of the photoresist that will affect the durability during subsequent processing and resolution of the photoresist.
- *Photoactive compound* is the photosensitive material that determines the sensitivity of the photoresist (mJ/cm^2) to the illumination needed to produce a chemical change.
- *Solvent* is the component affecting the viscosity of the resist that affects the application of photoresist, which is generally done by spinning the wafer and using centrifugal force to spread the photoresist to a uniform thickness. The solvent in the resist is then removed during the baking steps to make the material structurally rigid.

Fabrication Processes

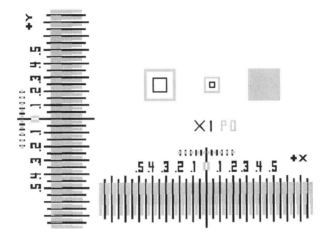

FIGURE 2.26 Example of a lithographic alignment target.

The lithographic process will transfer the image from the mask to photoresist on the wafer surface. Two types of photoresist can be used:

- *Negative resist.* The region of photoresist that has not been exposed to the illumination will dissolve during the development process and be removed.
- *Positive resist.* The region of photoresist that has been exposed to the illumination will dissolve during the development process and be removed. Positive resist has the best resolution and is more widely used.

After it is exposed and developed, the photoresist will be used as a physical mask during subsequent etching processes to transfer the pattern in photoresist on to the thin film of MEMS material beneath the photoresist. Photoresist is a key material in the lithographic processing sequence as well as the subsequent etch steps. It must have a diverse set of properties to enable the definition of the pattern and also maintain physical integrity during subsequent etching processes.

The aligner is the piece of mechanical equipment that supports the lithographic optical system and mask. It will align the masks relative to target patterns on the wafer; this will have the effect of aligning the masks and their subsequently etched patterns on the wafer with the mask and patterns utilized later in the fabrication sequence. Figure 2.26 shows a typical alignment target, which is typically specified by the lithographic system manufacturer. Two general types of aligners will be considered here:

- *Contact/proximity aligner.* The mask is held in contact or close proximity (a few microns) of the photoresist surface. These aligners utilize 1× masks and do not have pellicles due to the lack of available clearance between the mask and photoresist. The contact aligner actually

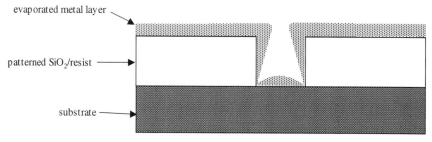

a. Evaporated metal layer on a patterned SiO$_2$ or resist layer

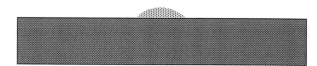

b. Strip the SiO$_2$ or resist layer leaving the metal on the substrate

FIGURE 2.27 Lift-off process schematic.

presses the mask under pressure against the photoresist, which will have the effect of degrading the mask under repeated use. This category of aligner is the least expensive, and has the lowest resolution capability. These aligners are generally used for research or limited production applications.

- *Projection aligner.* The mask and wafer are separated as dictated by the design of the optical system. This class of aligner can have high resolution that is only limited by optical system performance. These systems can be very expensive and are utilized in high-volume manufacturing.

2.6.2 Lift-Off Process

Lift-off is a patterning process frequently used in MEMS for patterning materials that do not possess a reliable process to etch them (e.g., noble metals). The lift-off process is accomplished via the use of an intermediate layer and deposition process, which has poor step coverage. Figure 2.27 is a schematic of a lift-off process that will deposit and pattern a material on a substrate or underlying layer. This process involves the following steps:

- Deposit and pattern a thick intermediate layer of a material that is easy to remove (e.g., SiO$_2$ or photoresist) and that will have a slightly reentrant profile.
- Deposit a layer of the material to be patterned utilizing a process that has poor step coverage (e.g., evaporation). The material thickness should be a fraction of the intermediate layer thickness.

Fabrication Processes

- Removal of the intermediate layer will cause the metal layer to fracture due to the stress concentration in the region of poor step coverage.

Alternatively, the lift-off process can involve a process that will explicitly form an undercut metal layer and not rely on the metal layer to fracture at the step. Figure 2.28 is a schematic of a process that will involve the explicit development of an undercut region:

- Deposit thick intermediate layer of a material (e.g., SiO_2).
- Deposit and pattern a layer of photoresist.
- Undercut the photoresist with a process such as wet chemical etching.
- Deposit a layer of the material to be patterned utilizing a process that has poor step coverage (e.g., evaporation). The material thickness should be a fraction of the photoresist and oxide layers.
- Remove the SiO_2 and photoresist, which will leave only the patterned metal layer.

(a) Substrate with an oxide layer and patterned resist

(b) Wet etch oxide layer to undercut the resist layer

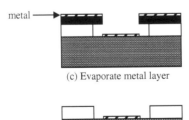

(c) Evaporate metal layer

(d) Strip resist

(e) Remove oxide

FIGURE 2.28 Lift-off process schematic with undercut metal layer.

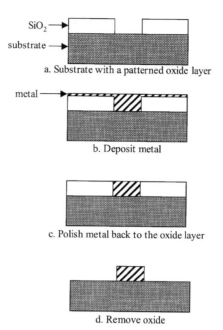

FIGURE 2.29 Damascene process schematic.

2.6.3 DAMASCENE PROCESS

The damascene process is an ancient process first developed in the Middle East [10] and utilized to inlay elaborate patterns on metal swords with various soft metals that could be easily polished. The concept of the damascene process (Figure 2.29) starts with forming a mold upon a substrate, depositing a metal which fills the mold and covers the surface. The surface is then polished so that only the material within the mold remains. Then the mold material can optionally be removed. The damascene process like the lift-off process is used to pattern materials that do not possess a reliable method of etching. For MEMS and microelectronics, these processes are used for patterning metal layers (e.g., copper, gold, etc.). The method frequently utilized for polishing the metal layer back to the mold is chemical mechanical polishing, which is discussed later in this chapter.

2.7 WAFER BONDING

Wafer bonding processes are used in packaging and to build up more complex structures. For example, bulk micromachined devices can be assembled into more complex structures by bonding multiple wafers together. Also, microfluidic channels can be formed by DRIE etching the channel in one wafer and bonding a wafer to seal the channel. The two categories of wafer bonding processes that will be discussed are silicon fusion bonding and anodic (electrostatic) bonding.

Fabrication Processes 51

The choice of bonding process will be influenced by the thermal budget of the devices and materials involved.

2.7.1 Silicon Fusion Bonding

At room temperature, two highly polished flat silicon wafers brought into contact will bond. The mechanism is believed to be hydrogen bonds between the surfaces. The bond can be converted into a stronger Si–O–Si bond at an elevated temperature (>800°C). The main concern with this bonding technique is voids in the bonded wafer due to surface defects, residues, and particulate on the surface. The processing sequence for silicon fusion bonding will generally include the following steps:

- Polishing of the wafer surfaces to be bonded
- Wafer surface wet cleaning processes to make the bond surface hydrophilic, which will facilitate the bonding mechanism
- Wafer surface inspection
- Precise alignment of bond surfaces in a clean environment
- Annealing of the bonded wafers

Other silicon-based materials such as polycrystalline silicon (polysilicon), silicon dioxide, and silicon nitride can be similarly bonded. However, yield of the bonded wafers due to voids will increase due to the different mechanical properties of the wafers.

2.7.2 Anodic Bonding

Anodic bonding is an electrostatic bonding technique for glass to silicon wafers or silicon wafers with a thin silicon dioxide layer between the wafers. Anodic bonding utilizes a heated chuck with an electrode capable of applying a DC voltage of up to 200 V. Pressure may be optionally applied to facilitate the bonding process. Figure 2.30 is a schematic of an anodic bonding process.

2.8 ANNEALING

Annealing is a process of elevating materials to a high temperature to achieve one of the following effects:

- *Reduction of residual stresses in a material.* The deposition process temperature will greatly influence the residual stress in a deposited film. A way to reduce the film residual stress is to anneal the film at a high temperature for a length of time (e.g., polysilicon residual stress can be reduced by annealing at >1100°C for several hours in an inert atmosphere, N_2, which minimizes oxidation).
- *Activation of dopants.* When dopants are implanted or diffused into a material to create active electronic devices or piezoresistors, the dopants

(a) Silicon wafers with hydrated surfaces.

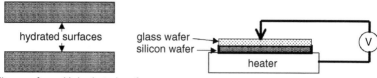

(b) Silicon wafers in contact and anneal.

FIGURE 2.30 Anodic bonding schematic.

are activated by an anneal process. This process facilitates the dopants' movement to an appropriate location within the material lattice structure. An anneal temperature of >600°C is typical for this application
- *Healing of material damage.* Implantation of dopants into a material will cause physical damage to the material through dislocations of the material atoms. The material damage is healed by annealing the material at an elevated temperature (e.g., >600°C).
- *Annealing bonded wafers to modify the wafer bond mechanism.* Silicon to silicon wafer bonds are initially hydrogen bonds that are transformed to a Si–O–Si bond through an annealing step that improves the wafer bond strength. An anneal temperature of >800°C is typical of this application.

Several anneal process issues are of concern, depending upon the application:

- Thermal stresses, particularly when dissimilar materials are involved
- Doping profile perturbation at elevated temperatures
- Melting of metal layers, which generally occurs at greater than 450°C

Annealing can be accomplished via an isothermal process in which the material is exposed to the elevated temperature over a long period of time. The isothermal annealing processes generally require a thermal controlled volume with a controlled atmosphere.

Rapid thermal annealing processes that anneal only a small distance into the material exist. This process will minimize doping profile perturbations and melting of metal layers within a device. This localized rapidly varying thermal cycling can be achieved within a fast ramp furnace, which can achieve thermal transients on the order of 75°C/min. A more rapid thermal transient can be achieved by moving the wafer within a furnace that has thermal gradients designed into the equipment. Figure 2.31 is a schematic of a rapid thermal anneal furnace in which

Fabrication Processes 53

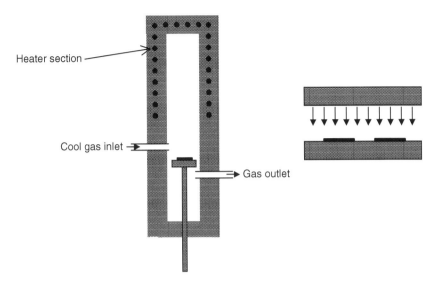

FIGURE 2.31 Rapid thermal anneal furnace utilizing a wafer elevator to control the temperature transient.

the wafer temperature is controlled via an elevator within the furnace. Thermal transients on the order of 100°C/sec can be achieved with this type of system.

2.9 CHEMICAL MECHANICAL POLISHING (CMP)

Chemical mechanical polishing (CMP) is a process originally developed in the microelectronics industry to planarize the layers of interconnect. CMP has also been adapted to MEMS processing in surface micromachining (see Figure 2.32). CMP has become essential to MEMS fabrication to remove the topography formed due to the repeated deposition, patterning, and etching of multiple thick film of material typical in MEMS surface micromachine processes. The topography generated in the MEMS processes make patterning with lithographic methods difficult due to depth of focus issues and poor photoresist coverage over the uneven surface. The topography also makes the design difficult as well. Figure 2.32 shows a comparison of a MEMS layer in a surface micromachine process with and without CMP in the process.

In Figure 2.33, CMP utilizes a slurry of silica particles and a dilute etching agent applied between the wafer and a pad that is mechanically moved (i.e., a combination of rotation and linear motion). The removal rate, typically on the order of 1000s of angstroms per minute, is a function of the pad pressure, relative velocity, and slurry chemistry. For surface micromachining processes, CMP can be applied to the sacrificial layers so that the structural layer deposited on top will not have topography. CMP can also be used in the damascene process discussed previously to polish the metal layer back to the mold.

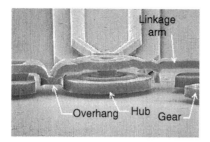
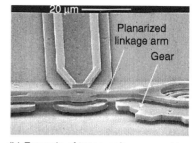

(a) Example of a conformable Layer

(b) Example of topography removed by Chemical Mechanical Polishing

FIGURE 2.32 SUMMiT™ (Sandia ultraplanar multilevel MEMS technology) polysilicon layer with and without CMP processing.

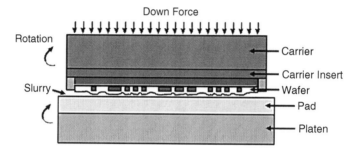

FIGURE 2.33 Schematic of a chemical mechanical polishing system.

2.10 MATERIAL DOPING

Dopants are impurities intentionally added to a semiconductor that can achieve several objectives. Active electronic devices are produced by adding dopants to a semiconductor material to create the necessary n- and p-junctions. Piezoresistors can be used in sensing applications and are created by doping a portion of a semiconductor material located in a region of high strain. Also, because doped silicon has different etch characteristics than single-crystal silicon, a doped layer can be used as an etch stop; this will facilitate bulk micromachining manufacture.

Figure 2.34 is a two-dimensional representation of a silicon lattice structure with a doping material included in the lattice. Silicon is in group IV of the periodic table and has four electrons in its outer shell; it shares these with four adjacent atoms to form a three-dimensional lattice structure. Covalent bonding is the sharing of electrons in the outer shell. For pure single-crystal silicon, all of the electrons in the outer shell of the silicon atoms are shared, thus producing a silicon crystal with no free electrons. If the silicon lattice has a small amount of a dopant, a free electron or a hole can be produced. Figure 2.34a shows silicon doped with phosphorus, a group V element with five electrons in the

Fabrication Processes

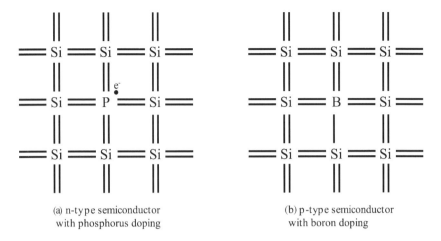

FIGURE 2.34 A two-dimensional schematic of a silicon lattice doped with phosphorous and boron to produce an n-type and a p-type semiconductor, respectively.

TABLE 2.7
Group III, IV, and V Elements Commonly Used in Semiconductors

Group III (three valance electrons-acceptors)	Group IV (four valance electrons)	Group III (five valance electrons-donors)
Boron (B)	Silicon (Si)	Phosphorus (P)
Aluminum (Al)		Arsenic (As)
Gallium (Ga)		Antimony (Sb)
Indium (In)		

outer shell. This will produce a free electron in the lattice and produce what is referred to as an n-type semiconductor. Similarly, a silicon lattice doped with a group III element, such as boron (with three electrons in the outer shell), will produce an unfulfilled covalent bond in the structure or a *hole*. This is a p-type semiconductor. Table 2.7 is a short list of the group III, IV, and V elements commonly used.

Two types of processes are utilized to introduce the dopants into a semiconductor material: diffusion and implantation. A diffusion will place the source material for the dopant atoms on the surface of the semiconductor material and allow the dopants to diffuse into the semiconductor under elevated temperature. Implantation will use a particle accelerator to implant the dopant atoms physically into the semiconductor. As a result of this physical implantation of the dopant atoms, the semiconductor lattice structure is damaged. The implant damage is then healed and the dopants activated via an anneal process step.

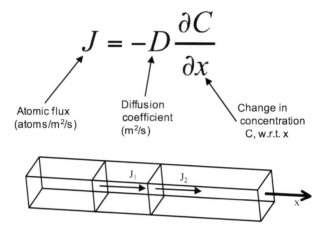

FIGURE 2.35 One dimensional schematic of Frick's laws.

2.10.1 Diffusion

The diffusion process is performed by placing the source material for the dopant atoms on the surface of a wafer and allowing the dopant atoms to diffuse into the semiconductor wafer under elevated temperature. Frick's law provides a mathematical model of the diffusion process (Equation 2.12), which is illustrated in a one-dimensional example in Figure 2.35.

$$\frac{\partial C}{\partial t} = D\nabla^2 C \qquad (2.12)$$

where
C = concentration of dopant atoms as a function of space and time
D = diffusion coefficient, which depends upon the dopant and condition of the diffusion process, such as temperature

The diffusion coefficient is the proportionality constant between the dopant atom flux across a surface, J, and the rate of change in dopant atom concentration across that interface, as shown in Equation 2.13. The negative sign indicates that the material will diffuse to decrease the concentration:

$$J = -D\frac{\partial C}{\partial x} \qquad (2.13)$$

Frick's law is a boundary value problem that can be solved analytically for the one-dimensional case to yield solutions that will correspond to physical implementations of the diffusion process used to dope semiconductors. Two instructive solutions of Frick's law will be examined.

Fabrication Processes

Case 1. In this case, the dopants constantly arrive at the surface of the wafer to maintain a constant dopant concentration, C_s, on the wafer surface (Equation 2.14). The dopant concentration deep within the wafer is 0, as shown in Equation 2.15:

$$C(x = 0, t) = C \tag{2.14}$$

$$C(x = \infty, t) = 0 \tag{2.15}$$

Equation 2.14 and Equation 2.15 are the problem boundary conditions. The initial condition at $t = 0$ is that the concentration is zero throughout the domain:

$$C(x, t = 0) = 0 \tag{2.16}$$

Frick's equation with these boundary and initial conditions can be solved to yield, Equation 2.17. $\sqrt{Dt}$ is called the *diffusion length* — a quantity that appears in the analytical solutions of Frick's equations. The solution of Frick's equation also involves the complementary error function, which is tabulated in a number of references [11]. Figure 2.36a plots this solution of Frick's equation, which shows the dopant concentration increasing within the wafer and penetrating deeper as time increases. The concentration remains constant at the wafer surface.

$$C(x, t) = C_s erfc \left[\frac{x}{2\sqrt{Dt}} \right] \tag{2.17}$$

Case 2. In this situation, a fixed amount of dopant, Q, has been initially introduced to the wafer, as shown in Equation 2.18. The initial distribution is approximated as a delta function at $x = 0$, and the initial concentration elsewhere in the wafer is 0 (Equation 2.19). There are two boundary conditions for this case. For the surface, $x = 0$, no additional dopants are added (Equation 2.20). This boundary condition is obtained from Frick's first law, Equation 2.13, which relates the dopant flux at a surface, J, and the rate of change of the concentration at the surface. The second boundary condition assumes that the wafer is thick; therefore, at $x = \infty$, the concentration is zero (Equation 2.21).

$$\int_0^\infty C(x, t) dx = Q = \text{constant} \tag{2.18}$$

$$C(x, t = 0) = 0, \quad x \neq 0 \tag{2.19}$$

$$J(x = 0, t) = \frac{dC(x = 0, t)}{dx} = 0 \tag{2.20}$$

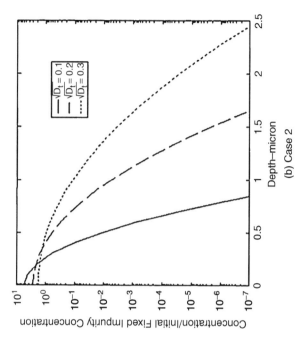

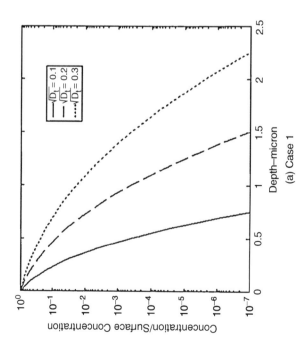

FIGURE 2.36 Solution of Frick's equation solutions for case 1 and case 2.

Fabrication Processes

$$C(x = \infty, t) = 0 \qquad (2.21)$$

The analytical solution for Frick's law with these initial and boundary conditions is a Gaussian function shown in Equation 2.22. The solution for this case is illustrated in Figure 2.36, which shows that the slope of the concentration is zero at $x = 0$ because there is no dopant flux at the $x = 0$ surface. Also, the concentration at the surface $x = 0$ decreases with time, which can be calculated from Equation 2.22:

$$C(x,t) = \frac{Q}{\sqrt{\pi D t}} e^{-\frac{x^2}{4Dt}} \qquad (2.22)$$

The implementation of the diffusion process can involve a gaseous, liquid, or solid source of dopants (Table 2.8). The wafer can be masked to allow the dopants to diffuse into selected locations (Figure 2.37). Silicon dioxide can be used for the masking material in the diffusion process because the diffusion coefficient for silicon dioxide is approximately 10^4 less than the diffusion coefficient of silicon at typical processing temperatures (Table 2.9). Two mechanisms of diffusion occur, as illustrated in Figure 2.38. Interstitial diffusion is the mechanism through which atoms (e.g., H_2, He, Na, O_2) that do not strongly interact with silicon diffuse through the silicon lattice. Lattice vacancy exchange is the mechanism used by the important group III and V dopants.

TABLE 2.8
Phosphorus, Boron, and Arsenic Chemical Sources for Diffusion

	Sources		
Dopant	Gaseous	Liquid	Solid
Arsenic	AsH_3, AsF_3	Arsenosilica	$AlAsO_4$
Phosphorus	PH_3, PF_3	$POCl_3$, phosphosilica	$NH_4H_2PO_4$, $(NH_4)_2H_2PO_4$
Boron	B_2H_6, BF_3, BCl_3	BBr_3, $(CH_3O)_3B$, borosilica	BN

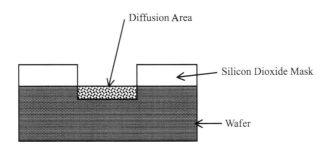

FIGURE 2.37 Diffusion with a silicon dioxide mask.

TABLE 2.9
Phosphorus and Boron Diffusion Coefficients in Silicon and Silicon Dioxide

Material	D_{Si} (cm²/s)	D_{SiO_2} (cm²/s)
B	2×10^{-15}	3×10^{-19}
P	1×10^{-14}	1×10^{-18}

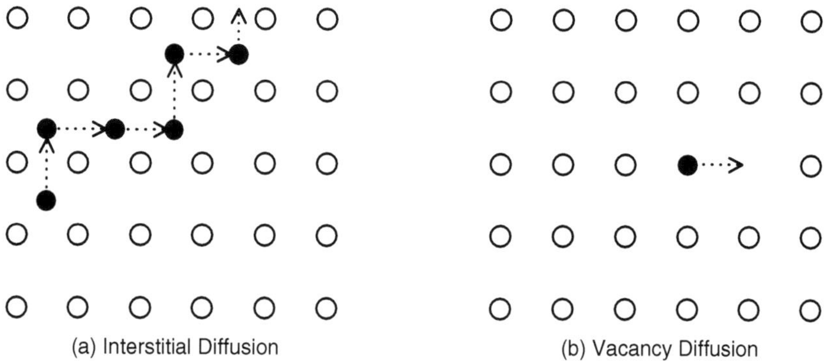

(a) Interstitial Diffusion   (b) Vacancy Diffusion

FIGURE 2.38 Diffusion mechanisms.

2.10.2 IMPLANT

Implantation is the most common method for doping a semiconductor used today. Implantation utilizes a particle accelerator to *implant* the ions of up to 200 keV physically into the wafer material. This physical implantation process will damage the lattice material, which is subsequently annealed to remove the damage and allow the implanted ion to fill vacancies in the silicon lattice (i.e., activate the dopants). Implantation is the most frequently used method of doping material because of several significant advantages:

- Precise dopant control
- Less lateral diffusion
- Variety of doping profiles attainable (Figure 2.39)

Figure 2.40 shows a schematic of an ion implantation system. The principal components are the ion source; magnetic analysis system; resolving aperture; accelerator; beam control system; mask; and Faraday cup. The ion source produces a flux of ions that is sent through the magnetic analysis system to select the desired ions, which produce a collumated beam after they pass through the resolving aperture. The ions are then accelerated to the desired energy level by the accelerator and the beam position controlled and scanned by the beam control system. The ion dose is measured by the Faraday cup where the wafer is posi-

Fabrication Processes

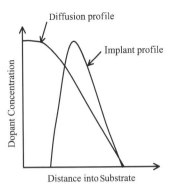

FIGURE 2.39 Comparison of doping profiles for diffusion and implantation.

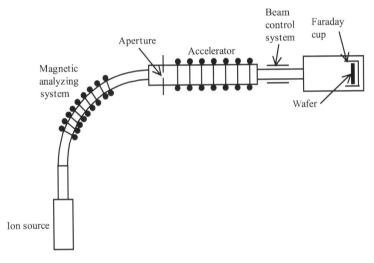

FIGURE 2.40 Schematic of an implantation system.

tioned. This type of implantation system can deliver a precise and pure dose of dopants to the wafer.

2.11 SUMMARY

This chapter provided an overview of the individual processes that can be used in MEMS fabrication technology. Most of the fabrication processes that have been presented are an outgrowth of the microelectronics industry. The fabrication processes can be combined to form a fabrication sequence or *technology*. *MEMS technologies* are discussed in the next chapter.

The fabrication process for starting material or substrates was presented in Section 2.2. MEMS fabrication frequently involves the deposition, patterning, and etching of a material upon a substrate (i.e., an additive process) or etching

62   Micro Electro Mechanical System Design

the substrate material (i.e., a subtractive process). The fabrication methods were presented for a single-crystal substrate (i.e., Czochralski growth, float zone) as well as for silicon on insulator (SOI) substrates (i.e., SIMOX, Unibond).

A fundamental process sequence will generally involve deposition, patterning, and etching. The deposition processes of physical vapor deposition (PVD) and chemical vapor deposition (CVD) and their variants were presented. The selection of deposition process and its parameters will control characteristics of the deposited film such as residual stress, and conformal or nonconformal layer. These characteristics can be very important to subsequent MEMS fabrication processes or the MEMS device.

Etching processes can be *isotropic* or *anisotropic* in nature. The selection of the specific etch process and material is important in determining these characteristics. The wet chemical etching, plasma etching, and ion milling processes were discussed. To a large degree, patterning is performed via lithographic methods. However, for materials that do not have a viable etch process, the lift-off or damascene process can be utilized.

A number of other processes used to achieve specific effects were presented. Annealing is utilized in a manner similar to macroscale manufacturing to relieve residual stresses. Chemical mechanical polishing (CMP) is used to remove topography in films or for patterning a film in a damascene process. The doping of a material for electrical purposes can be achieved by diffusion or implantation processes. Wafer bonding is a process that can be used to produce a specific structure for packaging or a MEMS device.

QUESTIONS

1. List the type of atomic bonds and describe them. Give an example of a material utilizing each type of bond.
2. What are Van der Waals forces and when are they important?
3. What are the three types of material structure?
4. Name and sketch the common variants of the cubic lattice structure.
5. Sketch a face-centered cubic lattice and the (1 0 0) plane. What is the plane equivalent to the (1 0 0) plane? Repeat for the (1 1 1) plane.
6. Give four examples of lattice defects. What is activation energy?
7. What is the maximum pull rate of a silicon seed crystal from the molten bath due to the heat transfer of the latent heat of fusion? (Hint: refer to Equation 2.1.) What are the parameters that could be used to control the boule diameter?
8. What are the majority carriers for single-crystal silicon doped with boron?
9. What are the dopants that could be used in an n-type silicon wafer?
10. What is meant by a conformable film? What process could deposit a conformable film?
11. What is meant by a nonconformable film? What process could be used to deposit a nonconformable film? Why would one want to do this?

Fabrication Processes 63

12. What is meant by step coverage? Give an example of why this is important.
13. What etchants can anisotropically etch single-crystal silicon? What crystal direction has the fastest etch rate? What crystal direction has the smallest etch rate?
14. What material could be used as a mask for a KOH etch of single-crystal silicon?
15. Describe the difference between contact and optical lithography. Which could produce a smaller dimension pattern in the photoresist?
16. What are the elements of a photoresist? Describe how photoresist is utilized in the patterning process.
17. Why would one use a lift-off or damascene process?
18. Describe the lift-off and damascene processes and their similarities.
19. Describe the annealing process for polysilicon. List two reasons for annealing a deposited film.
20. Why does the implant process provide a more accurate definition of the dopant profile and region?

REFERENCES

1. G.K. Teal, Single crystals of germanium and silicon — basic to the transistor and the integrated circuit, *IEEE Trans. Electron. Dev.* ED-23, 621, 1976.
2. W.R. Grove, On the electrochemical polarity of gases, *Philos. Trans. Faraday Soc.* 87, 1852.
3. R.A. Morgan, *Plasma Etching in Semiconductor Fabrication*, Elsevier, Amsterdam, 1985.
4. D.M. Manos and D.L. Flamm, *Plasma Etching, an Introduction*, Academic Press, Boston, 1989.
5. R.A. Morgan, VCSEL — a new twist in semiconductor-lasers, *Photonics Spectra*, 24(12), 89, 1990.
6. M.A. Michalicek, V.M. Bright, Flip-chip fabrication of advanced micromirror arrays, *Sensors Actuators A*, 95, 152–157, 2002.
7. S. Ohki, S. Ishihara, An overview of x-ray lithography, *Microelectron. Eng.*, 30(1–4), 171–178, January 1996.
8. R. DeJule, E-beam lithography, the debate continues, *Semiconductor Int.*, 19, 85, 1996.
9. M.V. Klein, *Optics*, John Wiley & Sons, New York, 1970.
10. J.D. Verhoeven, The mystery of Damascus blades, *Sci. Am.*, 74–79, January 2001.
11. M. Abramowitz, I.A. Stegun, *Handbook of Mathematical Functions with Formulas, Graphs, and Mathematical Table*, National Bureau of Standards Applied Mathematics Series 55, June 1964, 10th printing, December 1972.

3 MEMS Technologies

For all but the simplest of applications, the large assortment of individual fabrication processes discussed in Chapter 2 needs to be combined synergistically to form a *technology* that can produce useful MEMS devices. Three dominant MEMS fabrication technologies — philosophically different in their approach — are currently in use:

- LIGA (**Li**thographie, **G**alvanoformung, **A**bformung)
- Bulk micromachining
- Sacrificial surface micromachining

Figure 3.1 illustrates the basic concepts of each of the three fabrication approaches. Bulk micromachining and sacrificial surface micromachining are most frequently silicon based and are generally very synergistic to the microelectronics industry because they tend to use common tool sets.

Bulk micromachining (BMM) utilizes wet- or dry-etch processes to produce an isotropic or anisotropic etch profile in a material. Bulk micromachining can create large MEMS structures (tens of microns to a millimeter thick), which can be used for applications such as inertial sensing or fluid flow channels. Commercial applications of bulk micromachining have been available since the 1970s. These applications include pressure sensors, inertial sensors, and ink-jet nozzles.

Sacrificial surface micromachining (SSM) is a direct outgrowth of the fabrication processes of the microelectronics industry. SSM technology also has a path toward the integration of electronics with the MEMS structures, which will facilitate control or sensing functions. SMM technology has had several commercial successes in the last decade (Figure 3.2), including optical mirror arrays and inertial sensors. Both of these applications include integrated microelectronics for sensing and control functions. This technology is generally limited to individual film thicknesses of 2 to 6 μm with an overall device thickness of <15 μm. The resulting devices are assembled as they are fabricated, thus relieving a difficult task; this gives SSM technology a large advantage for applications involving large arrays of devices.

LIGA technology was first demonstrated in the1980s. This technology can fabricate devices with small critical dimensions and high aspect ratios (i.e., thickness/width) with electroplated metallic materials. The metallic LIGA parts can be used directly or as a die for an injection molding process to produce plastic parts. This gives this technology the advantage in applications requiring a broad

65

66 Micro Electro Mechanical System Design

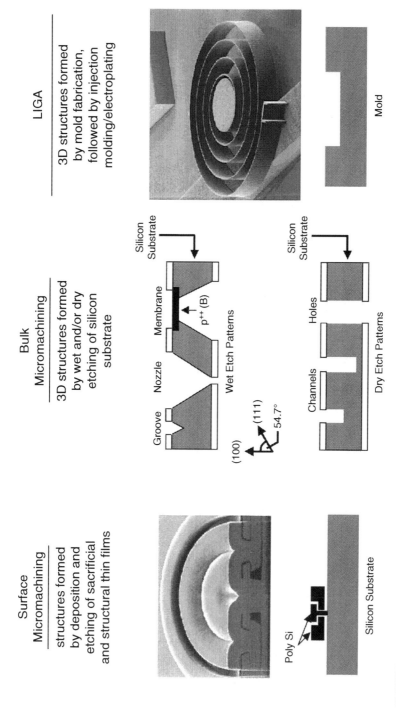

FIGURE 3.1 Types of MEMS fabrication technologies. (Courtesy of Sandia National Laboratories.)

MEMS Technologies

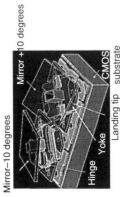
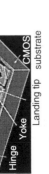

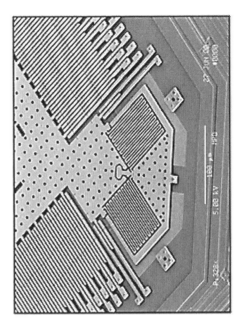

FIGURE 3.2 Surface micromachining commercial applications.

set of materials. However, assembly of large numbers or arrays of devices that require assembly is an issue that has spurred research in microassembly methods (see Section 9.1.1.4).

The evaluation and selection of a fabrication process appropriate for an application requires the assessment of a number of factors:

- Feature size: the smallest dimension that can be fabricated with the technology
- Device thickness: influences the mass and stiffness of the MEMS device; this is related to the aspect ratio capability of the fabrication technology
- Lateral dimension: how large a device can be made by the fabrication technology
- Precision: the technology precision (i.e., dimensional accuracy/nominal device dimension) that can be achieved
- Materials: the materials that can be utilized in the fabrication technology
- Assembly requirements: assembly required to produce a functioning MEMS device (i.e., piece part vs. assembled system)
- Scalability: whether the fabrication technology can produce large quantities of devices if required
- Ability to be integrated with microelectronics: the ability of the technology to be integrated directly with microelectronic circuitry on the same die

A comparison of these capabilities for the MEMS fabrication technologies as well as conventional machining for reference purposes is presented in Table 3.1. The impacts of the fabrication process capabilities on the capabilities of a MEMS device are summarized in Table 3.2. MEMS device capabilities include actuation method, mass, capacitance, out-of-plane stiffness, etc. These are the issues through which the MEMS design engineer must sort to select a fabrication process suitable for the device of interest.

3.1 BULK MICROMACHINING

Bulk micromachining technologies can use a combination of wet and dry etching methods to achieve isotropic and anisotropic etches of features in materials. In order to manufacture items of practical interest, a number of different aspects of the etch processes need to be considered. Petersen [1] and Kovacs et al. [2] present detailed information on etch rates and selectivity of crystallographic orientation and materials. Pister [3] offers a very instructive fold-up model of silicon and its etching rates for various orientations:

- Masking
- Etch selectivity due to crystallographic orientation or materials
- Etch stop and/or end-point detection

TABLE 3.1
Comparison of the Capabilities of MEMS Fabrication Technologies and Conventional Machining

Capability	LIGA	Bulk micromachining	Surface micromachining	Conventional machining
Feature size	~3–5 μm	~3–5 μm	1 μm	~10–25 μm
Device thickness	>1 mm	>1 mm	13 μm	Very large
Lateral dimension	>2 mm	>2 mm	2 mm	>10 m
Relative tolerance (see Figure 4.15)	~10^{-2}	~10^{-2}	~10^{-1}	>10^{-3}
Materials	Electroplated metals or injection molded plastics	Very limited material suite	Very limited material suite	Extremely large material suite
Assembly requirements	Assembly required	Assembly required	Assembled as fabricated	Assembly required
Scalability	Limited	Limited	Yes	Yes
Microelectronic integratability	No	Yes for SOI bulk processes	Yes	No
Device geometry	Two-dimensional high-aspect ratio	Two-dimensional high-aspect ratio	Multilayer; two-dimensional	Very flexible three-dimensional
Processing	Parallel processing at the wafer level	Parallel processing at the wafer level	Parallel processing at the wafer level	Serial processing

TABLE 3.2
Comparison of MEMS Device Capabilities within the Three Types of MEMS Fabrication Technologies

Device capability	Bulk micromachining	Surface micromachining	LIGA
Type of actuation	Electrostatic	Electrostatic	Electromagnetic
Mass	Large	Small	Large
Capacitance	>1 pF	<1 pF	>1 pF
Out-of-plane stiffness	Large	Small	Large
Range of motion	Restricted to the plane of fabrication	Three-dimensional motion capability	Restricted to the plane of fabrication
Large arrays of devices	No (electrical interconnect limited)	Yes (10^6 devices demonstrated)	No (electrical interconnect and assembly limited)
Integral on-chip microelectronics	Yes (on SOI wafers)	Yes	No

3.1.1 WET ETCHING

Wet etching is purely a chemical process that is an isotropic etch in an amorphous material such as silicon dioxide; it can be directional in crystalline materials such as silicon. Contaminants and particulate in this type of process are purely a function of the chemical purity and chemical system cleanliness. Agitation of the wet chemical bath is frequently used to aid in the movement of reactants and by-products to and from the surface. Agitation will also aid etch uniformity because the by-products may form solids or gases that must be removed. A modern wet chemical bench will usually have agitation, temperature, and time controls as well as filtration to remove particulate. For example, etching of SiO_2 utilizing a 6:1 by volume water to HF mixture is a common wet-etch process employed in a surface micromachining release etch or the production of isotropic features.

Wet-etching methods can also be used on crystalline materials such as silicon to achieve anisotropic directional etches. For example, a common directional wet etchant for crystalline silicon is potassium hydroxide (KOH). KOH etches 100 times faster in the (100) plane than the (111) plane. Patterned silicon dioxide or silicon nitride, which etches very slowly in KOH, can be used as an etch mask for these types of etches. Very directional etches can be achieved with these techniques, as illustrated in Figure 3.3. Note the angular features (54.7°) that can be etched in silicon due to the etch selectivity of the (100) vs. the (111) crystal plane. Table 2.6, Petersen [1], and Kovacs et al. [2] list some of the common etchants for crystalline silicon and their selectivity.

If there are no etch stops in a wet etching process, the two options available to the process engineer to achieve a specific etch depth are a timed etch or etching completely through the material. A timed etch is difficult to control accurately

MEMS Technologies

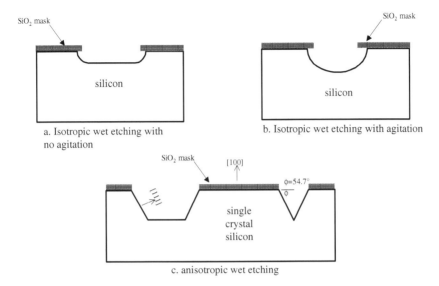

FIGURE 3.3 Wet etching of crystalline silicon.

due to the many other variables in a wet-etch process, such as temperature, chemical agitation, purity, and concentration. If this is not satisfactory, etch stops can be used in wet etching to define a boundary on which the etch can stop. Several etch stops methods can be utilized in wet etching:

- p+ (boron diffusion or implant) etch stop
- Material-selective etch stop
- Electrochemical etch stop

For example, the etch rate of boron-doped silicon (p-silicon) by KOH or EDP can be up to 100 times less than the etch rate in undoped silicon [1–3]. Therefore, boron-doped regions produced by diffusion or implantation have been used to form features or as an etch stop (Figure 3.4).

A material-selective etch stop can be produced by a thin layer of a material such as silicon nitride, which has a greatly reduced etch rate in etchants such as KOH, EDP, and TMAH (Table 2.6). For example, a thin layer silicon nitride can be deposited on a silicon device to form a membrane on which the etch will stop.

An electrochemical etch stop can also be used (Figure 3.5). Silicon readily forms a silicon oxide layer that will impede etching of the bulk material (Table 2.6). The formation of the oxide layer is a reduction oxidation reaction, which can be impeded by a reversed biased p–n junction that prevents the current flow necessary for the reduction oxidation reaction to occur. The p–n junction can be formed on a p-type silicon wafer with an n-type region diffused or implanted with an n-type dopant (e.g., phosphorus or arsenic) to a prescribed depth. With the p–n junction reverse biased, the p-type silicon will be etched

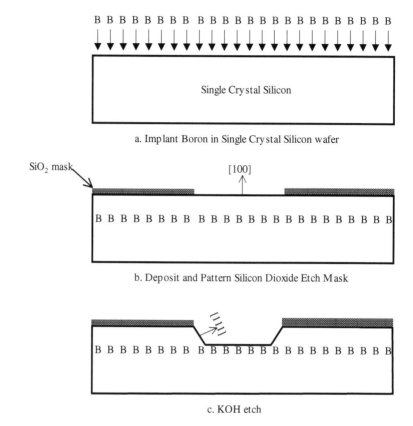

FIGURE 3.4 Boron-doped silicon used to form features or an etch stop.

because a protective oxide layer cannot be formed, and the etch will stop on the n-type material.

3.1.2 Plasma Etching

Plasma etching offers a number of advantages compared to wet etching, including:

- Easy to start and stop the etch process
- Repeatable etch process
- Anisotropic etches
- Few particulates

Plasma etching includes a large variety of etch processes and associated chemistries [5] that involve varying amounts of physical and chemical attach. The plasma provides a flux of ions, radicals, electrons, and neutral particles to the surface to be etched. Ions produce physical and chemical attack of the surface, and the radicals contribute to chemical attack. The details and types of etch

MEMS Technologies

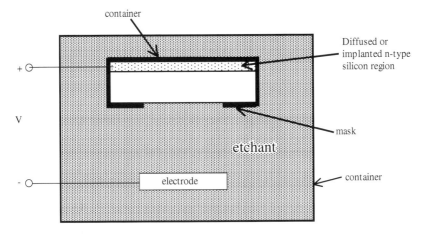

a) Electrochemical etch schematic

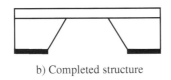

b) Completed structure

FIGURE 3.5 Electrochemical etch stop process schematic.

chemistries involved in plasma etching are varied and complex and beyond the scope of this book.

The anisotropy of the plasma etch can be increased by the formation of nonvolatile fluorocarbons that deposit on the sidewalls (Figure 3.6). This process is called *polymerization* and is controlled by the ratio of fluoride to carbon in the reactants. The side wall deposits produced by polymerization can only be removed by physical ion collisions. Etch products from the resist masking are also involved in the polymerization.

End-point detection of the etch is important in controlling the etch depth or minimizing the damage to underlying films. This detection is accomplished by analysis of the etch effluents or spectral analysis of the plasma glow discharge to detect.

Types of plasma etches include reactive ion etching (RIE); high-density plasma etching (HDP); and deep reactive ion etching (DRIE). RIE etching utilizes a low-pressure plasma. Clorine (Cl)-based plasmas are commonly used to etch silicon, GaAs, and Al. RIE etching may damage the material due to the impacts of the ions. However, this damage can be removed by annealing at high temperatures. HDP etches utilize magnetic and electric fields to increase dramatically the distance that free electrons can travel in the plasma. HDP etches have good selectivity of Si to SiO_2 and resist. The DRIE etch cycles between the etch

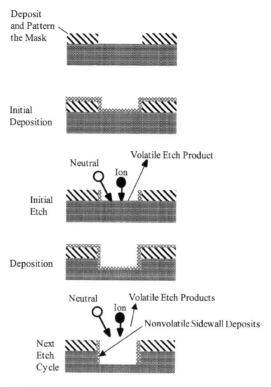

FIGURE 3.6 Schematic of sidewall polymerization to enhance anisotropic etching.

chemistry and deposition of the sidewall polymer; this enables the high aspect ratio and vertical side walls attainable with this process [6]. Figure 3.7 shows two sample applications of bulk micromachining utilizing deep reactive ion etching to produce deep channels and an electrostatic resonator.

3.1.3 Examples of Bulk Micromachining Processes

This section will present two examples of bulk micromachining processes used to make various devices for teaching and research purposes. The *SCREAM* and *PennSOIL* technologies utilize the fabrication processes discussed in Chapter 2 and Chapter 3 to produce two unique methods of bulk micromachine fabrication. The SCREAM process is developed around etching of single-crystal silicon to produce complex high aspect ratio devices. The PennSOIL process starts with an SOI wafer and uses anisotropic plasma etching and wet etching of single-crystal silicon to produce the device of interest. SCREAM and PennSOIL are very capable bulk micromachining processes with advantages for MEMS devices, depending on the device requirements. From a device design perspective, bulk micromachining provides the capabilities of large capacitance, mass, and out-of-plane stiffness, as listed in Table 3.2.

MEMS Technologies

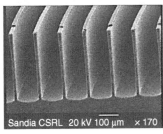
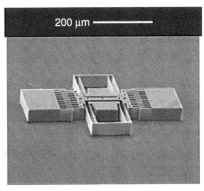

(a) Channels (b) Resonator

FIGURE 3.7 Bulk micromachined channels and resonator. (Courtesy of Sandia National Laboratories.)

3.1.3.1 SCREAM

The SCREAM (single-crystal reactive etching and metallization) process [7,8] is a bulk micromachining process that uses anisotropic plasma etching of single-crystal silicon to fabricate suspended single-crystal silicon (SCS) structures. High-capacitance actuators and sensors, such as accelerometers and vibratory gyroscopes, can be fabricated in this process. The fabricated structures may flex in the plane of fabrication. The SCREAM process yields millimeter-scale SCS structures greater than 100 μm deep and 1.5 μm minimum feature sizes (beam widths and separations). This results in a process capable of producing devices with an aspect ratio > 66.6. Devices have been fabricated with suspension space greater than 5 mm.

SCREAM process outline (Figure 3.8):

1. Start with a clean silicon wafer. (100) and (111) wafers with highly doped n-type (arsenic) or moderately doped p-type boron wafers have been used.
2. Deposit mask oxide. PECVD deposition of 1 to 2 μm oxide is used because of high deposition rate and low temperature (~240°C).
3. Pattern and etch mask oxide. Etch is accomplished by an RIE process.
4. Strip resist. This is an O_2 plasma strip.
5. Deep silicon etch I. The mask oxide is used to transfer the pattern into the substrate. Depending on the structure height to be obtained, 4 to 20 μm may be accomplished. An anistropic BCl_3/Cl_2 RIE etch is utilized. Process details are given in Shaw et al. [7].
6. Sidewall oxide deposition. Deposit ~0.3-μm conformal oxide layer with PECVD process. This oxide protects the sidewall during release.
7. Remove floor oxide. An RIE (CF_4/O_2) etch [7] is used to remove 0.3 μm of oxide from mesa top and trench bottom. This etch will leave the sidewall oxide largely undisturbed.

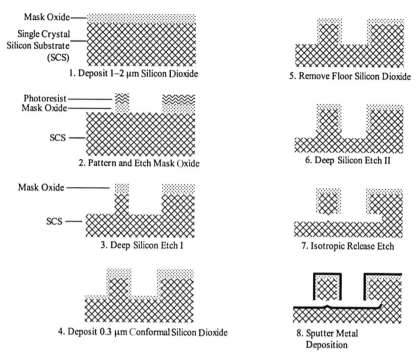

FIGURE 3.8 SCREAM (single-crystal reactive etching and metallization) process flow.

8. Deep silicon etch II. Use RIE to etch silicon floor down another 3 to 5 μm below the sidewall oxide. This exposed silicon on the sidewalls below the sidewall oxide will be removed via a subsequent release etch.
9. Isotropic release etch. The release is an isotropic SF_6 RIE etch [7] that removes the silicon at the bottom of the trench to produce a suspended structure. This etch is highly selective to oxide (i.e., several microns of silicon are etched with only a nominal erosion of the oxide coating).
10. Metal sputter deposition. Sputter deposition of a 0.1- to 0.3-μm aluminum layer is made. This produces a uniform coating.

NOTE: A thin silicon dioxide or silicon nitride passivation layer (50 nm) may be deposited to prevent electrical shorting of the electrode.

3.1.3.2 PennSOIL

PennSOIL (University of **Penn**sylvania **s**ilicon-**o**n-**i**nsulator **l**ayer) [9,10] is a silicon bulk micromachining process developed to pursue research on electrothermal-compliant (ETC) microdevices; this is an embedded actuation technique. ETC devices are compliant mechanisms that elastically deform due to constrained thermal expansion under joule heating. The shapes of ETC devices are designed so that the joule heating induced by the application of voltage between two points

MEMS Technologies

creates a nonuniform temperature distribution that causes the desired deformation pattern due to the material thermal expansion.

The qualities required of a fabrication process to pursue ETC research are:

- The ability to produce any two-dimensional shape
- Adequate out-of-plane stiffness
- Ability to etch through large depths with good dimensional control
- The ability to change the resistivity of the structure selectively by masked doping
- Released structures that can be mechanically anchored in desired locations
- Electrical insulating layer beneath the mechanical anchors of the device

The PennSOIL process utilizes silicon-on-insulator (SOI) wafers in which the handling wafer is KOH etched from the bottom and the epitaxial single-crystal silicon layer is plasma etched to define the shape of the ETC device. The buried oxide layer is etched with HF, which releases the device. The epitaxial layer can be selectively doped in specific locations to modify the resistivity. The PennSOIL process is described next and illustrated in Figure 3.9. Figure 3.10 contains some examples of ETC devices fabricated in the PennSOIL process.

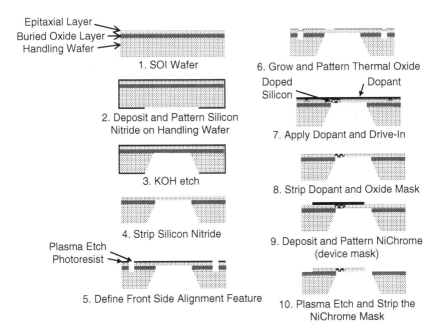

FIGURE 3.9 PennSOIL (University of **Penn**sylvania silicon-on-insulator layer) process flow.

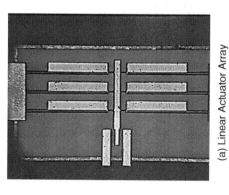

FIGURE 3.10 Devices fabricated with the PennSOIL process. (Courtesy of Dr. G.K. Ananthasuresh, University of Pennsylvania.)

(a) Linear Actuator Array (b) Three Degree of Freedom Platform (c) A Single Thermal Actuator

MEMS Technologies

PennSOIL process outline:

1. The SOI wafers that have been used in the PennSOIL process are 525 to 550 μm thick. The buried silicon dioxide layers (BOX) used in various runs have been 0.4, 2, and 3 μm, and the epitaxial layer thicknesses have been 10, 12, and 15 μm.
2. A thin layer of silicon nitride is deposited and patterned on the handling wafer side of the SOI wafer to define the membrane opening.
3. Conduct a KOH etch to form the membrane opening. This opening provides thermal isolation for the devices defined on the epitaxial layer.
4. Strip the nitride layer with an HF etch; this removes the exposed silicon dioxide as well.
5. Define a front (epitaxial) side alignment feature in the epitaxial layer. Apply and pattern photoresist. Perform a shallow plasma etch on the epitaxial layer to form the alignment features.
6. Grow and pattern silicon dioxide on the epitaxial layer to form a doping mask.
7. Apply dopant and drive in dopant with high temperature.
8. Strip dopant and oxide.
9. Deposit and pattern NiChrome to form the device mask on the epitaxial layer.
10. Conduct a plasma etch of the epitaxial layer to form the ETC device.
11. Strip the NiChrome mask.

3.2 LIGA

The LIGA (**Li**thographie, **G**alvanoformung, **A**bformung) process [11] is capable of making complex structures of electroplateable metals with very high aspect ratios with thicknesses up to millimeters. This process utilizes x-ray lithography, thick resist layers, and electroplated metals to form complex structures. Because x-ray synchonotron radiation is used as the exposure source for LIGA, the mask substrate is made of materials transparent to x-rays (e.g., silicon nitride or polysilicon). An appropriate mask opaque layer is a high atomic weight material such as gold, which will block x-rays.

The LIGA fabrication sequence shown schematically in Figure 3.11 starts with the deposition of a sacrificial material used for separating the LIGA part from the substrate after fabrication. The sacrificial material should have good adhesion to the substrate, yet be readily removed when desired. An example of a sacrificial material for this process is polyimide. A thin seed layer of material is then deposited; this will enable the electroplating of the LIGA base material. A frequently used seed material would be a sputter-deposited alloy of titanium and nickel. Then, a thick layer of the resist material, polymethylmethacrylate (PMMA), is applied. A synchrotron provides a source of high-energy collimated x-ray radiation, which is needed to expose the thick layer of resist material. The exposure system of the mask and x-ray synchrotron radiation can produce vertical

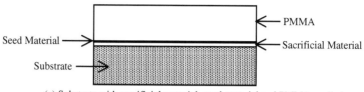

(a) Substrate with sacrificial material, seed material and PMMA applied

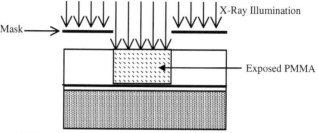

(b) Exposing PMMA with x-ray synchrotron radiation

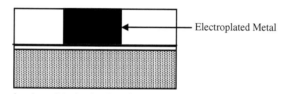

(c) Electroplated metal in the developed PMMA mold

FIGURE 3.11 LIGA fabrication sequence.

sidewalls in the developed PMMA layer. The next step is the electroplating of the base material (e.g., nickel) and polishing the top layer of the deposited base material. Then the PMMA and sacrificial material are removed to produce a complete LIGA part.

LIGA has the advantage of producing metal parts that enable magnetic actuation. However, the assembly of LIGA devices for large-scale manufacturing is a challenging issue (Section 9.1.1). Figure 3.12 shows an assembled LIGA mechanism. Alternatively, LIGA can fabricate an injection mold made of metal, which is then used to form the desired part typically made of plastic (see the next section).

3.2.1 A LIGA Electromagnetic Microdrive

New applications in medicine, telecommunications, and automation require powerful microdrive systems. Speeds up to 100,000 rpm and torques in the micronewton-meter range with a diameter of a few millimeters are typical requirements. Microdrive applications include a microdrive-equipped catheter that will enhance

MEMS Technologies

FIGURE 3.12 Assembled LIGA fabricated mechanism. (Courtesy of Sandia National Laboratories.)

the capability of minimally invasive surgery, automated assembly of miniaturized components, and use in small appliances such as a camcorder.

The Faulhaber Group [12] and the Institute for Microtechnology, Mainz, Germany [13], have jointly developed an electromagnetic motor with an outer diameter of only 1.9 mm [14–16]. Figure 3.13 shows the 1.9-mm motor and an exploded view of its components. For flexibility in application, these micromotors must be combined with microgear heads of the same outer diameter. The development of this system illustrates the development of a mesoscale device (>2 mm) that contains components fabricated with microscale fabrication technology (LIGA).

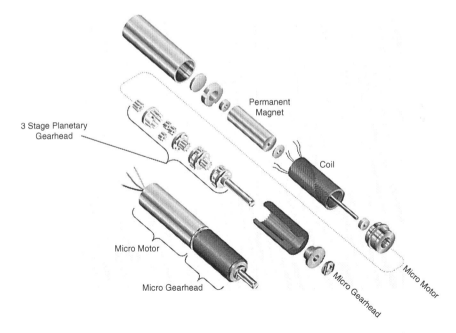

FIGURE 3.13 An exploded view of the 1.9-mm electromagnetic micromotor. (Courtesy of Dr. Fritz Faulhaber, GmbH & Co. KG.)

A synchronous motor design was utilized for the design to avoid the need for a mechanical commutator; this precludes a long operational lifetime. The synchronous motor consists of a permanent magnet (neodymium–iron–boron) coated with a very thin gold layer for corrosion protection mounted on a 240-μm diameter shaft with a microcoil mounted in the motor casing.

The microcoil is produced by winding enameled copper wires that have two different coatings. After the wires are wound, they are heated; this melds the outer coating to connect the separate wires mechanically. The winding process is optimized for an outer diameter of 1.6 mm, which allows the microcoil to fit within the motor casing.

A sleeve bearing was selected for the micromotor. Miniature ball bearings and jewel bearing used in the watch industry were considered. However, a high rotor speeds up to 100,000 rpm; the losses in a sleeve bearing are lower. Due to the manufacturing tolerances, the relative play is high and hydrodynamic gliding starts at speeds between 10,000 and 20,000 rpm.

Gear ratios from 50 to 1000 are required for the microgear head to convert the power of the micromotor to lower speeds and higher torques. For this application, a planetary gear system with involute tooth profile was found to be the most suitable. The advantages of a planetary gear system in this application include:

- High gear ratios attainable in one stage
- High-power density allowed by splitting the torque to the three planetary wheels
- Planetary gears supported by the sun gear, which eliminates the need for planetary gear bearings
- Multistage gear system realizable in a compact form

All the components of the microgear head (Figure 3.14), except the output shaft (steel) and output sleeve bearings (brass), are produced by microinjection molding in LIGA [17]-made molds. The microgears are made of the polymer POM (polyacetal polyoxymethylene) with a tooth-face width of 300 μm. The tip diameter of the planetary wheels is 560 μm and the axles fixed in the carrier have a 180 μm diameter. The frame is divided into two parts with rigid connections between the planetary wheels. The sun gear of the following stage or the output shaft is fixed to the upper part of the frame. The output shaft diameter is 500 μm.

The development of the microdrive system was accomplished with manual assembly techniques for the microgear head. The assembly must take place in a class 100 to 1000 clean-room environment. The assembly was accomplished with tweezers, vacuum pipettes, and specially designed tools and fixtures. The assembly was visually guided with a stereo microscope with variable magnification. Mass production is possible only with automated assembly whose development was guided by the experiences of manual assembly during the development phase.

The microdrive system, which consists of the micromotor and microgear head, has been developed and is commercially available. The micromotor is 5.5 mm long × 1.9 mm diameter and can produce 7.5 μN-m torque. The maximum

MEMS Technologies

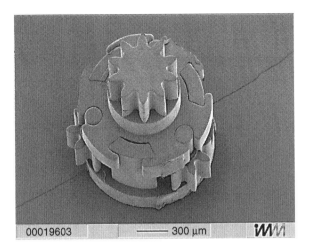

FIGURE 3.14 One stage planetary gear head composed of POM (polyacetal, polyoxymethylene) microinjection molded gears and planet carrier. (Courtesy of Dr. Fritz Faulhaber, GmbH & Co. KG.)

output torque of the microdrive, which combines the micromotor and a 47:1 gear head, is 150 µN-m in continuous operation. Operation times of 1500 h with a motor speed of 12,000 rpm have been demonstrated.

3.3 SACRIFICIAL SURFACE MICROMACHINING

The basic concept of surface micromachining fabrication process has its roots as far back as the 1950s and 1960s with electrostatic shutter arrays [18] and a resonant gate transistor [19]. However, it was not until the 1980s that surface micromachining utilizing the microelectronics tool set received significant attention. Howe and Muller [20,21] provided a basic definition of polycrystalline silicon surface micromachining; Fan et al. [22] illustrated an array of mechanical elements such as fixed-axle pin joints, self-constraining pin joints, and sliding elements. Pister et al. [23] demonstrated the design for microfabricated hinges that enable the erection of optical elements.

Surface micromachining is a fabrication technology based upon the deposition, patterning, and etching of a stack of materials upon a substrate. The materials consist of alternating layers of a *structural material* and a *sacrificial material*. The sacrificial material is removed at the end of the fabrication process via a *release etch*, which yields an assembled mechanical structure or mechanism. Figure 3.15 illustrates the fabrication sequence for a cantilever beam fabrication in a surface micromachine process with two structural layers and one sacrificial layer.

Surface micromachining uses the planar fabrication methods common to the microelectronics industry. The tools for depositing alternating layers of structural and sacrificial materials and photolithographically patterning and etching the

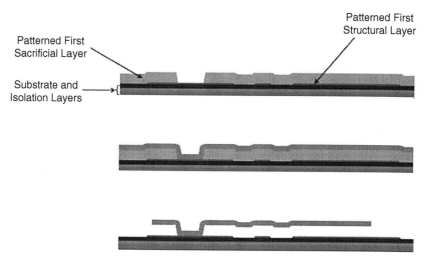

FIGURE 3.15 Surface micromachined cantilever beam with underlying electrodes showing the effect of topography induced by conformal layers.

layers have their roots in the microelectronics industry. The etches of the structural layers define the shape of the mechanical structure, and the etching of the sacrificial layers defines the anchors of the structure to the substrate and between structural layers. Deposition of a low-stress structural layer is a key goal in a surface micromachine process. From a device design standpoint, it is preferable to have a slightly tensile average residual stress with minimal or zero residual stress gradient. A small tensile residual stress alleviates the design consideration of device structure buckling. The stress in a thin film is a function of deposition conditions such as temperature. A postdeposition anneal is frequently used to reduce the layer stress levels. For polysilicon, the anneal step can require several hours at 1100°C in an inert atmosphere such as N_2.

Polycrystalline silicon (polysilicon) and silicon dioxide are common sets of structural and sacrificial materials, respectively, used in surface micromachining. The release etch for these materials is hydroflouric acid (HF), which readily etches silicon dioxide but minimally attacks the polysilicion layers. A number of different combinations of structural, sacrificial materials and release etches have been utilized in surface micromachine processes. Table 3.3 summarizes a sample of surface micromachine material systems utilized in commercial and foundry processes. The selection of the material system depends on several issues, such as the structural layer mechanical properties (e.g., residual stress, Young's modulus, hardness, etc.) or the thermal budget required in the surface micromachine processing, which may affect additional processing necessary to develop a product.

Even though surface micromachining leverages the fabrication processes and tool set of the microelectronics industry, several distinct differences and challenges exist. The surface micromachine MEMS devices are generally larger (>100 μm vs. <1 μm) and they are composed of much thicker films than microelectronic

MEMS Technologies

TABLE 3.3
Example of Surface Micromachining Technology Material Systems

Structural	Sacrificial	Release	Application
PolySi	SiO_2	HF	SUMMiT V
SiN	PolySi	XeF_2	GLV
Al	Resist	Plasma etch	TI DMD
SiC	PolySi	XeF_2	MUSIC

Notes: SUMMiT — Sandia ultraplanar, multilevel MEMS technology; GLV — grating light valve (silicon light machines); DMD — digital mirror device (Texas Instruments); MUSIC — multi user silicon carbide (FLX micro).

devices are (2 to 6 μm vs. <<1 μm). The repeated deposition and patterning of the thick films used in surface micromachining will produce topography of increasing complexity as more layers are added to the process. Figure 3.15 shows the topography induced on an upper structural by patterning of lower levels, which is caused by the conformal films deposited by processes such as chemical vapor deposition (CVD). Figure 3.16 shows a scanning electron microscope image of this effect in an inertial sensor made in a two-level surface micromachine process.

In addition to the topography induced in the higher structural levels by the patterning of lower structural and sacrificial layers, two other significant process difficulties can be encountered. The first difficulty results from the anisotropic plasma etch used for the definition of layer features to attain vertical sidewalls. The topography in the layer will inhibit the removal of material in the steps of the topographical features. This is illustrated in Figure 3.17, which shows an increased vertical layer height at the topographical steps that prevents removal of material at these discontinuities. This will give rise to the generation of small

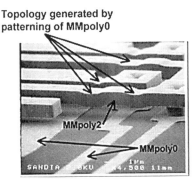

FIGURE 3.16 Scanning electron microscope image of topography in a two-level surface micromachine process. (Courtesy of Sandia National Laboratories.)

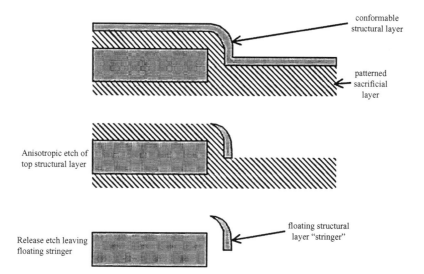

FIGURE 3.17 Illustration of stringer formation at a topographical discontinuity.

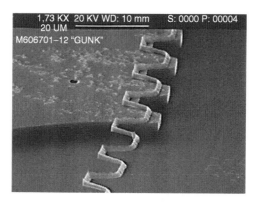

FIGURE 3.18 Scanning electron microscope image of a stringer that was formed and floated to another location on the die after release.

particles of material, *stringers*, that can be attached to the underlying layers or float away during the release etch (Figure 3.18). Stringers can cause a MEMS device to function improperly due to mechanical interference or electrical shorting. The second process difficulty is the challenge of photolithographic definition of layers with severe topography. The photoresist coating is difficult to apply and the *depth of focus* will lead to a decreased resolution of patterned features.

The application of chemical mechanical polishing (CMP) to a surface micromachine MEMS process directly addresses the issues of topography (Figure 3.19). CMP was originally utilized in the microelectronics industry for global planarization [24], which is needed as the levels of electrical interconnect increase. CMP planarization was first reported in MEMS by Nasby et al. [25,26]. Figure 3.19

MEMS Technologies

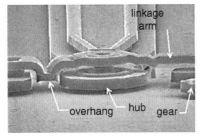

(a) Example of a conformable layer

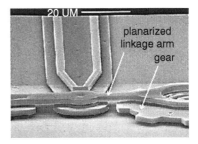

(b) Example of topography removed by chemical mechanical polishing

FIGURE 3.19 Example of a linkage fabricated in SUMMiT™ with and without CMP. (Courtesy of Sandia National Laboratories.)

shows a linkage that has been fabricated in a surface micromachined process, SUMMiT™, before and after CMP was included in the process. In addition to solving the fabrication issues of topography, the use of CMP also aids in realizing designs without the range of motion and interference constraints imposed by the topography. CMP will also aid in the development of MEMS optical devices [27] by enhancing the optical quality of the surface micromachined MEMS mirrors.

The release etch is the last step in the surface micromachine fabrication sequence. For a surface micromachine process such as SUMMiT, the release etch would involve a wet etch in an HF to remove the silicon dioxide sacrificial layers. The removal of the sacrificial layers will yield a mechanically free device capable of motion. For very long or wide structures, etch release holes are frequently incorporated into the structural layers to provide access for the HF to the sacrificial silicon dioxide in underlying layers. This will reduce the etch release process time. Because the MEMS device is immersed in a liquid during the release etch, an issue is the adhesion and stiction of the MEMS layers upon removal from the liquid release etchant [28]. Polysilicon surfaces are hydrophilic and the removal of liquids from the MEMS device can be problematic. Surface tension of the liquid between the MEMS layers will produce large forces pulling the layers together. Stiction of the MEMS layers after the release etch can be addressed in several ways:

- Make the MEMS device very stiff to resist the surface tension forces.
- Fabricate a bump (i.e., dimple) on the MEMS surfaces that will prevent the layers from coming into large area contact.
- Use a fusible link to hold the MEMS device in place during the release etch; this can be mechanically or electrically removed subsequently [29].
- Use a release process that avoids the liquid meniscus during drying, such as supercritical carbon dioxide drying [30] or freeze sublimation [31].
- Use a release process that will make the surface hydrophobic; this is accomplished via the use of self-assembled monolayer (SAM) coatings [32]. It has been reported that SAM coatings also have the effect of reducing friction and wear.

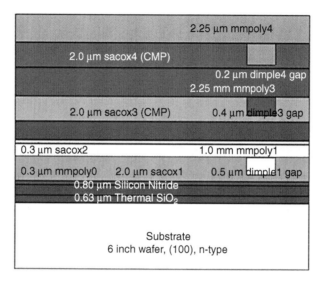

FIGURE 3.20 SUMMiT V™ layers and features. (Courtesy of Sandia National Laboratories.)

3.3.1 SUMMiT™

SUMMiT (Sandia ultraplanar, multilevel MEMS technology) is a state-of-the-art surface micromachine process developed by Sandia National Laboratories [33,34] that utilizes standard IC processes optimized for the thicker films (e.g., 2 to 6 μm) required in MEMS applications. Low-pressure chemical vapor deposition (LPCVD) is used to deposit the polycrystalline silicon (polysilicon) and silicon dioxide films. Optical photolithiography is utilized to transfer the designed patterns on the mask to the photosensitive material applied to the wafer (e.g., photoresist or resist). Reactive ion etches are used to etch the defined patterns into the thin films of the various layers. A wet chemical etch is also used to define a hub feature as well as the final release etch of the SUMMiT process. Figure 3.20 schematically shows the layers and features in the SUMMiT V™ surface micromachine process. This process uses 14 photolithography steps and masks to define the required features. Table 3.4 lists the layer and mask names and a summary of their use. To illustrate the SUMMiT V fabrication sequence, Figure 3.21 and Figure 3.22 show the masks and fabrication process at several intermediate stages.

The SUMMiT fabrication process begins with a bare n-type, (100) silicon wafer. A 0.63-μm layer of silicon dioxide (SiO_2) is thermally grown on the bare wafer. This layer of oxide acts as an electrical insulator between the single-crystal silicon substrate and the first polycrystalline silicon layer (MMPOLY0). A 0.8-μm thick layer of low-stress silicon nitride (SiN_x) is deposited on top of the oxide layer. The NITRIDE layer is also an electrical insulator, but acts as an etch stop as well, protecting the underlying oxide from wet etchants during processing. The NITRIDE layer can be patterned with the NITRIDE_CUT mask to establish

TABLE 3.4
SUMMiT V Layer Names, Mask Names, and Purposes

SUMMiT V layer	Mask	Purpose
NITRIDE	NITRIDE_CUT	Electrical contact to the substrate
MMPOLY0	MMPOLY0	Electrical Interconnect
SACOX1	DIMPLE1_CUT SACOX1_CUT	Dimple
		Anchors
MMPOLY1	MMPOLY1	Structural layer definition
	PIN_JOINT	Hub formation
SACOX2	SACOX2	Hub formation
MMPOLY2	MMPOLY2	Structural layer
SACOX3	DIMPLE3_CUT SACOX3_CUT	Anchors
SACOX3		Dimple
MMPOLY3	MMPOLY3	Structural layer definition
SACOX4	DIMPLE4_CUT SACOX4_CUT	Dimple sacrificial layer definition
MMPOLY4	MMPOLY4	Structural layer definition

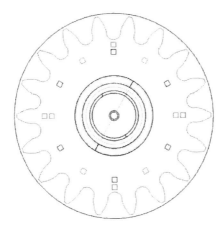

FIGURE 3.21 SUMMiT V layout for a multi-layered gear with substrate connection.

electrical contact with the substrate. A 0.3-μm thick layer of doped polycrystalline silicon known as MMPOLY0 is deposited on top of the nitride layer. MMPOLY0 is not a structural layer, but it is usually patterned and is used as a mechanical anchor, electrical ground, or electrical wiring layer.

Following MMPOLY0 deposition, the first sacrificial layer of 2 μm silicon dioxide (SACOX1) is deposited. SACOX1 is a *conformable* layer that will reflect any patterning of the underlying MMPOLY0 layer. Upon deposition of the SACOX1 layer, dimples are patterned and etched into the oxide. The dimples (primarily used for antistiction purposes) are formed in the MMPOLY1 (the next polysilicon deposition) by filling the holes etched into the SACOX1 layer. The dimple depth is controlled via timed 1.5-μm deep etch.

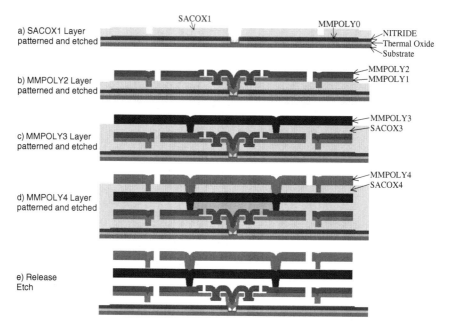

FIGURE 3.22 SUMMiT V fabrication sequence at several intermediate stages for a multi-layered gear with substrate connection.

Following the dimple etches, the SACOX1 layer is patterned again with the SACOX1_cut mask and etched to form anchor sites through the depth of SACOX1 to the MMPOLY0 layer. Figure 3.22a shows the deposited and patterned SACOX1 layer. MMPOLY1 deposited over the SacOx1 layer will be anchored or bonded to MMPOLY0 at the SACOX1 cuts as well as to the substrate at the nitrade cuts. This will act as electrical connections between MMPOLY0, MMPOLY1 and the substrate. With the anchor sites defined, a 1-µm thick layer of doped polysilicon (MMPOLY1) is deposited.

The MMPOLY1 layer can be patterned with the MMPOLY1 mask to define a pattern in the polysilicon layer, or the PIN_JOINT_CUT mask to define a feature used in the formation of a rotational hub/pin-joint structure. The hub/pin-joint is defined at the PIN_JOINT_CUT site by the combination of an anisotropic reactive ion etch and a wet etch to undercut the MMPOLY1 layer. This feature will be used to form a *captured rivet head* for the hub/pin-joint.

A 0.3-µm layer of silicon dioxide, SACOX2, is then deposited and patterned with the SACOX2 mask. The SACOX2 is deposited by an LPCVD process that is comfortable and will deposit on the inside wall of the hub structure. The thickness of SACOX2 defines the clearance of the hub structure. SACOX2 can also be used as a hard mask to define MMPOLY1 using the subsequent etch that also defines MMPOLY2.

Upon completion of the SACOX2 deposition, pattern, and etch, a 1.5-µm thick layer of doped polysilicon, MMPOLY2, is deposited. Any MMPOLY2 layer mate-

MEMS Technologies

rial that is deposited directly upon MMPOLY1 (i.e., not separated by SACOX2) will be bonded together. Following the MMPOLY2 deposition, an anisotropic reactive ion etch is performed to etch MMPOLY2 and composite layers of MMPOLY1 and MMPOLY2 (laminated together to form a single layer approximately 2.5 μm thick). The MMPOLY2 etch will stop on silicon dioxide; thus, MMPOLY1 will be protected by any SACOX2 on top of MMPOLY1 and the SACOX2 layer can be used as a hard mask to define a pattern in MMPOLY1. Figure 3.22b shows the SUMMiT V fabrication after the MMPOLY2 etch has been completed.

At this point in the SUMMiT V process, all the layers have been conformable (i.e., assume the shape of the underlying patterned layers). To enable the addition of subsequent structural and sacrificial levels without the fabrication and design constraints of the conformable layers, chemical mechanical polishing (CMP) is used to planarize the sacrificial oxide layers. With the MMPOLY2 etch complete, approximately 6 μm of TEOS (tetraethoxysilane) silicon dioxide (SACOX3) is deposited. CMP is used to planarize the oxide to a thickness of about 2 μm above the highest point of MMPOLY2. Following planarization, SacOx3 is patterned and etched to provide dimples and anchors to the MMPOLY2 layer using the DIMPLE3_CUT and SACOX3_CUT masks, respectively. The DIMPLE3_CUT etch is performed by etching all the way through the SACOX3 layer and stopping on MMPOLY2. Then 0.4 μm of silicon is deposited to backfill the dimple hole to provide the 0.4-μm stand-off distance. The processing of the DIMPLE3 feature will provide a repeatable stand-off distance.

A 2-μm thick layer of doped poly (MMPOLY3) is deposited on the CMP planarized SACOX3 layer. The MMPOLY 3 layer will be flat and not have any of the topography due to the patterning of the underlying layers (Figure 3.22c). This will ease design constraint on the higher levels and enhance the use of MMPOLY3 and MMPOLY4 layers as mirror surfaces in optical applications. The MMPOLY3 layer is patterned and etched using the MMPOLY3 mask.

The processing for the SACOX4 and MMPOLY4 layers proceeds using the SACOX4_CUT, DIMPLE4_CUT, and MMPOLY4 mask in an analogous fashion to the SACOX3 and MMPOLY3 layers, except that the DIMPLE4 stand-off distance is 0.2 μm (Figure 3.22d).

Release and drying of the SUMMiT V die are the final fabrication steps (Figure 3.22e). The device is released by etching all the exposed oxides away with a 100:1 HF:HCl wet etch. Following the wet release etch, a drying process can be employed using simple air evaporation, supercritical CO_2 drying [30], or CO_2 freeze sublimation [31]. The choice of the drying process will depend upon the design of the particular devices. Very stiff structures will be less sensitive to the surface tension forces, and they can be processed by simple air drying. Supercritical CO_2 drying processing for large devices would be a better option.

The SUMMiT V sacrificial surface micromachine fabrication process is capable of fabricating complex mechanisms and actuators (Figure 3.23 and Figure 3.24). The ability to make a low-clearance hub enables rotary mechanisms and gear reduction systems, as well as hinges that can be used to fabricate moveable mirrors (Figure 3.25). Figure 3.26 shows a vertically erected mirror

FIGURE 3.23 Rack and pinion drive, gear reduction system. (Courtesy of Sandia National Laboratories.)

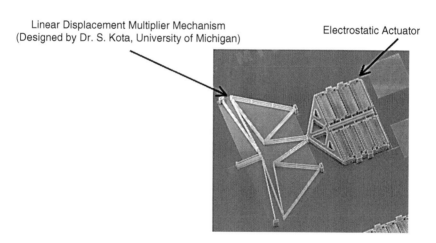

FIGURE 3.24 Electrostatic drive with linear displacement multiplying mechanism. (Courtesy of Sandia National Laboratories.)

MEMS Technologies

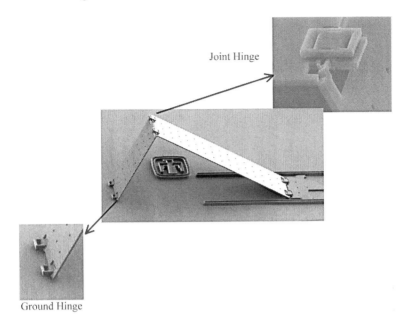

FIGURE 3.25 Pop-up mirror with ground and joint hinges. (Courtesy of Sandia National Laboratories.)

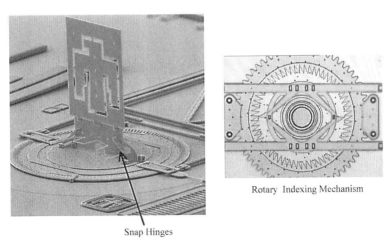

FIGURE 3.26 Rotary indexing device and vertically erected mirror with snap hinges. (Courtesy of Sandia National Laboratories.)

94 Micro Electro Mechanical System Design

held in place by elastic snap hinges. The vertical mirror is mounted upon a rotationally indexing table driven by an electrostatic comb drive actuator. SUM-MiT V has also been used to fabricate arrayed devices; this is possible because surface micromachined devices are assembled when they are fabricated.

3.4 INTEGRATION OF ELECTRONICS AND MEMS TECHNOLOGY (IMEMS)

The integration of electronics for control and sense circuitry and MEMS technology becomes essential for sensing applications, which require increased sensitivity (e.g., Analog Devices ADXL accelerometers [35]), or actuation applications that require the control of large arrays of MEMS devices (e.g., TI DMD™ [36]). For sensor applications, the packaging integration of a MEMS device and an electronic ASIC becomes unacceptable when the parasitic capacitances and wiring resistances have an impact on sensor performance (i.e., RC time constants of the integrated MEMS system are significant). For actuation applications such as a large array of optical devices that require individual actuation and control circuitry, a packaging solution becomes untenable with large device counts.

Of the three MEMS fabrication technologies previously discussed, surface micromachining is the most amenable to integration with electronics to form an IMEMS process. The development of an IMEMS process faces some challenges:

- *Large vertical topologies.* Microelectronic fabrication requires planar substrates due to the use of precision photolithographic processes. Surface micromachine topologies can exceed 10 μm due to the thickness of the various layers.
- *High-temperature anneals.* The mitigation of the residual stress of the surface micromachine structural layers can require extended periods of time at high temperatures (such as several hours at 1100°C for polysilicon). This would have adverse effects due to a thermal budget of microelectronics that is limited because of dopant diffusion and metalization.

There are three strategies for the development of an IMEMS process [37]:

- *Microelectronics first.* This approach overcomes the planarity restraint imposed by the photolithographic processes by building the microelectronics before the nonplanar micromechanical devices (Figure 3.27). The need for extended high-temperature anneals is mitigated by the selection of MEMS materials (e.g., aluminum, amorphous diamond [38]) and/or selection of the microelectronic metallization (e.g., tungsten instead of aluminum); these make the MEMS and microelectronic processing compatible. Examples of this IMEMS approach include an all-tungsten CMOS process developed by researchers at Berkeley Sensor and Actuator Center [39], and the Texas

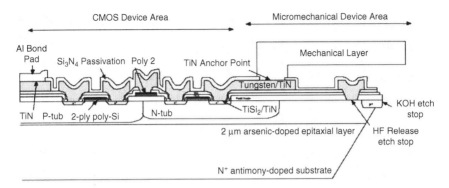

FIGURE 3.27 Microelectronics first approach to MEMS–microelectronic process integration.

Instruments' process [36] used to fabricate the DMD (Figure 3.2), which utilizes aluminum and photoresist as the device structural layer and sacrificial layer, respectively.

- *Interleave the microelectronics and MEMS fabrication.* This approach may be the most economical for large-scale manufacturing because it optimizes and combines the manufacturing processes for MEMS and microelectronics. However, this requires extensive changes to the overall manufacturing flow in order to accommodate the changes in the microelectronic device or the MEMS device. Analog Devices has developed and marketed an accelerometer and gyroscope that illustrate the viability and commercial potential of the interleaving integration approach [35].
- *MEMS fabrication first.* This approach fabricates, anneals, and planarizes the micromechanical device area before the microelectronic devices are fabricated, thus eliminating the topology and thermal processing constraints. The MEMS devices are built in a trench that is then refilled with oxide, planarized, and sealed to form the starting wafer for the CMOS processing (Figure 3.28). This technology was targeted for inertial sensor applications. Figure 3.29 shows prototypes designed in this technology by the University of California, Berkeley Sensor and Actuator Center (BSAC) and by Sandia National Laboratories.

3.5 TECHNOLOGY CHARACTERIZATION

The design of MEMS devices requires adequate knowledge of the material properties of the technology in which the device is built. Many classical methods for obtaining bulk material properties are available; however, the thin-film material properties, which are difficult to obtain accurately because of size-scale issues, are required for MEMS device design. These properties are unique to the specific technology used to build a MEMS device due to the variety of processing steps

96   Micro Electro Mechanical System Design

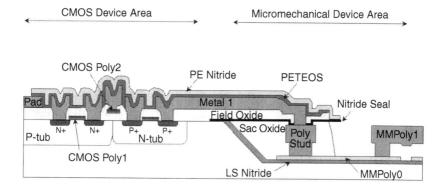

FIGURE 3.28 MEMS first approach to MEM–microelectronics process integration.

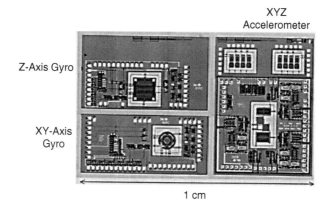

FIGURE 3.29 Inertial measurement unit fabricated in the MEMS first approach to MEMS–microelectronics process integration method. Designed by University of California, Berkley, Berkley Sensor Actuator Center; fabricated by Sandia National Laboratories.

that can be utilized and the sequence in which they are utilized. Appendix E provides a list of common MEMS material properties extracted from a number of sources. Reference 40 provides a broad catalog of MEMS material properties that is frequently updated from a variety of sources.

The development of technology characterization capability is essential to process monitoring which ensures the technology processing is consistent from run to run. Also, *in-situ* monitoring of some parameters may be quite beneficial to design development and qualification. Figure 3.30 shows some representative *in-situ* technology characterization devices.

Technology characterization requires the measurement of a number of parameters. The measured parameters that relate directly to MEMS device function are *process dimensional control* and *process material properties*. Examples of process dimensional control parameters are:

MEMS Technologies

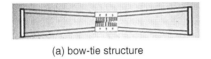
(a) bow-tie structure

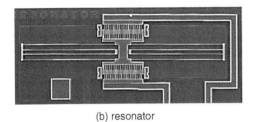
(b) resonator

(c) fixed-fixed beam array

FIGURE 3.30 SEM of some representative *in-situ* mechanical property test structures.

- *Etching profiles.* The shape of the sidewall of a MEMS structure depends upon the control of the etch process. These variations can result in stiffness or natural frequency perturbations in the MEMS device.
- *Registration errors.* Multilayer processes such as surface micromachining can have up to 14 masks, which must be aligned relative to one another to produce the MEMS device. Registration errors can cause structural errors in the device.
- *Resolution errors.* The lithography process, which transfers the image on the mask to photoresist on the wafer surface, and the etch process, which etches the resulting photoresist image onto the MEMS material, define the shape (clearances, widths) of the MEMS device. These types of errors can cause variations in device operation and possibly affect the device's ability to operate.
- *Layer thickness and uniformity.* Errors of this type will also affect the device's operation.

The principal process material properties important to MEMS device design are Young's modulus (E), residual stress/strain (σ_r, ε_r), fracture strength (σ_f), and electrical resistance.

Because the methods for process dimension control methodology tend to be specific to technology or fabrication process equipment, techniques for material property characterization will be the focus here. This section will discuss examples of material property characterization methods for common useful material parameters. The structures for this characterization will be presented generically. A specific implementation of the device for a particular technology will be required for implementation.

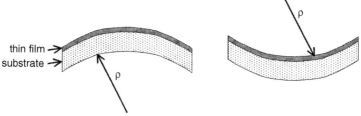

(a) Thin Film Compressive Residual Stress   (b) Thin Film Tensile Residual Stress

FIGURE 3.31 Thin films with compressive or tensile residual stress will cause the wafer to warp.

3.5.1 RESIDUAL STRESS

The behavior of MEMS devices is considerably influenced by residual stress of the materials. Residual stresses can cause significant change of shape and "warpage" of a device, as well as change the device's operating parameters such as stiffness or natural frequencies. Fracture or delaminating of the material is also a possibility.

Residual stress in a thin film deposited on a substrate is caused by a *thermal expansion coefficient mismatch* of the film, α_{film}, and substrate material, α_{sub}, and the *deposition conditions*. A mismatch in the thermal expansion coefficients will produce a residual strain in the film, ε_{film}, if the temperature of deposition, T_{dep}, is different from room temperature, T_0 (Equation 3.1). Residual stress due to deposition conditions is less well understood and is influenced by a number of factors such as substrate temperature, deposition rate, and film thickness. Generally, low-temperature deposition will result in lower residual stress.

$$\varepsilon_{film} = \frac{1}{1-\upsilon_{film}} \int_{T_0}^{T_{dep}} (\alpha_{film} - \alpha_{sub}) dT \qquad (3.1)$$

Residual stress can be compressive or tensile. A thin film with compressive residual stress that is deposited upon a wafer will want to expand, causing the wafer to bow, as shown in Figure 3.31a. Conversely, a thin film with tensile residual stress that is deposited upon a wafer will want to contract, causing the wafer to bow, as shown in Figure 3.31b. If the thickness of the thin film, t_{film}, and substrate, t_{sub}, are known, the *Stoney equation* (Equation 3.2) [41] can be used to calculate the residual stress of the thin film, σ_{film}, by measuring the wafer initial and final radius of curvatures, ρ_i and ρ_f, respectively.

$$\sigma_{film} = \frac{E_{sub} t_{sub}^2}{6(1-\upsilon_{sub})t_{film}} \left(\frac{1}{\rho_i} - \frac{1}{\rho_f} \right) \qquad (3.2)$$

MEMS Technologies

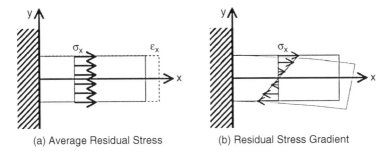

FIGURE 3.32 Material layer with average residual stress and residual stress gradient components.

Residual stress may vary through the thickness of a MEMS material layer (Figure 3.32). The portion of the residual stress distribution that is constant through the material thickness is called the average residual stress. Alternatively, the portion of the residual stress that varies through the material thickness is the residual stress gradient. The average residual stress will cause deflections primarily in the plane of the material layer except where the constrained boundary condition of a structural member may cause buckling and deflect out of plane. The residual stress gradient will cause an internal bending moment in the material that will directly result in out-of-plane deflections.

The simplest test structure for compressive residual stress [42] is an array of fixed–fixed beams (Figure 3.33a). When the underlying material (sacrificial layer) is removed, the compressive residual stress of the material will cause expansion. Buckling will occur on beams greater than a critical length as defined by Euler column theory (Section 6.1.7). Buckling of the beams can be observed by inspection using a microscope, interferometer, or SEM. The fixed–fixed beam test structure is a "proof" test device that establishes that the residual stress is greater than a particular value by inspection of the array. The residual stress, σ_r, can be calculated with Equation 3.3 if the beam thickness, t, and Young's modulus, E, are known for the shortest buckled beam of length, L. The negative sign in Equation 3.3 indicates compressive residual stress. Tensile residual stress will produce no observable effect on this test structure.

$$\sigma_r = -\frac{\pi^2 t^2 E}{3L^2} \qquad (3.3)$$

The ring and beam test structure [43–45] shown in Figure 3.33b is also a "proof" test device fabricated in an array of various sizes that establish that the tensile residual stress is greater than a particular value by inspection of the array. When the underlying material (sacrificial layer) is removed, the tensile residual stress of the material will cause the ring structure to contract; however, the ring is anchored at a particular diameter. The ring diameter at a 90° position away

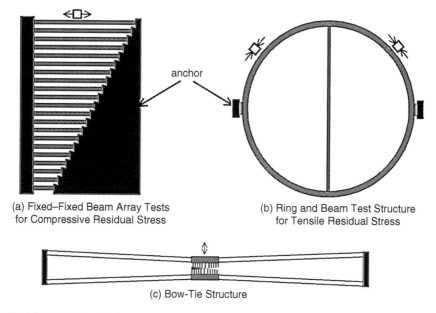

FIGURE 3.33 Residual stress test structures.

from the anchors will contract causing the beam structure to compress. If the beam length is greater than a critical length specified by Euler column theory, the beam will buckle, which is determined by optical inspection. The residual stress, σ_r, can be calculated with Equation 3.4 if the beam thickness, t, and Young's modulus, E, are known for the smallest buckled ring-beam structure of radius, R. G is a geometry parameter that is a function of the inner and outer ring radius and $G \leq 0.918$.

$$\sigma_r = \frac{\pi^2 t^2 E}{12 R^2 G} \tag{3.4}$$

A bent-beam test structure [46,47] will produce a quantitative measurement of residual strain for tensile and compressive residual strain. This test structure (Figure 3.33c) consists of a pair of opposed vernier scales; each is supported by two beams at the center with a shallow angle of bend. The opposed vernier scales allow for a 2× amplified quantitative readout of tensile and compressive residual strains. The relationships between residual stress and vernier deflections are given by the detailed analyses in Gianchandani and Najafi [46] and Zavracky [47]. Lin et al. [48] present another test structure that utilizes mechanical amplification of displacements and can also produce a quantitative vernier readout for compressive and tensile residual stress.

The residual stress gradient depicted in Figure 3.32 is due to an internal bending moment of the film whose magnitude is due to the integration of the

MEMS Technologies

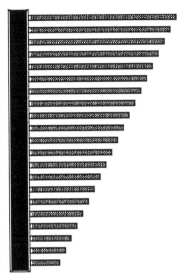

(a) Cantilever Beam Array

(b) Archimede Spirals with inner and outer anchors

FIGURE 3.34 Residual stress gradient test structures.

stress through the film thickness (Equation 3.5). The simplest test structure to assess the residual stress gradient is an array of cantilever beams (Figure 3.34a). The out-of-plane deflection due to the internal bending moment can be simply calculated [49]; however, the quantitative measurement of the out-of-plane deflection requires an SEM or interferometer. Figure 3.34b shows another test structure used to measure residual stress gradients, the Archimedes spiral [50]. The spiral will expand or contract upon release from the substrate; three response variables (endpoint height, endpoint rotation, and lateral contraction) may be related to the residual stress gradient. This gradient can be estimated from just one of the variables; two of the variables can be simply obtained with an optical microscope, which is advantageous. However, the Archimedes spiral may need to be large to obtain the required sensitivity.

$$M = \int_{-\frac{1}{2}}^{\frac{t}{2}} \sigma_x(y) y \, dy \tag{3.5}$$

3.5.2 Young's Modulus

Young's modulus, E, (Section 6.1.1), which is the proportionality between stress, σ, and strain, ε, and the essential parameter for calculation of the stiffness of structures, is necessary for design. This modulus may be obtained by directly

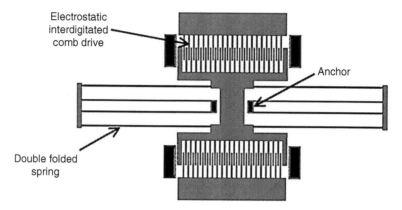

FIGURE 3.35 Electrostatic resonator test structure.

testing the thin-film material using specialized devices such as a nanoindenter, which plunges a diamond tip into the material and measures the deformation.

Alternatively, a lateral electrostatic resonator (Figure 3.35) may be used to extract the value of Young's modulus. The lateral resonator moves parallel to the substrate and thus minimizes damping effects and allows observation with an optical microscope. The resonator structure is driven by opposed interdigitated electrostatic comb drives. The resonator is suspended by a pair of folded beams that minimize the effect of residual stress. The stiffness of the suspension can be calculated using the equations in Appendix F. Resonance is the frequency, f, at which the resonator obtains its largest amplitude of motion; this is observed via a microscope. The resonance frequency is a function of the resonator mass, M, and spring stiffness, K. The mass of the resonator is readily obtained by the dimension of the moving structure and density of the material. Young's modulus is estimated from the spring stiffness equations of Appendix F.

$$f = \frac{1}{2\pi}\sqrt{\frac{K}{M}} \tag{3.6}$$

3.5.3 Material Strength

The traditional method for obtaining material strength for a bulk material is a pull test of a tensile specimen until failure occurs. This has been attempted with thin-film materials [52] with specialized instruments such as a nonoindenter or atomic force microscope. Figure 3.36 shows two thin-film test structures for material strength measurement. Figure 3.36a [53,54] is a structure moved with a probe; the movement of the shuttle brings several beams fixed to the shuttle in contact with a fixed post. The beams are deflected until the material fails. Nonlinear beam theory can extract the material strength, σ_f, when given data collected by observation with an optical microscope system.

MEMS Technologies

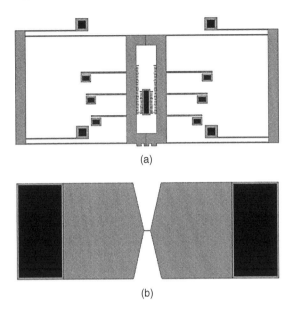

FIGURE 3.36 Material strength test structure.

Figure 3.36b shows a structure similar in intent to a bulk material tensile specimen. The wide portion of material can produce sufficient force via residual stress or electrostatic force [55] to fracture the small material specimen in the narrow portion of the structure. Many other kinds of strength measurement devices have been proposed. One comprises T- and H-shaped structures [56,57] and deflects due to tensile residual strain that ultimately fractures the material. The movement at the top of the T- or H-structure is measured to provide data for the ultimate strength, σ_f, calculation.

3.5.4 Electrical Resistance

Electrical resistance is a quantity that must be known for device design. The several ways in which resistance can be expressed (resistance, resistivity, sheet resistance) need to be explained. Figure 3.37 shows a slab of material with a specified thickness (t), width (W), and length (L) that is part of an electrical circuit. Equation 3.7 states that *resistance*, R, which is measured in ohms is a product of the *resistivity*, ρ — a characteristic of the material with units of ohms-meter and a geometric term. Equation 3.7 shows that resistance varies directly with the length of the slab and inversely with the slab cross-section area ($A = Wt$). In most MEMS and microelectronic technologies, the layers have a fixed thickness, and the resistivity is a characteristic of the material and doping that is also fixed for a specific technology.

Grouping these terms together, the *sheet resistance*, R_s, which is a constant for a particular layer in a MEMS or microelectronic technology, is defined in

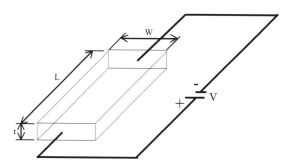

FIGURE 3.37 A slab of material within an electrical circuit.

Equation 3.8. Equation 3.9 states that the resistance of a slab of material is the product of the sheet resistance, which has units of *ohms per square* and the length-to-width ratio, which has units of *squares*. This ratio is defined as the *number of squares*, N_s. The unit "square" is, of course, dimensionless and it is frequently denoted symbolically by □.

The use of the sheet resistance concept enables an easy method for calculation of the resistance of run of material. For example, a run of a material ten units long by one unit wide has ten squares of material, N_s; therefore, the resistance of the run of material is $10 \times R_s$. If the run of material is doubled in width, the number of squares, N_s, is five. This means that the resistance of this wider run of material is $5 \times R_s$, which is half of what it was before.

$$R = \rho \frac{L}{A} = \rho \frac{L}{Wt} \tag{3.7}$$

$$R_s = \frac{\rho}{t} \tag{3.8}$$

$$R = R_s \frac{L}{W} = R_s N_s \tag{3.9}$$

The sheet resistance can be measured in a number of ways. The simplest is the four-point probe method (Figure 3.38a). In this method, current is passed between the two outer probes and voltage is measured across the inner pair of probes. The sheet resistance is the ratio of the voltage drop to the forced current time — a geometric factor that depends upon the probe geometry [61].

The second method is the van der Pauw method [62] (Figure 3.38b). Current is forced between one pair of electrodes and voltage is measured across the other pair of electrodes. To improve accuracy, the measurement is repeated three times by rotating the probe configuration 90° and repeating the measurement. The

MEMS Technologies

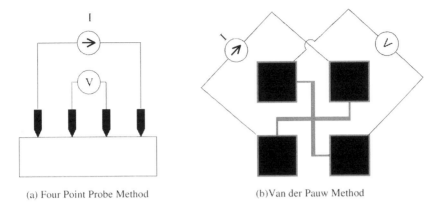

(a) Four Point Probe Method   (b) Van der Pauw Method

FIGURE 3.38 The four-point probe method and van der Pauw methods for determining sheet resistance.

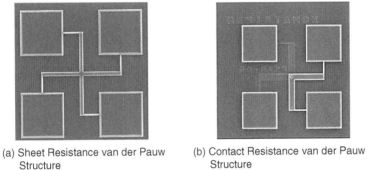

(a) Sheet Resistance van der Pauw Structure   (b) Contact Resistance van der Pauw Structure

FIGURE 3.39 Example of van der Pauw test structures.

measured resistance is then averaged. The calculation of sheet resistance also involves a geometrical correction factor. Figure 3.39 shows examples of van der Pauw structures. The measurement of *thermal sheet resistances* for thin films can also be measured with a van der Pauw type of test structure [63].

3.5.5 Mechanical Property Measurement for Process Control

The test structure discussed in previous sections utilized some combination of proof test structure arrays, optical microscope obtainable data, and mechanical probing of devices to obtain data to extract the material parameters. In recent years, it has been determined that a combination of mechanical analysis, high-precision optical measurements (interferometry), and data extraction methods is needed to obtain material properties of an accuracy necessary for process control. The literature shows two major approaches, M-TEST [58] and IMap [59,60], to this difficult problem, which is essential to MEMS process control and MEMS design.

M-TEST is a set of electrostatically actuated MEMS test structures and analysis procedures utilized for MEMS process monitoring and property measurement. M-TEST uses electrostatic pull-in of three sets of test structures (cantilever beams, fixed-fixed beams, and clamped circular diaphragms) followed by the extraction of two intermediate quantities, S and B parameters, that depend on a combination of material properties and test structure geometry. The test structure geometry, such as beam width and gap, is obtained with high accuracy with a profilometer.

The IMaP (interferometry for **ma**terial **p**roperty measurement in MEMS) uses a set of test structures that are electrostatically actuated to obtain the full voltage vs. displacement relationship. Values for the material properties and nonidealities of the test structure such as support post compliance are extracted to minimize the error between the measured and modeled deflections. It is clear that, for MEMS process control and material property information, automation of detailed measurement procedures such as M-TEST or IMAP will be required.

3.6 ALTERNATIVE MEMS MATERIALS

3.6.1 SILICON CARBIDE

Silicon carbide (SiC) has outstanding mechanical properties, particularly at high temperatures. Silicon is generally limited to lower temperatures due to a reduction in the mechanical elastic modulus above 600°C and a degradation of the electrical p–n junctions above 150°C. Silicon carbide is a wide bandgap semiconductor (2.3 to 3.4 ev) suggesting the promise of high-temperature electronics [64]. SiC has outstanding mechanical properties of hardness, elastic modulus, and wear resistance [66] (Table 3.5). SiC does not melt but sublimes above 1800°C and also has excellent chemical properties. Therefore, SiC is an outstanding material for harsh environments [65].

TABLE 3.5
Comparative Properties of Silicon, Silicon Carbide, and Diamond

Property	3C-SiC	Diamond	Si
Young's modulus E (GPa)	448	800	160
Melting point (°C)	2830 (sublimation)	1400 (phase change)	1415
Hardness (kg/mm^2)	2840	7000	850
Wear resistance	9.15	10.0	<<1

Note: Properties obtained from a number of sources, such as MEMS and Nanotechnology Clearinghouse Web site, material database, http://www.memsnet.org/material/40; G.L. Harris, 1995; and G.R. Fisher and P. Barnes, *Philos. Mag.*, B.61, 111, 1990.

MEMS Technologies

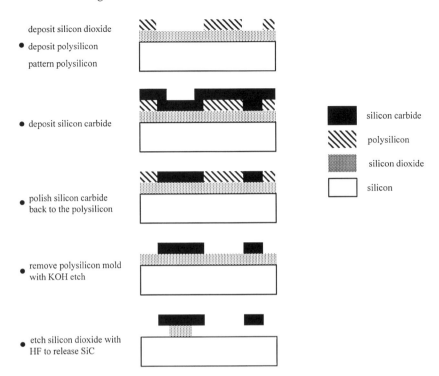

FIGURE 3.40 Example of a single-layer micromolding process for silicon carbide.

SiC has a large number (>250) of crystal variations [67], polytypes. Of these polytypes, 6H-SiC and 4H-SiC are common for microelectronics and 3C-SiC are attractive for MEMS applications. Technology exists for the growth of high-quality 6H-SiC and 4H-SiC 50-mm wafers. Single-crystal 3C-SiC wafers have not been produced, but 3C-SiC can be grown on (100 to 150 mm) Si wafers. However, polycrystalline 3C-SiC wafers are available.

The chemical inertness of SiC or polycrystalline SiC presents challenges for micromachining of these materials. Uses of conventional RIE techniques for SiC result in relatively low etch rates compared to polysilicon surface micromachining and the etch selectivity of SiC to Si or SiO_2 is poor; these characteristics make them inadequate etch stop materials.

An alternative approach for micromachining of SiC is a *micromolding* technique (damascene process) to pattern the SiC films [68]. An example of a single-layer SiC micromolding process is shown in Figure 3.40 and outlined next.

1. Deposit a 2-µm SiO_2 layer on a silicon wafer.
2. Deposit and pattern a 2-µm polysilicon layer to form the mold.
3. Deposit poly-SiC so that the mold and its surface are covered. Poly-SiC is deposited with atmospheric pressure chemical vapor deposition (APCVD) in which hydrogen is the carrier gas. Silane and propane

108 Micro Electro Mechanical System Design

are the precursor gases for the chemical reactions involved in this CVD process.

4. Polish the wafer with a diamond slurry to remove the poly-SiC from the top surface of the mold and planarize the wafer.

5. Remove the polysilicon mold with KOH. Poly-SiC is inert to most acids; however, it can be etched by alkaline hydroxide bases such as KOH at elevated temperatures (>600°C).

6. The SiO_2 is not etched by the KOH in the previous step. The patterned poly-SiC can now be released by removing the SiO_2 with hydrofluoric acid (HF) and partially undercutting the base of the poly-SiC to form an anchored region.

The micromolding process for SiC is able to bypass the RIE etch rate and selectivity issues for SiC mentioned earlier and yields a planarized wafer amenable to multilayer processing. However, control of the in-plane stress and stress gradients of SiC is still under development. SiC micromachining technologies have been used to fabricate prototype devices [69] required to operate under extreme conditions of temperature, wear, and chemical environments.

3.6.2 SILICON GERMANIUM

Polycrystalline silicon–germanium alloys (poly-$Si_{1-x} Ge_x$) have been extensively investigated for electronic devices and also present some attractive features as a MEMS material [71]. Poly-$Si_{1-x} Ge_x$ has a lower melting temperature than silicon and is more amenable to low-temperature processes, such as annealing, dopant activation, and diffusion, than silicon is. Poly-$Si_{1-x} Ge_x$ offers the possibility of a MEMS mechanical material with properties similar to polysilicon; however, the fabrication processing can be accomplished as low as 650°C. This will make poly-$Si_{1-x} Ge_x$ an attractive micromachining material for monolithic integration with microelectronics, which requires a low thermal budget [72].

Also, a surface micromachining process can be implemented utilizing poly-$Si_{1-x} Ge_x$ as the structural film and poly Ge as the sacrificial film with a release etch of hydrogen peroxide when x < 0.4. Poly Ge can be deposited as a highly conformable material and thus enables many MEMS structures.

3.6.3 DIAMOND

Diamond and hard amorphous carbon are a promising class of materials with extraordinary properties that would enable MEMS devices. The various amorphous forms of carbon, such as amorphous diamond (aD) tetrahedral amorphous carbon (ta-C) and diamond-like carbon (DLC), have hardness and elastic modulus properties that approach crystalline diamond, which has the highest hardness (~100 GPa) and elastic modulus (~1100 GPa) of all materials [73]. The appeal of this class of materials for MEMS designers is the extreme wear resistance,

MEMS Technologies 109

hydrophobic surfaces (i.e., stiction resistance), and chemical incrtncss. Recent progress has been achieved in the area of surface micromachining and mold-based processes [74,75] and a number of diamond MEMS devices have been demonstrated [76,77]. The use of diamond films in MEMS is still in the research stages. Recent progress in stress relaxation of the diamond films at 600°C [78,79] has been essential to the development of diamond as a MEMS material.

3.6.4 SU-8

EPON SU-8 (from Shell Chemical) is a negative, thick, epoxy–photoplastic, high-aspect-ratio resist for lithography [80]. SU-8 is a UV-sensitive resist that can be spin-coated in a conventional spinner in thicknesses ranging from 1 to 300 μm. Up to 2-mm thicknesses can be obtained with multilayer coatings. SU-8 has very suitable mechanical and optical properties and chemical stability; however, it has the disadvantages of adhesion selectivity, stress, and resist stripping. SU-8 adhesion is good on silicon and gold, but for materials such as glass, nitrides, oxides, and other metals the adhesion is poor. The thermal expansion coefficient mismatch between SU-8 and silicon or glass is large. SU-8 has been applied to MEMS fabrication [80,81] for plastic molds or electroplated metal micromolds. SU-8 MEMS structures have also been used for microfluidic channels and biological applications [82].

3.7 SUMMARY

Three categories of micromachining fabrication technologies have been presented: bulk micromachining, LIGA, and sacrificial surface micromachining. Bulk micromachining is primarily a silicon-based technology that employs wet chemical etches and reactive ion etches to fabricate devices with high aspect ratio. Control of the bulk micromachining etches with techniques such as etch stops and material selectivity is necessary to make useful devices. Commercial applications utilizing bulk micromachining, such as accelerometers and ink-jet nozzles, are available.

LIGA is a fabrication technology utilizing x-ray synchrotron radiation, a thick resist material, and electroplating technology to produce high-aspect-ratio metal-lic devices. Surface micromachining uses thick films and processes from the microelectronic industry to produce devices. This technology employs a sacrificial material and a structural material in alternating layers. A release process removes the sacrificial material in the last step in the process; this produces free-function structural devices. Surface micromachining enables large arrays of devices because no assembly is required. It can also be integrated with microelectronics for sensing and control. Two notable commercial applications of surface micro-machining are Texas Instruments' digital mirror device (DMD) [36] and Analog Devices' ADXL accelerometers [35].

QUESTIONS

1. Research a commercial MEMS application (e.g., accelerometer, pressure sensor, optical device, etc.). Discuss how the device was fabricated. Why was that fabrication approach selected for this application?
2. What are the difficulties involved in integrating microelectronics with MEMS?
3. Research a commercial MEMS application that has integrated microelectronics and MEMS. Why was integrated microelectronics needed for this application? How was the integration accomplished?
4. Why is it important to characterize the mechanical and electrical properties of a MEMS technology? What are the difficulties in obtaining these properties?
5. What is a micromolding process? Why is micromolding utilized in the process outlined in Section 3.6.1?

REFERENCES

1. K.E. Petersen, Silicon as a mechanical material, *Proc. IEEE*, 70(5), 420–457, May 1982.
2. G.T.A. Kovacs, N.I. Maluf, K.E. Petersen, Bulk micromachining of silicon, *Proc. IEEE*, 86(8), 1536–1551, August 1998.
3. K.S.J. Pister, Foldup illustration of silicon lattice planes and etch rates, http://www-bsac.eecs.berkeley.edu/~pister/crystal.pdf.
4. K.R. Williams, R.S. Muller, Etch rates for micromachining processing, *JMEMS*, 5(4), December 1996.
5. R.J. Shul, S.J. Pearton, *Handbook of Advanced Plasma Processing Techniques*, Springer–Verlag, New York, 2000.
6. U.S. Patent 5,501,893: Method of anisotropically etching silicon, F. Laermer, A. Schlp, Robert Bosch GmbH, issued March 26, 1996.
7. K.A. Shaw, Z.L. Zhang, N.C. MacDonald, SCREAM I: a single mask, single-crystal silicon, reactive ion etching process for microelectromechanical structures, *Sensors Actuators A*, 40, 63–70, 1994.
8. N.C. MacDonald, SCREAM microelectromechanical systems, *Microelectron. Eng.*, 32, 49–73, 1996.
9. J. Li, G.K. Ananthasuresh, A quality study on the excimer laser micromachining of electro-thermal-compliant micro devices, *J. Micromech. Microeng.*, 11, 38–47, 2001.
10. T. Moulton, G.K. Ananthasuresh, Micromechanical devices with embedded electro-thermal-compliant actuation, *Sensors Actuators A*, 90, 38–48, 2001.
11. E.W. Becker, W. Ehrfeld, P. Hagmann, A. Maner, D. Muchmeyer, Fabrication of microstructures with high aspect ratios and great structural heights by synchrotron radiation lithography, galvanoforming, and plastic molding (LIGA process), *Microelectron. Eng.*, 4, 35, 1986.
12. FAULHABER Group: http://www.faulhaber.com/.
13. Institute for Microtechnology, Mainz, Germany: http://www.imm-mainz.de/.

MEMS Technologies

14. C. Thürigen, W. Ehrfeld, B. Hagemann, H. Lehr, F. Michel, Development, fabrication and testing of a multi-stage micro gear system, *Proc. Triboloby Issues Opportunities MEMS*, Kluwer Academic Publishers, Dordrecht, 1998.
15. C. Thürigen, U. Beckord, R. Bessey, F. Michel, Design rules and manufacturing of micro gear systems, *Actuator98 6th International Conference on New Actuators*, 575–578, 1998.
16. C. Thürigen, U. Beckord, R. Bessey, F. Faulhaber, Construction and manufacturing of a micro gearhead with 1.9 mm outer diameter for universal application, *Proc. SPIE*, 3680, 526–533, March–April 1999.
17. W. Ehrfeld, H. Lehr, Deep x-ray lithographie for the production of three-dimensional microstructures from metals, polymers and ceramics, *Radiation Phys. Chem.*, 45(3), 349–365, 1995
18. U.S. Patent 2,749,598: Method of preparing electrostatic shutter mosaics, filed Feb. 1, 1952, issued June 12, 1956.
19. H.C. Nathanson, W.E. Newell, R.A. Wickstrom, J.R. Davis, The resonant gate transistor, *IEEE Trans. Electron Devices* ED-14, 117–133, 1967.
20. R.T. Howe, R.S. Muller, Polycrystalline silicon micromechanical beams, *J. Electrochem. Soc.: Solid-State Sci. Technol.*, 130(6), 1420–1423, June 1983.
21. R.T. Howe, Surface micromachining for microsensors and microactuators, *J. Vac. Sci. Tech.* B 6 (6), 1809–1813, November/December 1988.
22. L-S. Fan, Y-C Tai, R.S. Muller, Integrated movable micromechanical structures for sensors and actuators, *IEEE Trans. Electron Devices*, 35(6), 724–730, 1988.
23. K.S.J. Pister, M.W. Judy, S.R. Burgett, R.S. Fearing, Microfabricated hinges, *Sensors Actuators A*, 33, 249–256, 1992.
24. W. Patrick, W. Guthrie, C. Standley, P. Schiable, Application of chemical mechanical polishing to the fabrication of VLSI circuit interconnections, *J. Electrochem. Soc.*, 138(6), 1778–1784, June 1991.
25. R.D. Nasby, J.J. Sniegowski, J.H. Smith, S. Montague, C.C. Barron, W.P. Eaton, P.J. McWhorter, D.L. Hetherington, C.A. Apblett, J.G. Fleming, Application of chemical mechanical polishing to planarization of surface micromachined devices, Solid State Sensor Actuator Workshop, Hilton Head Is., SC, 48–53, June 1996.
26. U.S. Patent 5,804,084, issued September 8, 1998, Use of chemical mechanical polishing in micromachining, R.D. Nasby, J.J. Sniegowski, P.J. McWhorter, D.L. Hetherington, C.A. Apblett.
27. A. Yasseen, S. Smith, M. Mehregany, F. Merat, Diffraction grating scanners using polysilicon micromotors, Proc. *8th Int. Conf. Solid-State Sensors Actuators Eurosensors IX*, Stockholm, Sweden, 1, 206–209, 1995.
28. R. Legtenberg, J. Elders, M. Elwenspoek, Stiction of surface microstructures after rinsing and drying: model and investigation of adhesion mechanisms, *Proc. 7th Int. Conf. Solid State Sensors Actuators*, 198–201, 1993.
29. G.K. Fedder, R.T. Howe, Thermal assembly of polysilicon microstructures, *Proc. Microelectromechanical Syst. '89*, 63–68.
30. G.T. Mulhern, D.S. Soane, R.T. Howe, Supercritical carbon dioxide drying of microstructures, *Proc. Int. Conf. Solid-State Sensors Actuators (Transducers '93)*, Yokohama, Japan, 296–299, 1993.
31. H. Guckel, J.J. Sniegowski, T.R. Christenson, S. Mohney, T.F. Kelly, Fabrication of micromechanical devices from polysilicon films with smooth surfaces, *Sensors Actuators*, 20, 117–122, 1989.

32. M.R. Houston, R. Maboudian, R.T. Howe, Self assembled monolayer films as durable anti-stiction coatings for polysilicon microstructures, *Proc. Solid-State Sensor Actuator Workshop*, Hilton Head Island, SC, 42–47, 1996.
33. U.S. Patent 6,082,208, Method for fabricating five level microelectromechanical structures and five level microelectromechanical transmission formed, M.S. Rodgers, J.J. Sniegowski, S.L. Miller, P.J. McWhorter, issued July 4, 2000.
34. SUMMiT™ (Sandia ultra-planar, multi-level MEMS technology), Sandia National Laboratories, http://mems.sandia.gov.
35. K.H. Chau, R.E. Sulouff, Technology for the high-volume manufacturing of integrated surface-micromachined accelerometer products, *Microelectron. J.*, 29, 579–586, 1998.
36. P.F. Van Kessel, L.J. Hornbeck, R.E. Meier, M.R. Douglass, A MEMS-based projection display, *Proc. IEEE*, 86(8), August 1998.
37. R. Howe, Polysilicon integrated microsystems: technologies and applications, *Proc. Transducers 95*, 43–46, 1995.
38. J.P. Sullivan, T.A. Friedmann, M.P. de Boer, D.A. LaVan, R.J. Hohlfelder, C.I.H. Ashby, M. Mitchell, R.G. Dunn, Developing a new material for MEMS: amorphous diamond, 2000 Fall MRS Meeting, Nov. 27–Dec 1. 2000, Boston, MA.
39. W. Yun, R. Howe, P. Gray, Surface micromachined, digitally force balanced accelerometer with integrated CMOS detection circuitry, *Proc. IEEE Solid-State Sensor Actuator Workshop'92*, 126, 1992.
40. MEMS and Nanotechnology Clearinghouse Web site, material database, http://www.memsnet.org/material/.
41. G.G. Stoney, The tension of metallic films deposited by electrolysis, *Proc. R. Soc. London*, Ser. A, 82, 172, 1909.
42. H. Guckel, T. Randazzo, D.W. Burns, A simple technique for the determination of mechanical strain in thin films with applications to polysilicon, *J. Appl. Phys.*, 67(5), March 1985.
43. H. Guckel, G.W. Burns, C.C.G. Viser, H.A.C. Tilmans, D. Deroo, Fine-grained polysilicon films with built-in tensile strain, *IEEE Trans. Electron Dev.*, 35(6), 800–801, 1988.
44. H. Guckel, G.W. Burns, C.C.G. Viser, H.A.C. Tilmans, D. Deroo, C.R. Rutigliano, Mechanical properties of fine grained polysilicon the repeatability issue, *Tech. Dig. IEEE Solid State Sensor Actuator Workshop*, Hilton Head Island, SC, 96, 1988.
45. H. Guckel, G.W. Burns, Polysilicon thin film process, U.S. Patent No. 4,897,360 1990.
46. Y.G. Gianchandani, K. Najafi, Bent-beam strain sensors, *JMEMS*, 5(1), 52–58, March 1996.
47. P.M. Zavracky, G.G. Adams, P.D. Aquilino, Strain analysis of silicon-on-insulator films produced by zone melting recrystallization, *JMEMS*, 4(1), 42–48, March 1995.
48. L. Lin, A.P. Pisano, R.T. Howe, A micro strain gauge with mechanical amplifier, *JMEMS*, 6(4), 313–321, December 1997.
49. R.J. Roark, *Formulas for Stress and Strain*, McGraw-Hill Book Company, New York, 1989.
50. L.S. Fan, R.S. Muller, W. Yun, R.T. Howe, J. Huang, Spiral microstructures for the measurement of average strain gradients in thin films, *Proc. IEEE Conf. Micro Electro Mechanical Systems*, 177, 1990.

MEMS Technologies

51. T.P. Weihs, S. Hong, J.C. Bravman, W.D. Nix, Mechanical deflection of cantilever microbeams: a new technique for testing the mechanical properties of thin films, *J. Mater. Res.*, 3, 931, 1988.
52. P.T. Jones, G.C. Johnson, R.T. Howe, Fracture strength of polycrystalline silicon, *Mat. Res. Soc. Proc.*, 518, 197–202.
53. P.T. Jones, G.C. Johnson, R.T. Howe, Micromechanical structures for fracture testing of brittle thin films, *Int. Mechanical Eng. Conf. Exposition*, DSC-59, 325–330, 1996.
54. M.P. de Boer, B.D. Jensen, F. Bitsie, A small area *in-situ* MEMS test structure to measure fracture strength by electrostatic probing, *Proc. SPIE*, 3875, 97–103, Sept. 1999.
55. M.G. Allen, M. Mehregany, R.T. Howe, S.D. Senturia, Microfabricated structures for the *in situ* measurement of residual stress, Young's modulus, and ultimate strain of thin films, *Appl. Phys. Lett.*, 51(4), 241–243, July 1987.
56. M.G. Allen, M. Mehregany, R.T. Howe, S.D. Senturia, Novel structures for the *in situ* measurement of mechanical properties of thin films, *J. Appl. Phys.*, 62(9), 3579–3584, November 1987.
57. P.M. Osterberg, S.D. Senturia, M-TEST: a test chip for MEMS material property measurement using electrostatically actuated test structures, *JMEMS*, 6(2), 107–118, June 1997.
58. B.D. Jensen, M.P. de Boer, S.L. Miller, IMAP: interferometry for material property measurement in MEMS, Proc. Int. Conf. Modelling Simulation Microsystems, Semiconductors, Sensors Actuators, 206–209, April 1999.
59. M.B. Sinclair, M.P. de Boer, N.F. Smith, B.D. Jensen, S.L. Miller, Method and system for automated on-chip material and structural certification of MEMS devices, U. S. Patent 6,567,715, issued May 20, 2003.
60. M. Yamashita, M. Agu, Geometrical correction factor of semiconductor resistivity measurement by four point probe method, *Jpn. J. Appl. Phys.* 23, 1499, 1984.
61. L.J. van der Pauw, A method of measuring specific resistivity and Hall effect of discs of arbitrary shape, *Philips Res. Rep.*, 13, 1–9, 1958.
62. O. Paul, P. Ruther, L. Plattner, H. Baltes, A thermal van der Pauw test structure, *IEEE Trans. Semiconductor Manuf.*, 13(2), 159–166, May 2000.
63. M. Mehregany, C.A. Zorman, SiC MEMS: opportunities and challenges for applications in harsh environments, *Thin Solid Films*, 355–356, 518–524, 1999.
64. M. Mehregany, C.A. Zorman, N. Rajan, C.H. Wu, Silicon carbide MEMS for harsh environments, *Proc. IEEE*, 86(8), 1594–1609, August 1998.
65. G.L. Harris, Properties of silicon carbide, IEEE, Stevenage, U.K., 1995.
66. G.R. Fisher, P. Barnes, Toward a unified view of polytypism in silicon carbide, *Philos. Mag.*, B.61, 2, 217–236, February 1990.
67. A.A. Yasseen, C.A. Zorman, M. Mehregany, Surface micromachining of polycrystalline SiC films using microfabricated molds of SiO_2 and polysilicon, *J. Microelectromech. Syst.*, 8(3), 237–242, September 1999.
68. N. Rajan, M. Mehregany, C.A. Zorman, S. Stefanescu, T.P. Kicher, Fabrication and testing of micromachined silicon carbide and nickel fuel atomizers for gas turbine engines, *J. Microelectromech. Systems*, 8(3), 251–257, September 1999.
69. G.T.A. Kovacs, *Micromachined Transducers*, McGraw-Hill, New York, 1998.
70. S. Sedky, P. Fiorini, M. Caymax, S. Loreti, K. Baert, L. Hermans, R. Mertens, Structural and mechanical properties of polycrystalline silicon germanium for micromachining applications, *J. Microelectromech. Syst.*, 7(4), 365–372, December 1998.

71. A.E. Franke, J.M. Heck, T.J. King, Polycrystalline silicon-germanium films for integrated microsystems, *J. Microelectromech. Syst.*, 12(2), 160–171, April 2003.
72. J.P. Sullivan, T.A. Friedmann, M.P. de Boer, D.A. LaVan, R.J. Hohlfelder, C.I.H. Ashby, M.T. Dugger, M. Mitchell, R.G. Dunn, A.J. Magerkurth, Developing a new material for mems: amorphous diamond, *Mat. Res. Soc. Symp. Proc.*, 657, 2001.
73. H. Bjorkman, P. Rangsten, P. Hollman, K. Hjort, Diamond replicas from microstructured silicon masters, *Sensors Actuators*, 73, 24–29, 1999.
74. R. Ramesham, Fabrication of diamond microstructures from microelectromechanical systems (MEMS) by a surface micromachining process, *Thin Solid Films*, 340, 1–6, 1999.
75. H. Bjorkman, P. Rangsten, P. Hollman, K. Hjort, Diamond microstructures for optical micro electromechanical systems, *Sensors Actuators*, 78, 41–47, 1999.
76. T. Shibata, Y. Kitamoto, K. Unno, E. Makino, Micromachining of Diamond Film for MEMS Applications, *J. Microelectromech. Syst.*, 9(1), 47–51, March 2000.
77. T.A. Friedmann, J.P. Sullivan, J.A. Knapp, D.R. Tallant, D.M. Follstaedt, D.L. Medlin, P.B. Mirkarimi, Thick stress-free amorphous-tetrahedral carbon films with hardness near that of diamond, *Appl. Phys. Lett.*, 71, 3820, 1997.
78. T.A. Friedmann, J.P. Sullivan, Method of forming a stress relieved amorphous tetrahedrally coordinated carbon film, U.S. Patent no. 6,103,305, issued Aug. 15, 2000.
79. E.H. Conradie, D.F. Moore, SU-8 thick photoresist processing as a functional material for MEMS applications, *J. Micromech. Microeng.* 12, 368–374, 2002.
80. H. Lorenz, M. Despont, N. Fahrni, J. Brugger, P. Vettiger, P. Renaud, High-aspect-ratio, ultrathick, negative-tone near-UV photoresist and its applications for MEMS, *Sensors Actuators* A 64, 33–39, 1998.
81. Y. Choi, R. Powers, V. Vernekar, A.B. Frazier, M.C. LaPlaca, M.G. Allen, High aspect ratio SU-8 structures for 3-D culturing of neurons, 2003 ASME International Mechanical Engineering Congress, IMECE 42794, 2003.

4 Scaling Issues for MEMS

Mankind has been driven over hundreds of years toward miniaturization of devices for various reasons. Some of these reasons are merely aesthetic and some are to attain increased functionality. However, only during the late 20th century, when the investigation of the engineering and physics of systems involved a very large size scale decrease of more than three orders of magnitude (i.e., 0.001), has the issue of size had significant impact on the "relevant" physical phenomena. These large-scale reductions have come about via the engineering of microelectronic and MEMS devices and on to the extreme scale reduction of quantum mechanics at the atomic level. The issue of relevance can arise in a number of ways.

- Entering different physics regimes at a particular scale
- Physical phenomena scaling at different rates, which changes their relative importance

These are important issues for the MEMS design engineer to consider because the intuition attained in the engineering experience of macroscale devices does not directly transfer to the microscale in many ways. Figure 4.1 shows the diversity of size encountered in the macro, micro, and nano domains.

This chapter will explore a number of aspects of how things change with scale. Things that will be considered range from simple geometric effects; the behavior of physical systems of interest (e.g., mechanics, fluidics, electrical, etc.); new physical regimes (e.g., Brownian motion, electron tunneling, etc.); fabrication tolerances; material issues; and even computational issues.

4.1 SCALING OF PHYSICAL SYSTEMS

4.1.1 GEOMETRIC SCALING

In order to evaluate the effect on a system due to size reduction, it is necessary first to look at the system geometry and define a framework to make that evaluation. For the discussions in this chapter, an isomorphic scaling of the system (i.e., all dimensions scaled equally) will be considered. A dimension of length, X_o, can be scaled to a smaller dimension, X_s, by a scale factor, S. Because we are studying the effect of scale reduction, $0 < S \leq 1$. The geometry of length, area, and volume scale by decreasing powers of S is shown in Equation 4.1 through Equation 4.3.

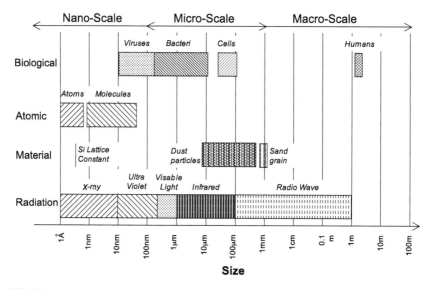

FIGURE 4.1 A perspective of size in the macro, micro, and nano domains.

$$X_s = S X_o \quad (4.1)$$

$$A_s = X_s Y_s = S^2 X_o Y_o = S^2 A_o \quad (4.2)$$

$$V_s = X_s Y_s Z_s = S^3 X_o Y_o Z_o = S^3 V_o \quad (4.3)$$

Figure 4.2 shows the effect that various powers of the scaling parameter, S, will have on the scaled variable. The effect of MEMS device scaling would not affect heuristic macroscale engineering expectations if all the scaling is linear. If other scalings arise, this will change the relative importance of phenomena or give rise to new phenomena.

The mass of an object is directly proportional to the object volume. Therefore, because the object size is reduced by 1000 (i.e., $S = 10^{-3}$), the volume and mass are reduced by 10^{-9}. The mass is a significant variable in numerous engineering phenomena and calculations.

Various geometric ratios such as the area to volume ratio are significant in many engineering fields — especially fluid dynamics and heat transfer. The area to volume ratio scales as the inverse of S (Equation 4.4). Therefore, as systems are reduced to the MEMS scale the area to volume ratio increase, which implies that physical phenomena sensitive to the area–volume ratio will change from heuristic macroworld expectations.

$$A_s/V_s = 1/S \, (A_o/V_o) \quad (4.4)$$

Scaling Issues for MEMS 117

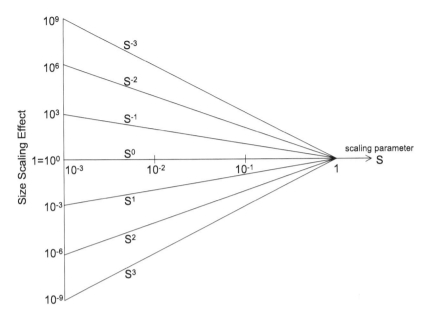

FIGURE 4.2 Effect of scaling for various powers of the scaling parameter S.

4.1.2 Mechanical System Scaling

The two most fundamental parameters describing a mechanical system are the mass and stiffness. The mass of a linear translating object is simply the volume times the density. Therefore, the mass scales the same as volume, S^3. As the system scale decreases by S, the mass and volume are decreasing even more rapidly (i.e., S^3).

$$M_s = \rho S^3 V_o = S^3 M_o \qquad (4.5)$$

The stiffness of a mechanical system is a fundamental quantity describing the ability of a mechanical system to resist applied forces. Stiffness, K, is the ratio of the force applied to a mechanical member to the resulting deflection. The stiffness of a mechanical suspension may consist of a combination of beams and rods. Figure 4.3 shows a schematic of a simple circular beam and rod. A beam will resist deflection due to transverse bending and a rod will resist axial deflection. The details of beams and rods will be discussed in more detail in Chapter 7.

Equation 4.6 and Equation 4.7 show the basic proportionality of stiffness for a circular cross-section beam and rod, respectively, to the material property for stiffness, E (i.e., Young's modulus), and the geometric quantities of area, A, and area moment of inertial, I. In these equations, the area, A, and area moment of inertia, I, have been converted to their underlying dimensional definitions and the scaling parameters inserted. These proportionality equations show that the mechanical stiffness decreases linearly in proportion to the system scaling.

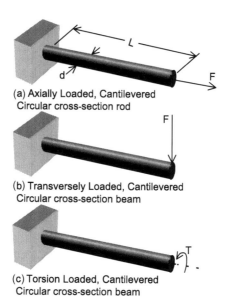

FIGURE 4.3 Schematic of a circular beam and rod.

$$K_{bending} \propto \frac{EI}{L^3} \propto \frac{Ewt^3}{L^3} \propto S \tag{4.6}$$

$$K_{axial} \propto \frac{EA}{L} \propto \frac{Ewt}{L} \propto S \tag{4.7}$$

Example 4.1

Problem: Given the mechanical member shown in Figure 4.3c, how does its torsion stiffness vary with scale?

Solution: The stiffness of a torsion bar is given below. The details of torsion stiffness are discussed in Chapter 7.

$$K_{torsion} = \frac{Torque}{angle} = \frac{GJ}{L}$$

where
    G = modulus of rigidity
    J = area polar moment of inertia
    L = bar length

For a circular bar, the area polar moment of inertia is given by

Scaling Issues for MEMS

$$J = \frac{\pi d^4}{32}$$

In terms of geometrical dimensions,

$$K_{torsion} \propto \frac{d^4}{L}$$

Use the scaling parameter to determine the effect on stiffness:

$$K_{torsion} \propto \frac{\left(Sd_0\right)^4}{SL_0} \propto S^3$$

The torsion stiffness scales as S^3, which is different than the bending and axial stiffness that scales as S, due to the units involved in torsion stiffness (i.e., N–M/radian). However, the net effect on more descriptive parameters such as natural frequency is consistent with previous results for mechanical systems.

Mass and stiffness are fundamental quantities of a mechanical system, but the actual values of these quantities for comparing one mechanical system or design to another are not very insightful. The natural frequency is a parameter that relates the elastic and inertial forces to define directly the frequency of oscillation of a mechanical system. This frequency of oscillation is a measure of the relative stiffness of a mechanical system and is encountered in the design of mechanical oscillators and filters.

The natural frequency of a one degree of freedom translational mechanical system is defined as the square root of the stiffness divided by the mass (Equation 4.8). The scaling parameters for mass and stiffness that have been developed previously have been inserted to show that, as a system reduces in scale, S, the natural frequency scales by $1/S$. This means that the natural frequency increases for system scale reductions. The effect is due to the stiffness decreasing more slowly than the mass for a reduction in size.

$$f_n = \frac{1}{2\pi} \sqrt{\frac{K}{M}} \propto \sqrt{\frac{K}{M}} \propto \sqrt{\frac{S}{S^3}} \propto \frac{1}{S} \tag{4.8}$$

It has been noted that MEMS scale devices are more rugged in mechanical shock and vibration environments [8,9] than their macro world counterparts. MEMS inertial sensors have been shown to survive shock environments of tens of thousands of g's (gravitational acceleration) [5,7]. A MEMS reliability study [6] has shown that a packaged MEMS device was tested to 40,000 g, which caused failure of the package; however, the MEMS device was still operational (Figure 4.4).

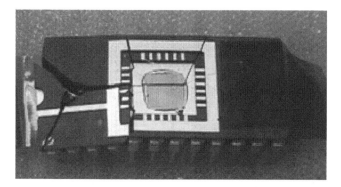

FIGURE 4.4 Packaged MEMS device after shock testing. (Courtesy of Sandia National Laboratories.)

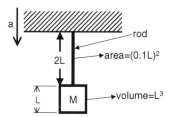

FIGURE 4.5 A concentrated mass, M, and rod subjected to an acceleration, a.

Example 4.2

Problem: In a system shown in Figure 4.5, the cubic mass of dimension, L, is supported by a square rod of cross-section dimension, $0.1L$, which is $2L$ long. This system is subjected to a step acceleration of a, which causes the rod to fail due to stress, σ_s. If this system is isomorphically scaled by a factor of $S = 0.001$, what acceleration level would be achieved when the rod fails due to stress?

Solution: The stress is related to a dynamic loading factor, D, times the inertial force (Ma) divided by the cross-section area, A, which, upon substitution for the problem dimensions, yields the following equation. The dynamic loading factor, D, accounts for the type of dynamic loading (i.e., impulse, step, etc.) for an acceleration or force [10]. For a step load, the dynamic loading factor is equal to 2 ($D = 2$).

$$\sigma_s = \frac{D\,Ma}{A} = \frac{D\rho L^3 a}{(0.01L)^2}$$

In this case, the density, ρ, is constant as the system scales in size; therefore, the failure stress, σ_s, is proportional to the product of length and the input acceleration, a.

Scaling Issues for MEMS

$$\sigma_s \propto La$$

At the reduced scale, S, the failure stress, σ_s, is given by

$$\sigma_s \propto SLa$$

This means that the acceleration, a, required to achieve failure due to stress scales by $1/s$, as shown next.

$$a \propto \frac{\sigma_s}{SL} \propto \frac{1}{S}$$

Therefore, when this system is isomorphically scaled by a factor of $S = 0.001$ the acceleration required for a stress failure of the rod increases by 1000.

Reduction in scale in mechanical systems has some negative consequences. For example, the deflection of an inertial mass in a sensing device will require a more sensitive detection system, because the inertial mass and the resulting inertial force scale as S^3 but the stiffness scales as S. This different scaling causes the mass to deflect less as smaller size scales.

4.1.3 THERMAL SYSTEM SCALING

The effect of scaling a thermal system can readily be determined by analyzing the basic heat transfer relationships [11]. The thermal energy storage capability for an object is determined by Equation 4.9. The thermal energy storage of an object is the product of the mass, m, constant pressure specific heat, c_p, and the temperature change, ΔT, in the material; it is a measure of the temperature increase in an object due to thermal input. The thermal mass for any specific system will involve the volume of the system, V, and it will therefore scale with size by S^3:

$$mc_p \Delta T = \rho V c_p \Delta T \propto S^3 \tag{4.9}$$

Heat can be transferred within a system by the mechanisms of conduction, convection, and radiation (see Figure 4.6 and Equation 4.10 through Equation 4.12, respectively). Conduction is heat transfer through the material of an object due to a temperature gradient, ∇T. Convection is a form of heat transfer through a liquid or gas to an object, due to a temperature difference between the object surface or wall, T_w, and the temperature of the bulk fluid or gas, T_∞. K and h are the conduction and convection heat transfer coefficients, respectively, and are functions of the medium through which the heat is transferred. The convection heat transfer coefficient, h, also involves complex relationships of the medium fluid flow.

Radiation heat transfer between objects in space is driven by the temperatures of the objects to the fourth power, where σ is the Boltzmann constant. These

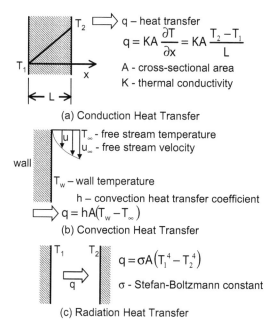

FIGURE 4.6 Heat transfer mechanisms.

three heat transfer mechanisms all involve the area, A, through which the heat is transferred. Therefore, the heat transfer mechanisms scale with area by S^2. Any specific system can involve any combination of these heat transfer mechanisms and may be modeled via an *effective* heat transfer coefficient, K_{eff}.

$$q = KA \nabla T \propto S^2 \tag{4.10}$$

$$q = hA(T_w - T_\infty) \propto S^2 \tag{4.11}$$

$$q = A\sigma T^4 \propto S^2 \tag{4.12}$$

A significant question remaining to be answered is how the heat transfer rate differs between the macro- and microscale. The preceding discussion showed that heat storage of a system is scaled by S^3, and heat transfer in/out of a system is scaled by S^2.

Now consider a thermal system, shown in Figure 4.7, that involves a system with internal heat generation, q_{gen}, thermal storage within the system, and heat transfer in/out of the system. These effects can be modeled by the first-order differential Equation 4.13. This equation can be rearranged into a form in which a time constant, τ, of the system will become apparent. The time constant, τ, can

Scaling Issues for MEMS

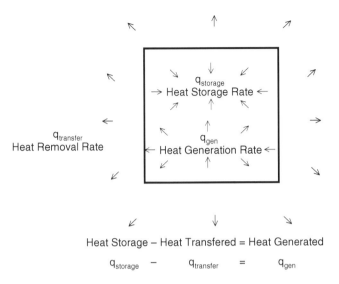

FIGURE 4.7 Thermal system.

be shown to consist of a term, α, that involves the material and heat transfer mechanisms of the system and the volume to area ratio. Alpha is known as the thermal diffusivity constant. Therefore, the thermal time constant scales by S. As the system scale decreases, the heat storage capacity decreases by S^3 and the heat transfer decreases by S^2. This means that the thermal time constant decreases by S (i.e., the thermal system responds more quickly as scale decreases) and that thermal systems will respond quickly at the microscale.

$$mc_p \dot{T} + KAT = q_{gen} \tag{4.13}$$

$$\tau \dot{T} + T = \frac{q_{gen}}{KA} \tag{4.14}$$

$$\tau = \frac{mc_p}{KA} = \left(\frac{\rho c_p}{K}\right)\left(\frac{V}{A}\right) \propto S \tag{4.15}$$

$$\alpha = \left(\frac{\rho c_p}{K}\right) \tag{4.16}$$

Thermal and fluid mechanics involve complicated governing equations with many parameters. Dimensionless ratios of these parameters have been used frequently to access system response or relate to empirical data. Two dimen-

124 Micro Electro Mechanical System Design

sionless ratios that imply heat transfer effects due to reduction in scale will now be discussed.

The Biot number (Bi) (Equation 4.17) is the ratio of the heat convection coefficient, h, and characteristic length, L, product divided by the thermal conductivity of a solid, K. Bi is the ratio of the convection heat transfer coefficient to the conduction heat transfer coefficient for a solid body immersed in a liquid or gaseous substance. The Biot number indicates the ability of a body to come to thermal equilibrium rapidly without setting up significant internal thermal stresses. For example, an ice cube immersed in hot water will frequently crack. For $Bi <$ 1, the internal conduction heat transfer coefficient of the object is greater than the convection heat transfer coefficient of the surrounding fluid. This means that internal thermal gradients and stresses are not likely to be developed in the object. Equation 4.17 shows that, as the size is decreased, the Bi number is also decreased.

$$Bi = \frac{hL}{K} \propto S \qquad (4.17)$$

The Grashof number (Gr), Equation 4.18, is a dimensionless group frequently used to access or empirically determine the free convection heat transfer coefficient. Gr is the ratio of the buoyancy forces (i.e., numerator of equation) to the viscous forces. Free convection heat transfer is due to the circulation or induced flow of the fluid medium due to fluid expansion due to temperature change and buoyancy of the expanded less dense medium. The Grashof number is a comparison of the buoyancy forces inducing convective fluid motion vs. the viscous forces impeding the convective fluid motion. The scaling effects on the Gr are apparent in Equation 4.18, which yields an S^3 dependency on scale; this means that natural convection is less effective as size is reduced due to fluid motion being impeded.

$$Gr = \frac{g\beta\left(T_w - T_\infty\right)L^3}{\upsilon^2} \propto S^3 \qquad (4.18)$$

where

$\quad g$ = gravitational acceleration
$\quad \beta$ = coefficient of thermal expansion of the fluid
$\quad T_w$ = wall/surface temperature of the object
$\quad T_\infty$ = bulk temperature of the fluid medium
$\quad L$ = characteristic length
$\quad \upsilon$ = kinematic viscosity

4.1.4 FLUIDIC SYSTEM SCALING

The equations governing fluid mechanics, Navier–Stokes equations, are nonlinear and complex and involve many forces, such as gravity, inertia, viscosity, and

Scaling Issues for MEMS

surface tension. The difficulty in solving these equations has caused engineers to utilize nondimensional ratios of these quantities to relate the various physical regimes of fluid mechanics for different systems of varying complexity and scale. To evaluate the effect of scale on the fluid mechanics of MEMS systems, three of these dimensionless quantities will be examined: Reynolds number, Knudsen number, and the Weber number.

The Reynolds number, Re, is a dimensionless ratio relating inertia forces to viscosity forces in a fluidic system. Equation 4.19 defines Re with density, ρ, characteristic length, L, characteristic velocity, V, and viscosity, μ. The Reynolds number is indicative of the regime of the fluid flow (i.e., laminar flow vs. turbulent flow). Laminar flow occurs when $Re < 2000$, and the velocity distribution across the flow channel is parabolic. The viscous forces are dominant in laminar flow. Turbulent flow occurs when $Re > 4000$, where the inertial forces dominate and the viscous forces cannot dampen disturbances caused by roughness of the flow channel walls. The velocity profile in turbulent flow is more uniform, with eddies in the flow field, which enhances a mixing action. A transition region occurs $2000 < Re < 4000$, where the nature of the flow field depends upon the surface roughness of the flow channel.

$$\text{Re} = \frac{\rho VL}{\mu} \propto S \qquad (4.19)$$

The Reynolds number scales by S, which will decrease the Reynolds number, indicating that the viscous forces are dominating and microscale fluid flow will be laminar. The laminar flow regime will make fluidic mixing at the microscale difficult.

The Knudsen number (Kn), Equation 4.20, is the ratio of the molecular mean free path, λ, to a characteristic length of the flow field, L. Kn is used as a measure for gas flow field characteristics. Gas flows occur in a number of MEMS applications such as microactuators, inertial sensors (i.e., accelerometers, gyroscopes) as well as in MEMS fabrication tools such as CVD reactors. Based upon the Knudsen number, gas flow can be classified [12,13] as follows:

- $Kn < 0.01$: continuum
- $0.01 < Kn < 0.1$: slip
- $0.1 < Kn < 10$: transition
- $Kn > 10$: free molecular

Table 4.1 lists several gas flow examples and their Knudsen number regime. An exercise at the end of this chapter will discuss some of these examples in more detail.

For a very small size gas flow field, which occurs in MEMS devices and the clearance in a gas bearing (i.e., ~2 μm), there are only a few tens of mean free paths (mfps) across the clearance. In a macroflow of gas in a pipe, there may be

TABLE 4.1
Knudsen Number Regimes in a Few Gas Flow Examples

Application	Mean free path λ	Characteristic length	Knudsen regime	Ref.
Pipe flow	~0.1 μm	>1 mm	Continuum	
MEMS device	~0.1 μm	2 μm	Slip	14
CVD reactor	~0.1 μm–100 μm @STP–1 torr respectively	0.1–2 μm	Transition-free molecular	2
Disk drive heads	65 nm	15 nm	Free molecular	13

many 1000s of mfps across the diameter of the pipe, thus putting this in the continuum regime. At the other extreme of the Knudsen regime is the free molecular flow that can be seen to occur in a disk drive head with a clearance of approximately 15 nm. The various Knudsen regimes are important to the operation of the physical system involved and indicative of the assumptions required to model the phenomena properly.

The fluid mechanics that occurs in the microdomain are one of the least understood physical phenomena. The work of Dohner et al. [14] is just one instance of the inconsistencies of current fluid mechanic theories at the microscale that must be addressed in the future to aid MEM design engineers to understand the physics with which they are dealing.

$$Kn = \frac{\lambda}{L} \tag{4.20}$$

The Weber number, We, is a nondimensional ratio relating inertia forces and surface tension in a fluid system. Equation 4.21 defines the Weber number with density, ρ, characteristic length, L, characteristic velocity, V, and surface tension, σ (N/M). We scales with S, which is indicative of the increasing influence of surface tension forces as the system scale, S, decreases. An example of the significance of surface tension at the microscale is the release process in surface micromachining and the resulting issues of stiction. During the drying phase of the release process, it is possible for a meniscus to form in which the surface tension force can possibly pull the mechanical layers together.

Surface tension forces can also be used to advantage in MEMS applications. These forces can be used to assemble a MEMS structure as demonstrated in Syms et al. [15]. Figure 4.8 shows a MEMS steam engine in which the meniscus is used to seal the space between the piston and cylinder.

$$We = \frac{\rho V^2 L}{\sigma} \propto S \tag{4.21}$$

Scaling Issues for MEMS

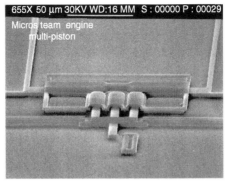

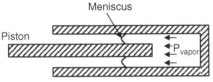

FIGURE 4.8 MEMS "steam" engine. (Courtesy of Sandia National Laboratories.)

Example 4.3

Problem: Given a cube of dimension, d, on a side and density, ρ_c, floating in a liquid of surface tension, σ. (a) Calculate the cube dimension, d at which surface tension force is greater than the cube weight. (b) The cube is made of silicon (ρ_c = 2300 kg/M^3) and the liquid is water (σ =0.072 N/M). What is the cube dimension from part a?

Solution: (a) The surface tension force is the liquid surface tension, σ, around four sides of length d. The cube weight is the product of cube volume, d^3, cube density, ρ_c, and gravity, g:

$$F_{surface\_tension} > W_{cube}$$

$$4\sigma d > \rho_c d^3 g$$

$$\sqrt{\frac{4\sigma}{\rho_c g}} > d$$

(b) Substituting the preceding values with g = 9.8 M/s^2, yields d = 3.575 mm. A cube of this dimension or smaller will float in the water due to surface tension effects only.

Example 4.4

Problem: (a) Evaluate the distributed force per unit length on the beam shown in Figure 4.9 for the forces of gravity, surface tension. (b) Find the voltage required

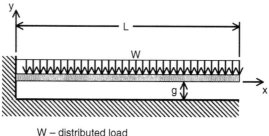

FIGURE 4.9 Distributed load on a cantilever beam.

to produce an electrostatic distributed force of the same magnitude as the gravity and surface tension forces. (c) Find the scaling relationships between the distributed forces of gravity, surface tension, and electrostatics.

Solution: (a) The relationships for the distributed forces per unit length on the beam of gravity and surface tension are:

$$W_{gravity} = \frac{Ma_{gravity}}{L} = \frac{\rho V a_{gravity}}{L} = \frac{\rho L t b a_{gravity}}{L}$$

$$W_{SurfTension} = \frac{2L\gamma}{L}$$

Using the values shown in Figure 4.9 yields the following values for the distributed force due to gravity and surface tension:

$$W_{gravity} = 4.5 \times 10^{-7} \quad N$$

$$W_{SurfTension} = 0.144 \quad N$$

(b) The electrostatic distributed force for the beam is:

$$Wes = \frac{1}{2}\frac{\varepsilon A V^2}{g^2}\frac{1}{L} = \frac{1}{2}\frac{\varepsilon (bL) V^2}{g^2}\frac{1}{L}$$

Scaling Issues for MEMS

The voltage required to produce an equivalent electrostatic distributed force to the distributed forces produced by gravity and surface tension is:

$$V = 0.2 \ V \implies W_{es} = W_{gravity}$$

$$V = 114.1 \ V \implies W_{es} = W_{SurfTension}$$

(c) The scaling for the three types of distributed forces is shown in the following equations. As the scale factor, S, decreases, the only force that increases is electrostatic. However, the surface tension force is strong as evidenced by the high voltage to offset it in part (b). To reduce the required voltage may require an impractically small gap.

$$W_{gravity} = \rho t b a_{gavity} \propto S^2$$

$$W_{SurfTension} = \frac{2L\gamma}{L} \propto S^0$$

$$Wes = \frac{1}{2}\frac{\varepsilon A V^2}{g^2}\frac{1}{L} = \frac{1}{2}\frac{\varepsilon b V^2}{g^2} \propto S^{-1}$$

4.1.5 Electrical System Scaling

The effect of scale for electrical systems will be considered by accessing the impact of scale on the basic electrical circuit elements (i.e., resistor, capacitor, inductor), shown in Figure 4.10, and the actuation capabilities of electric and magnetic fields.

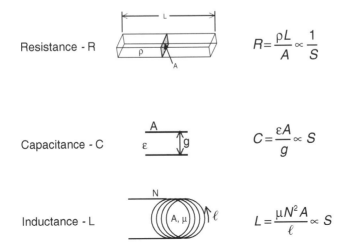

FIGURE 4.10 Electrical circuit element scaling.

The resistance, R, of a piece of material is a function of the resistivity, ρ, of the material and its geometric size. The geometry enters the equation via the ratio of the length of material, l, in direction of current flow and the cross-section area, A. This results in the resistance of an object increasing by $1/S$ when the object is isomorphically reduced in size, S.

$$R = \rho \frac{l}{A} \propto \frac{1}{S} \tag{4.22}$$

The capacitance of a capacitor, C, is defined by the permittivity, ε, of the dielectric material between two plates of area, A, separated by a gap, g, as shown next. Therefore, the capacitance will be reduced by S when isomorphically scaled.

$$C = \varepsilon \frac{A}{g} \propto S \tag{4.23}$$

The inductance, L, of a N loops of material encompassing material of permeability, μ, can be calculated by Equation 4.24. The inductance, L, similar to a capacitor, will scale by S upon isomorphic scaling.

$$L = \mu N \frac{A}{l} \propto S \tag{4.24}$$

Electrical circuit elements of inductance and capacitance will decrease in proportion to the reduction in scale, but resistance will increase inversely with scale. CMOS microelectronics has capabilities of fabricating transistors, resistors, and capacitors; in a simple view, CMOS transistors are just capacitance controlled switches. This means that a simple reduction of size of an existing MEMS or CMOS design containing some combination of these circuit elements cannot be done because the different circuit elements scale differently.

Knowledge of the scaling effects on the electrical circuit elements is useful in the design of many MEMS devices that may contain electrical circuit elements on the same die or as part of the mechanical structure for the MEMS device. For example, a MEMS inertial sensing device (e.g., accelerometer, gyroscope) will contain a suspended mass and use a variable capacitor to enable capacitive sensing of the displacement of the suspended mass.

Electrical scaling also has importance for the design of actuated MEMS devices. In the macroworld, electromagnetic actuation dominates. With extremely few exceptions, electrical to mechanical (i.e., electric motors) or mechanical to electrical (i.e., electric generators) utilize magnetic fields. As the sizes of devices are reduced to the MEMS scale, it is necessary to evaluate whether this still remains the best choice [21].

Scaling Issues for MEMS

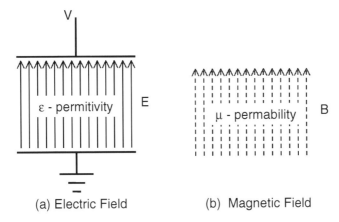

FIGURE 4.11 Electric and magnetic fields in a region of space.

One way to make this assessment of electric vs. magnetic fields for actuation is to consider the energy density of an electric, $U_{electric}$, and a magnetic, $U_{magnetic}$, field for a region of space at the appropriate operational condition (Figure 4.11).

Equation 4.25 and Equation 4.26 define the electric and magnetic field density, respectively, where ε is the permittivity and μ is the permeability of the region that contains the electric field, E, and the magnetic field, B. For purposes of this assessment, the free space permittivity, $\varepsilon_0 = 8.84 \times 10^{-12}$ F/M, and the free space permeability, $\mu_0 = 1.26 \times 10^5$ H/M will be used. The maximum value of the electric field, E, and magnetic field, B, will be limited by the maximum obtainable operational values.

The maximum obtainable electric field is at the point just before electrostatic breakdown. This breakdown occurs when the electrons or ions in an electric field are accelerated to a sufficient energy level so that, when they collide with other molecules, more ions or electrons are produced, resulting in an avalanche breakdown of the insulating medium; high current flow is produced. For air at standard temperature and pressure, the electric field at electrostatic breakdown in macroscopic scale gaps between electrodes (i.e., > ~10 μm) is $E_{max} = 3 \times 10^6$ V/M.

$$U_{electric} = \frac{1}{2} \varepsilon E^2 \tag{4.25}$$

$$U_{magnetic} = \frac{1}{2} \frac{B^2}{\mu} \tag{4.26}$$

The maximum obtainable magnetic field energy density is limited by the saturation of the magnetic field flux density in magnetic materials. In materials, the spin of an electron at the atomic level will produce magnetic effects. In many

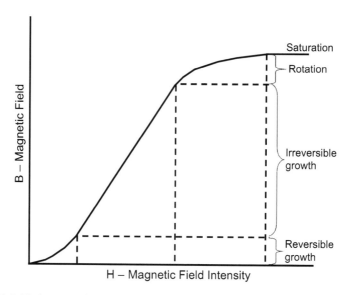

FIGURE 4.12 An example a magnetization curve.

materials, these atomic level magnetic effects are canceled out due to their random orientation. However, in ferromagnetic materials, adjacent atoms have a tendency to align to form a magnetic domain in which their magnetic effects collectively add up. Each magnetic domain can be from a few microns to a millimeter in size [17], depending upon the material and its processing and magnetic history. However, the domains are randomly oriented and the specimen exhibits no net external magnetic field. If an external magnetic field is applied, the magnetic domains will have a tendency to align with the magnetic field.

Figure 4.12 shows a plot of the magnetic flux density, B, vs. the magnetic field intensity, H, for a ferromagnetic material. The magnetic field intensity, H, is a measure of the tendency of moving charge to produce flux density (Equation 4.27). Figure 4.12 shows that, as H is increased, the magnetic flux density, B, increases to a maximum in which all the magnetic domains are aligned. For magnetic iron materials, the saturated magnetic flux, B_{sat}, is approximately 1 to 2 T. A B_{sat} of 1 T will be used for this assessment of magnetic field density.

$$H = \frac{B}{\mu} \tag{4.27}$$

Using the limiting values of E_{max} and B_{sat} discussed earlier to calculate the electric and magnetic field densities will yield the values shown next. These results indicate that the magnetic field energy density is 10,000 times greater than the electric field energy density. This calculation explains why electromagnetic actuation is dominant in the macroworld.

$$U_{electric} = \frac{1}{2}\varepsilon_0 E_{max}^2 = 3.98 \times 10^1 \ \frac{J}{M^3}$$

$$U_{magnetic} = \frac{1}{2}\frac{B_{max}^2}{\mu_0} = 3.96 \times 10^5 \ \frac{J}{M^3}$$

(4.28)

However, for MEMS scale actuators, the electrode spacing or gaps can be fabricated as close as 1 μm. MEM researchers [1,2,19] have noticed that the electric field, E, can be raised significantly above the breakdown electric field, E_{max} discussed earlier for macroscale gaps. This increased breakdown electric field for small gap sizes is predicted by Paschen's law [18], which was developed over 100 years ago. This law predicts that the electric field at breakdown, E_{max}, is a function of the electrode separation (d) – pressure (p) product. Figure 4.13 illustrates the basic functional dependence of Paschen's law, $E_{max} = f(p,d)$. Figure 4.13 shows that the separation-pressure product decreases to a minimum, which is the macroscopic breakdown electric field, E_{max}^{macro}.

However, as the separation-pressure product is decreased further, the breakdown electric field starts to increase. This increase in the electric field required for breakdown is because the gap is small and there are few molecules for ionization to occur. As the electrode separation becomes smaller, a fewer number of collisions occur between an electron or ion with a gas molecule because the mfp (mean free path) between collisions is becoming a greater fraction of the electrode separation distance. Decreasing the gas pressure also results in fewer collisions because decreasing the number of molecules increases the mfp length between collisions. This means that fewer collisions occur in a given electrode separation distance. The effect causes the breakdown electric field to increase

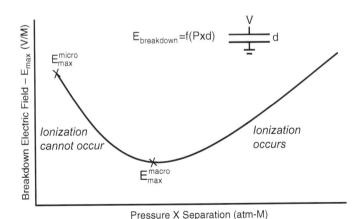

FIGURE 4.13 Paschen's law: breakdown electric field, E_{max} (V/M), vs. the electrode separation — pressure product (M-atm).

134   Micro Electro Mechanical System Design

with decreasing separation-pressure product up to a maximum, E_{max}^{micro}, for microscale electrode spacings. The electric field for small electrode separation distances in vacuum have been reported [20] to be

$$E_{max}^{micro} = 3.0 \times 10^8 \; \frac{V}{M}$$

Using this new value for E_{max} will change the comparison of the electric and magnetic field energy density calculation of Equation 4.29 as shown next. This results in a more favorable but neutral comparison of the energy density of electric and magnetic fields. However, the literature indicates that, for MEMS applications, electrostatics predominates. This is due to the added fabrication and assembly complexity of fabricating MEMS scale permanent magnets, coils of wire, and the associated resistive power losses with their use.

$$U_{electric} = \frac{1}{2}\varepsilon_0 \left(E_{max}^{micro} \right)^2 = 3.98 \times 10^5 \;\; \frac{J}{M^3}$$

$$U_{magnetic} = \frac{1}{2}\frac{B_{max}^2}{\mu_0} = 3.96 \times 10^5 \;\; \frac{J}{M^3}$$

$$(4.29)$$

In another simple comparison of electric and magnetic fields, it can be seen that the magnetic field energy density, $U_{magnetic}$, does not change with size scaling because B_{sat} and μ are material properties that do not change appreciably with scaling to the microdomain. However, assuming that the applied voltage remains constant up to the limit of E_{max} at electrostatic breakdown shows that the electric field energy density, $U_{electric}$, varies with scale as shown in Equation 4.30. This gives electrostatic actuation increasing importance as devices are scaled to the microdomain.

$$U_{electric} = \frac{1}{2}\varepsilon_0 E^2 \propto \frac{1}{S^2}$$

$$U_{magnetic} = \frac{1}{2}\left(\frac{B}{\mu_0} \right)^2 \propto S^0$$

$$(4.30)$$

4.1.6 OPTICAL SYSTEM SCALING

Optical MEMS applications and research is an extremely active area, with MEMS devices developed for use in optical display, switching, and modulation applications. These MEMS scale optical devices [23,24] include LEDs, diffraction gratings, mirrors, sensors, and waveguides. Their operation can depend upon optical absorption or reflection for functionality.

Scaling Issues for MEMS

Optical absorption-based devices are governed by Beer's law (Equation 4.31), which can be seen to scale unfavorably to MEMS size because absorption depends on path length. This has spurred the development of folded optical path devices [22] to overcome this disadvantage, but this is ultimately limited by the reflectivity losses incurred with a large number of path folds.

$$A = \varepsilon CL \propto S \qquad (4.31)$$

where
    A = Optical absorption
    ε = molar absorptivity (wavelength dependent)
    C = concentration
    L = distance into the medium

Optical reflection-based MEMS devices are used for optical switching, display, and modulation devices. MEMS optical devices that have a displacement range from small fractions of a micron to several microns can be made. This corresponds to the visible light spectrum up to the near infrared wavelengths (Figure 4.1). Because electrostatic actuation is frequently used in MEMS devices, very precise submicron displacement accuracy is attainable. Also, very thin low-stress optical reflective coatings are possible. These attributes make a MEMS optical element very attractive.

4.1.7 CHEMICAL AND BIOLOGICAL SYSTEM CONCENTRATION

Miniaturization of fluidic sensing devices with MEMS technology has made miniature chemical and biological diagnostic and analytical devices possible [25,26]. To assess the effect that reduction in scale will have on these devices, the concentration of chemical or biological substances and how it is quantified must be studied.

Before the concentration of a chemical solution can be defined, a few preliminary definitions will be stated. A *mole* (mol) is a quantity of material that contains an Avogadro's number (N_A = 6.02 × 10²³) of molecules. The mass in grams of a mole of material is the molecular weight of the chemical substance in grams. The is known as the *gram molecular weight* (MW) and has units of grams per mole. Example 4.5 illustrates how the MW is calculated for salt.

Example 4.5

Problem: Calculate the gram molecular weight (MW) of common table salt (i.e., sodium chloride, NaCl). The atomic mass of sodium (Na) = 23.00. The atomic mass of chlorine (Cl) = 35.45. The molecular weight of NaCl = 58.45. The gram molecular weight of NaCl is MW = 58.45 g/mol.

Solution: The concentration, C, of a chemical in a solution is known as the *molarity* of the solution. A 1-*molar* solution (i.e., 1 M) is 1 mol of a chemical

136 Micro Electro Mechanical System Design

dissolved in 1 liter of solution. For example, a 1-M solution of NaCl consists of 58.45 g of NaCl dissolved in a liter of solution. This relationship is expressed in Equation 4.32.

$$W = MW \cdot C \cdot V$$
$$gram = \frac{gram}{mole} \cdot \frac{mole}{liter} \cdot liter \tag{4.32}$$

For chemical detection, the number of molecules, N, in a given sample volume, V, may be important to quantify. This relationship between number of molecules in a given concentration of solution, C, and volume of solution, V, is:

$$N = N_A \cdot C \cdot V$$
$$molecules = \frac{molecules}{mole} \cdot \frac{mole}{liter} \cdot liter \tag{4.33}$$

Figure 4.14 shows the relationship between concentration, C, and sample volume, V, as expressed by the preceding equation. The boundary for less than one molecule, N_1, of chemical or biological substance in a given sample volume is shown; this is an absolute minimum sample volume for analysis. The number of molecules required for detection, N_D, is some amount greater than N_1 (i.e., $N_D > N_1$). The required sample volume for analysis would be at the intersection of the N_D boundary with the concentration of the analyte available for analysis. Petersen et al. [26] have shown that the typical concentrations of chemical and biological material available for a few types of analyses are as shown in Table 4.2.

The miniaturization of chemical and biological systems has a few fundamental limits:

- The trade-off between sample volume, V, and the detection limit, N_D, for a given concentration of analyte, C, is illustrated in Figure 4.14.
- Further miniaturization may require increasing the concentration of analyte or increasing the sample volume.
- The use of small sample volumes requires increasingly sensitive detectors, which may be limited by other scaling issues (i.e., electrical, fluidic, etc.).
- The physical size limitation of biological sensing devices is limited by the size of the biological entity. A cell is approximately 10 to 100 µm, whereas DNA has a width of only ~2 nm but is very long.

Scaling Issues for MEMS 137

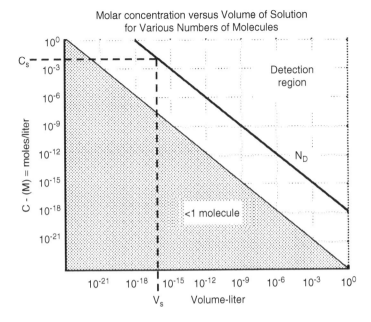

FIGURE 4.14 Concentration vs. sample volume.

TABLE 4.2
Typical Analyte Concentrations for Various Types of Analyses

Uses	Concentration (moles/liter)
Clinical chemistry assays	10^{-10}–10^{-4}
Immunoassays	10^{-17}–10^{-6}
Chemical, organisms, DNA analyses	10^{-22}–10^{-17}

4.2 COMPUTATIONAL ISSUES OF SCALE

The computational aspects of the scale of MEMS devices need to be considered because much of modern engineering design depends upon numerical simulation to achieve success. Due to fabrication challenges, long fabrication times, and experimental measurement difficulties, MEMS applications rely more upon simulation than their macroworld counterparts do. Therefore, time would be well spent in assessing the unique issues encountered in simulation of MEMS scale devices.

Engineering calculations are almost exclusively performed on digital computers in which the numbers representing the input data (i.e., mechanical and electrical properties, lengths, etc.) and the variables to be calculated are represented by a fixed number of digits. Due to this digital representation of numbers,

138   Micro Electro Mechanical System Design

the quantity known as *machine accuracy*, ε_m, is the smallest floating point number that can be represented on a given computer. The machine accuracy is a function of the design of the particular computer. Two types of errors arise in the calculations performed on digital computers [38]:

- *Truncation error* arises because numbers can only be represented to a finite accuracy (i.e., machine accuracy) on a digital computer.
- *Round-off error* arises in calculations, such as the solution of equations, due to the finite accuracy of the computer. Round-off error accumulates with increasing amounts of calculation. If the calculations are performed so that the errors accumulate in a random fashion, the total round-off error would be on the order of $\sqrt{N}\varepsilon_m$, where N is the number of calculations performed. However, if the round-off errors accumulate preferentially in one direction, the total error will be of the order $N\varepsilon_m$.

The topics of truncation and round-off error arise in regular macroscale engineering simulation; however, a unique aspect of computation for MEMS scale simulation needs to be addressed:

- *Convenient units scale of numbers for MEMS simulation.* The system of units typically used in engineering simulations (e.g., MKS) uses units of measure of quantities typically encountered for macroscale devices. For example, the MKS system of unit length measure is meters. However, MEMS devices are on a size scale of microns (i.e., 0.000001 m).
- *Numerically appropriate scale of unit for MEMS simulation.* Numerical simulations such as finite element analysis (FEM) [39,40] typically involve the solution of a large system of equations (e.g., 1,000 → 1,000,000). This system of equations will become ill conditioned when the quantities involved in the equations vary widely in magnitude. A large ill-conditioned system of equations can produce inaccurate results or may even be unsolvable. For example, ill conditioning can arise when a very small number is subtracted from a very large number; this will make the result unobservable due to the truncation and round-off errors of digital computation.

From a CAD layout perspective, the unit of length most appropriate for a MEMS scale device is a micron (i.e., 1 μm = 0.000001 m). This will allow the CAD design of the device to be done using reasonable multiples of a basic unit of measure.

From a numerical computation perspective, the system of units needed to express the basic quantities used in MEMS device simulation should be a numerically similar order of magnitude. This will avoid the ill conditioning of the numerical simulation problem. A system of units for MEMS simulation has been proposed [41] for finite element analysis. Appendix C provides the conversion

Scaling Issues for MEMS

factors between the MKS system and the μMKS system, which will be used in the design sections of this book. Several different permutations of an appropriate system of units are possible. However, a *consistent set of units* must be used in any simulation. This will maintain dimensional consistency for material properties and simulation problem parameters such as loads and boundary conditions.

4.3 FABRICATION ISSUES OF SCALE

To assess the fabrication issues unique for MEMS scale devices, it is necessary to put MEMS fabrication processes and technologies in perspective with manufacturing processes for other size scales. The size scales for manufacturing that will be discussed are large-scale construction, macroscale machining, MEMS fabrication, and integrated circuit (IC) and nanoscale manipulation. These are individually discussed next. These four size groups provide a wide spectrum that will enable the evaluation of any fabrication issues due to scale.

- Large-scale construction (>15 m). The fabrication of things in this size category includes civil structures, marine structures, and large aircraft. Manufacturing at this size scale involves a wide array of processes for materials such as wood, metal, and composite materials.
- Macroscale machining (2 mm to 15 m). Manufacturing at this scale includes a plethora of processes and materials. In many cases, the manufacturing processes and materials have been under development and improvement for an extended period. These manufacturing processes are mature and quite flexible. In most instances, more than one approach to the manufacture of a given item is available. Examples of items manufactured in this category include automobile or aircraft engines, pumps, turbines, optical instruments, and household appliances.
- MEMS scale fabrication (1 μm to 2 mm). MEMS fabrication includes the processes and technologies discussed in Chapter 2 and Chapter 3 to produce devices that range in size from 1 μm to 2 mm. This category of manufacturing has been under development for 30 years and has started to produce commercial devices within the last 10 years. To a large degree, the fabrication methods for MEMS are rooted in the IC infrastructure. As a result, the range of materials and the flexibility of the fabrication processes are more restrictive than in macroscale machining. Silicon-based materials are frequently used in surface and bulk micromachining. LIGA uses electroplateable materials (e.g., nickel, cooper, etc.). When LIGA molds are used with a hot embossing, plastic materials can be utilized to create devices.
- IC and nanoscale manipulation (<1 μm). The size scale for these fabrication technologies is 1 μm and below (i.e., <1 μm). IC fabrication technology has been under development and continuous improvement for 40 years [29] and relies on leading edge photolithography, CVD deposition, and etching techniques similar to those presented in Chap-

ter 2. The IC manufacture included in this category are state-of-the-art capabilities that are rapidly approaching 0.1 µm feature sizes and below. Nanoscale manipulation [32] is a recent demonstrated use of surface profiling tools [30,31] such as an atomic force microscope (AFM) and a scanning tunneling microscope (STM). These enable the individual manipulation of molecules. Nanoscale manipulation is a laboratory-based research capability as contrasted with IC manufacture, which is a mature large industrial capability.

The smallest feature that can be fabricated on a part is the *feature size*. From a design perspective, a more useful quantity to assess a fabrication capability is the *relative tolerance*. Relative tolerance is defined as the feature size divided by part size; this provides a measure of the precision with which a fabrication process can produce a part of any given size.

Figure 4.15 shows a graph of the relative tolerance vs. size over a considerable range. The four size categories defined earlier are noted in this figure, and the data for this graph are extracted from a number of sources [2,27,28,30–35]. Due to the extended size range and large number of fabrication processes that exist, the data in this graph should be viewed as a broad statement of the fabrication processes in a given size range rather than as indicative of any specific fabrication process or capability. Because of the large number and variety of macroscale fabrication processes, data were extracted [27,33] for some broad ranges of processes (e.g., grinding, milling, etc.) within this category. Figure 4.15 shows that macroscale fabrication has the smallest relative tolerance or precision, with the relative tolerance increasing as the size scale increases or decreases. This shows that MEMS scale fabrication has about the same precision as that of large-scale fabrication (i.e., MEMS devices have about the same level of precision as one's house!).

Due to the large variety and flexibility of macroscale fabrication processes, a number of categories of precision or relative tolerance have been defined [27,33]; these are shown in Figure 4.16 and Table 4.3. Ultraprecision machining is at the extreme level of precision and is reserved for only a few applications due to the time and expense necessary. Only a few instances, such as some large optical applications [36,37], require this level of precision. Figure 4.16 shows where these levels of precision lie relative to the MEMS-scale and nanoscale manipulation.

The fabrication issues of scale show that a MEMS designer is faced with fewer options and more restrictions than those faced by the macroworld design engineer. MEMS scale fabrication imposes the following concerns for the design engineer; they will need to be addressed in the device design:

- Limited material set availability
- Fabrication process restrictions upon design
- Reduced level of precision in the fabricated device

Scaling Issues for MEMS

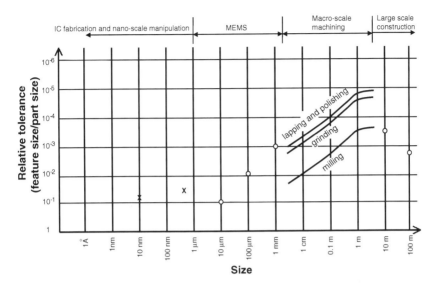

FIGURE 4.15 Manufacturing accuracy at various size scales.

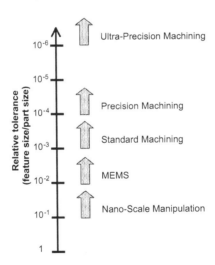

FIGURE 4.16 Relative tolerance levels.

4.4 MATERIAL ISSUES

As the size of a device is decreased, two general trends become evident:

- *The granularity of the solid or fluid materials* becomes increasingly apparent. This granularity can be expressed by quantities (see Table 4.4) such as the grain size of a material or the mfp in a gas. Does this

TABLE 4.3
Summary of Fabrication Methods, Size, and Relative Tolerances at Various Scales and Precisions

Fabrication scales	Methods	Size	Relative tolerance	Ref.
Large scale construction	Cutting, forging, forming processes, welding and fastening	>15 m	$<10^{-2}$	
Macromachining				
Ultraprecision machining	Single-point diamond turning, polishing, lapping	2 mm–15 m	$<10^{-6}$	33, 37
Precision machining	Grinding, lapping, polishing		$<10^{-4}$	35, 36
Standard machining	Milling, cutting processes, grinding		$<10^{-3}$	27, 28
MEMS	LIGA, bulk micromachining, surface micromachining.	1μm–2 mm	$<10^{-2}$	
IC	Photolithography, CVD, etching processes	1μm–100 nm	$<10^{-2}$	
Nanoscale manipulation	Focused ion beam, scanning tunneling microscope, atomic force microscope	<100 nm	~0.1	32

TABLE 4.4
Size Scale of Phenomena Relevant to MEMS

Physical entity	Approximate size
Mean free path of air @ STP	65 nm @ STP
Lattice constant	5.431Å for silicon
Material grain size	300–500 nm for polysilicon
Magnetic domains	25 μm

    violate the assumption of continuum mechanics frequently used in the macroworld to model engineering phenomena?

- *New physical phenomena* (e.g., Brownian motion, Paschen effect, electron tunneling current) become significant due to the reduced volume or spacing in MEMS devices.

    The classical engineering models used to design and simulate macroworld physics and devices are based upon continuum mechanics, which models the physics of interest with a set of partial differential equations. Table 4.5 shows a sampling of the array of physical phenomena modeled by such equations. These equations involve partial derivatives of the variable of interest, such as

Scaling Issues for MEMS

143

TABLE 4.5
Physical Phenomena Modeled by Continuum Mechanics

Physical phenomenon	Partial differential equation
Three-dimensional heat flow	$$\frac{\partial u}{\partial t} = c^2 \nabla^2 u$$ $$= c^2\left(\frac{\partial^2 u}{\partial x^2} + \frac{\partial^2 u}{\partial y^2} + \frac{\partial^2 u}{\partial z^2}\right)$$
Three-dimensional wave equation	$$\frac{\partial^2 u}{\partial t^2} = c^2 \nabla^2 u$$ $$= c^2\left(\frac{\partial^2 u}{\partial x^2} + \frac{\partial^2 u}{\partial y^2} + \frac{\partial^2 u}{\partial z^2}\right)$$
Elastic equations of equilibrium for solid mechanics	$$\frac{\partial \sigma_x}{\partial x} + \frac{\partial \tau_{yx}}{\partial y} + \frac{\partial \tau_{zx}}{\partial z} + F_x = 0$$ $$\frac{\partial \tau_{xy}}{\partial x} + \frac{\partial \sigma_y}{\partial y} + \frac{\partial \tau_{zy}}{\partial z} + F_y = 0$$ $$\frac{\partial \tau_{xz}}{\partial x} + \frac{\partial \tau_{yz}}{\partial y} + \frac{\partial \sigma_z}{\partial z} + F_z = 0$$
Maxwell's free space electromagnetic equations	$$\nabla \cdot \left(\varepsilon_0 \mathbf{E}\right) = \rho$$ $$\nabla \cdot \mathbf{B} = 0$$ $$\nabla \times \mathbf{E} = -\frac{\partial \mathbf{B}}{\partial t}$$ $$\nabla \times \frac{\mathbf{B}}{\mu_0} = J + \frac{\partial\left(\varepsilon_0 \mathbf{E}\right)}{\partial t}$$
Navier–Stokes equations for compressible fluid dynamics	$$\rho\left[\frac{\partial \mathbf{V}}{\partial t} + \left(\mathbf{V} \cdot \nabla\right)\mathbf{V}\right] = -\nabla P + \mathbf{F} \cdots$$ $$-\nabla \times \left[\mu\left(\nabla \times \mathbf{V}\right)\right] + \nabla\left[\left(\lambda + 2\mu\right)\nabla \cdot \mathbf{V}\right]$$

stress, displacement, or temperature, and some parameters (i.e., modulus of elasticity, heat transfer coefficients, speed of sound in a media) that model the domain that the set of equations govern. For these equations to be easily solved, the parameters must be known and the variable of interest smoothly varying over the domain of interest (i.e., differentiable). If a material is discrete or

144 Micro Electro Mechanical System Design

discontinuous (e.g., *granular*), it is more difficult to model the system with a continuum mechanics approach.

As one tries to design and model systems on smaller scales, a certain granularity of the physics is observed. In Chapter 2, the material structures of crystalline, polycrystalline, and amorphous were discussed. (Figure 2.2 illustrates these three material structures.) The spacing of atoms in crystalline and amorphous materials is at the atomic scale (i.e., <1 nm). The size of the individual crystals in a polycrystalline material are on the order of 100 to 500 nm, depending upon the material processing used. Many materials of engineering significance are polycrystalline. The physical parameters used to describe material behavior (e.g., Young's modulus, speed of sound) in a continuum mechanics model are statistical averages of the effects of the individual grains or molecules of material within a large object (relative to the grain size).

For example, for a macrodevice that is 2 cm wide with a 500 nm grain size, the statistically averaged property representing a parameter such as Young's modulus is adequate. However, a 2-μm wide microdevice contains only a few grains of material, and a statistically averaged approximation of a material property is not adequate. Research has been ongoing to measure microscale effects [42]; develop theories that apply at the microscale [43,44]; and incorporate these effects into simulations of the microscale phenomena [45].

The statistically averaged assumption also plays a role in the failure model of materials. The stress at which a material yields or fails is quantified by the parameters, yield strength, S_y, or failure strength, S_u. These parameters also have statistics in their origin. A material has a certain number of defects in the material structure (e.g., crystal lattice imperfections, corrosion products in the grain boundaries) that give rise to locations at which a material will yield or ultimately fail. These defects are assumed to be statistically distributed throughout the material. The *defect density* of a material and statistical process control is frequently used in the microelectronic community [46] in assessments and modeling of the yield (i.e., percentage of good devices manufactured) of their processes. A *potential advantage* of scaling devices down to densities approaching the defect density of the material is that devices could be produced with a low defect rate.

4.5 NEWLY RELEVANT PHYSICAL PHENOMENA

Several new phenomena are enabled or become relevant at the MEMS scale. The three briefly discussed next are examples of such phenomena, which gain importance because of the size of a MEMS device or the small gaps used in MEMS devices.

- *Brownian noise.* Also called *thermal noise* or *Johnson noise* for electrical systems, Brownian noise is a low-level noise present in electrical and mechanical systems. This thermal noise is present everywhere in the environment and is due to such things as the vibrations of atoms in the materials from which a device is made and the environment in

Scaling Issues for MEMS

which the device operates. This indicates that the thermal noise is a function of temperature of these materials. The mechanisms that couple these thermal vibrations to the mechanical or electrical device of interest are the energy dissipation mechanisms (i.e., damping for mechanical devices, resistance for electrical devices). As a device is reduced in size, these thermal noises or vibrations become significant for MEMS scale sensors. A detailed discussion of Brownian noise is in the chapter on MEMS sensors.

- *Paschen's effect.* The phenomenon that the *breakdown voltage* in a gap increases as the product of the pressure of the gas in the gap and gap spacing is reduced was discovered in 1889 [19]. This phenomenon is effective when the gap size is very small (<2 μm), which is typical of MEMS devices. This enables increased effectiveness of electrostatic actuation as discussed in detail in Section 4.1.5.

- *Electron tunneling current.* Quantum entities such as electrons can "tunnel" across a very small gap (on the order of nanometers) due to the uncertainty in the wave description of quantum mechanical entities. This especially appears to be strange due to the barrier of classical physics in which like charges repel. This phenomenon can be used in MEMS devices as a very sensitive displacement transduction method capable of resolving displacements on the order of 0.01 nm. A MEMS cantilever can be fabricated with a tip suitable for tunneling that is electrostatically brought within operating distance for this phenomenon to be effective. The tunneling phenomenon will be discussed in more detail in the chapter on MEMS sensors.

4.6 SUMMARY

A MEMS designer needs to be aware of a number of wide ranging issues and cannot rely solely on macroworld engineering experiences and training when considering the implementation of a MEMS design. System parameters will change in relative importance as the system scale is reduced. Table 4.6 shows four quantities that can be directly or indirectly related to actuation forces (i.e., gravity, surface tension, electrostatic, magnetic) in a device. If these forces all scaled in the same manner, heuristic macroworld intuition would be valid; however, these forces all scale differently.

Gravity forces become increasingly small with reduced size, and surface tension increases in importance. Surface tension forces can be used for assembly of devices; however, they can be a concern during MEMS fabrication release processes. Also, the table shows that the electric and magnetic fields and the forces derived from them scale differently, with the magnetic field forces not depending on scale. Table 4.7 summarizes a number of scaling effects for mechanical, fluidic, and thermal systems. The data in this table show that mechanical and thermal time constants are reduced for MEMS systems, and regimes of operation for thermal and fluidic systems are different at MEMS scale. The

TABLE 4.6
Scaling of Force-Generating Phenomena

Force-related quantities	Relationships	Scale factor	Trend as S ↓
Gravity force	$Ma_{gravity} = \rho V a_{gravity}$	$\propto S^3$	↓
Surface tension force	$4L\sigma$	$\propto S^3$	↓
Electric field energy density	$\dfrac{1}{2}\varepsilon E^2$	$\propto \dfrac{1}{S^2}$	↑
Magnetic field energy density	$\dfrac{1}{2}\varepsilon\left(\dfrac{B}{\mu}\right)^2$	$\propto S^0$	↔

discrete nature of solids and fluids (e.g., material grain size, mfp of a gas) also become apparent at MEMS scale.

Furthermore, new physical phenomena such as Paschen's effect, which greatly enables electrostatic actuation, become apparent for MEMS scale devices. Brownian motion and the tunneling effect also become significant at small size, which may cause concern in some instances (i.e., Brownian noise in sensors) or provide additional capability in others (i.e., electron tunneling sensors).

Scaling also has impact in calculations for MEMS devices. An appropriate set of units must be utilized to be convenient in CAD systems and reduce adverse numerical effect in large-scale calculations for MEMS devices.

TABLE 4.7
Summary of Mechanical, Fluidic, and Thermal Scaling

Trend as S

Quantity	Scaling	Interpretation
Mechanical		
Mass $= \rho V$	S^3	Mass of an object
Natural frequency $\omega_n = \sqrt{\dfrac{K}{M}}$	S^{-1}	Transfer function pole
Time constant $\tau = \dfrac{2\pi}{\omega_n}$	S	Mechanical system speed of response
Fluidic		
Reynolds number $\mathrm{Re} = \dfrac{\rho VD}{\mu}$	S	Inertia to viscous forces ratio; metric for fluid flow transition from laminar to turbulent
Weber number $We = \dfrac{\sigma V^2 L}{\sigma}$	S	Inertia to surface tension forces ratio

TABLE 4.7 (Continued)
Summary of Mechanical, Fluidic, and Thermal Scaling

Quantity	Scaling	Interpretation	Trend as S
Knudsen number $Kn = \dfrac{\lambda}{L}$	S^{-1}	Mean free path to characteristic dimension ratio	↑
		Thermal	
Biot number $Bi = \dfrac{hL}{K}$	S	Ratio of the convection and conduction heat transfer coefficients; indicative of the ability of a body to come to thermal equilibrium without thermal stresses	↓
Grashof number $Gr = \dfrac{g\beta\left(T_w - T_\infty\right)L^3}{\upsilon^2}$	S^3	Ratio of the buoyancy forces to the viscous forces in a convection thermal system; empirically related to the convection heat transfer coefficient	⇓
Thermal time constant $\tau = \left(\dfrac{\rho c_p}{K}\right)\left(\dfrac{V}{A}\right) = \alpha\left(\dfrac{V}{A}\right)$	S	Indicative of the thermal time response of the system	↓

Scaling Issues for MEMS

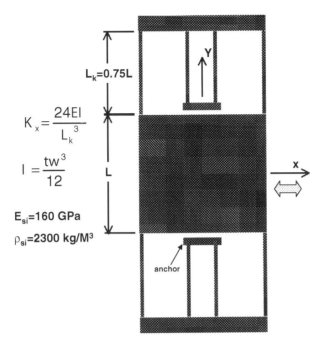

FIGURE 4.17 Double folded spring and mass resonator.

QUESTIONS

1. Explain the effect that scale factor reduction has on mechanical system parameters of mass, stiffness, and natural frequency.
2. Figure 4.17 shows a resonator made with a single level surface micromachine process that oscillates in the x axis. The layer thickness is $t = 2.5$ μm. The width of the springs is 2 μm. This system can be idealized as a lumped spring mass system, in which the total spring stiffness of the resonator can be calculated from the equation in Figure 4.17. I is the area moment of inertial of the spring (see Appendix G). Assume the mass of the springs is negligible and consider only the mass of the central oscillating plate. Calculate the natural frequency of the resonator for several spring lengths: $L = 10$ mm, 1 mm, and 100 μm. Does this follow the approximate scaling for natural frequency discussed in this chapter?
3. The spring mass system shown in Figure 4.18 will be actuated by an electrostatic force and have electrical contact on the opposite end. The switch is required to close repeatedly in 0.1 ms. Which of the spring lengths considered in question 2 is most appropriate?
4. The electrodes shown in Figure 4.19 are to be used to produce an actuation force of 10 μN with an applied voltage of less than 10 V. A gap of 1 μm is the smallest that can be manufactured. Plot the obtained

150   Micro Electro Mechanical System Design

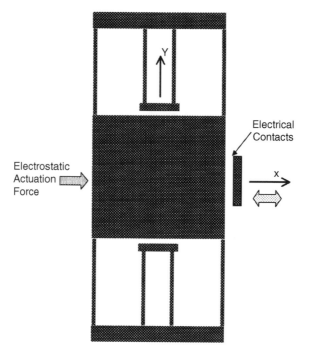

FIGURE 4.18 Actuated spring mass electrical relay contacts.

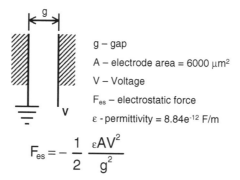

g – gap
A – electrode area = 6000 μm²
V – Voltage
F_{es} – electrostatic force
ε - permittivity = 8.84e⁻¹² F/m

$$F_{es} = -\frac{1}{2}\frac{\varepsilon A V^2}{g^2}$$

FIGURE 4.19 Electrostatic gap for actuation.

   force vs. the gap for 10 V applied. What gap size is recommended? If
   the gap cannot be made small enough, what are the possible alternatives?
5. Calculate the Reynolds number for flow in a square channel of length
   L on a side for a range of L = 10 mm, 1 mm, 100 μm, and 10 μm.
6. Calculate the Knudsen number and determine the gas flow regime for
   the following situations:
   a. A magnetic disk drive head with a "fly" height of 10 nm. Assume
      mfp of air at standard temperature and pressure.

Scaling Issues for MEMS   151

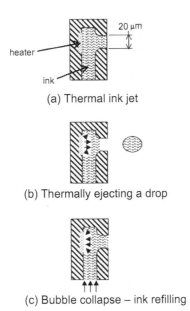

(a) Thermal ink jet

(b) Thermally ejecting a drop

(c) Bubble collapse – ink refilling

FIGURE 4.20 Thermal ink-jet print head.

    b. Gas flow over a MEMS feature (i.e., a 2-μm step) in a CVD reactor operating at low pressure with a gas mfp of 90 λm
    c. Air at STP flowing through a 50-μm MEMS channel
    d. Air at STP between the substrate and an oscillating MEMS structure (i.e., gap of 6 μm)
7. What will be the effect of increasing pressure of a gas have on the mean free path and the Knudson number?
8. Calculate the Reynolds number and the flow regime for the following situations.
    a. A bacteria (assume 2-μm size) moving at a velocity of 0.1 μm/s in water
    b. Water flowing 20 mm/s in a 2-mm pipe
    c. Water flowing at 10 μm/s in a 10-μm channel
9. Explain the effect of the volume/surface area ratio on the thermal characteristics of a system as the scale is reduced.
10. An ink-jet print head is schematically shown in Figure 4.20. The ink jet consists of a heating element, ink channels, and a nozzle. Assume the ink has the fluidic properties of water (Table 4.8). The ink is ejected due to bubble formation by heating the ink. When the bubble collapses, the ink channel refills with ink. The square ink channels in the print head are 20 μm. The ink jet ejects a 10-pL drop on each operating cycle. Calculate the following:
    a. The Reynolds number in the ink-jet nozzle when the 10-pL drop is ejected in 20 μs

TABLE 4.8
Water and Air Properties

Property			Water	Air @ STP
Mean free path	l	nm	—	0.65
Surface tension	s	N/m	72×10^{-3}	—
Dynamic viscosity	m	kg/(m s)	10^{-3}	1.85×10^{-5}
Kinematic viscosity	g	m²/s	10^{-6}	1.43×10^{-5}
Density	r	kg/m³	1000	—

    b. The Reynolds number in the ink jet when the ink channel is refilling upon the collapse of the bubble. The refilling operation takes 200 μs.

    c. From a thermal-scaling perspective, why are such rapid cycle times possible in the ink jet?

    d. If the ink-jet channels were decreased in size, would the thermal cycle time increase or decrease?

11. A chemical sample has a concentration of 10^{-6} mol/l. The detection system has a detection sensitivity of ten molecules per liter.

    a. What volume of sample is required?

    b. If that volume is too big, what should be done?

    c. What effect would increasing the sensitivity have on the required sample size?

REFERENCES

1. J.W. Judy, Microelectromechanical systems (MEMS): fabrication, design and applications, *Smart Mater. Struct.*, 10, 1115–1134, 2001.

2. M.J. Madou, *Fundamentals of Microfabrication, The Science of Miniaturization*, 2nd ed., CRC Press, Boca Raton, FL, 2000.

3. J.D. Meindel, Q. Chen, J.A. Davis, Limits on silicon nanoelectronics for terascale integration, *Science*, 293, 2044–2049, 14, September 2001.

4. W.T. Thompson, *Theory of Vibrations with Applications*, Prentice Hall, Inc., Englewood Cliffs, NJ.

5. A. Lawrence, *Modern Inertial Technology, Navigation, Guidance, and Control*, 2nd ed., Springer-Verlag, Heidelberg, 156, 1998.

6. D.M. Tanner, N.F. Smith, L.W. Irwin. W.P. Eaton, K.S. Helgesen, J.J. Clement, W.M. Miller, J.A. Walraven, K.A. Peterson, P. Tangyunyong, M.T. Dugger, S.L. Miller, MEMS reliability: infrastructure, test structures, experiments, and failure modes, SAND2000-0091, Sandia National Laboratories, January 2000.

7. N. Barbour, J. Connelly, J. Gilmore, P. Greiff, A. Kourepenis, M. Weinberg, Micromechanical silicon instrument and systems development at per laboratory, AAIA Guidance, Navigation Control Conf., San Diego, CA, 20–31 July 1996.

8. D.M. Tanner, J.A. Walraven, K.S. Helgesen, L.W. Irwin, D.L. Gregory, J.R. Stake, N.F. Smith, MEMS a vibration environment, *Proc. IRPS*, 139–145, 2000.

Scaling Issues for MEMS

9. D.M. Tanner, J.A. Walraven, K. Helgesen, L.W. Irwin, F. Brown, N.F. Smith, N. Masters, MEMS reliability in shock environments, *Proc. IRPS*, 129–138, 2000.

10. C.M. Harris, C.E. Crede, *Shock and Vibration Handbook*, 2nd ed., McGraw-Hill Book Company, New York, 1976.

11. J.P. Holman, *Heat Transfer*, 6th ed., McGraw–Hill Book Company, New York, 1986.

12. G.E. Karniadakis, A. Beskok, *Micro Flows: Fundamentals and Simulation*, Springer, New York, 2002.

13. P. Bahukudumbi, A. Beskok, A phenomenological lubrication model of the entire Knudsen regime, *J. Micromech., Microeng.*, 13, 873–884, 2003.

14. J.L. Dohner, M. Jenkins, T. Walsh, K. Klody, Anomalies in the theory of viscous energy losses due to shear in rotational MEMS resonators, Sandia National Laboratories Report, SAND2003-4314, December 2003.

15. R.A. Syms, E.M. Yeatman, V.M. Bright, G.M. Whitesides, Surface tension-powered self-assembly of microstructures — the state of the art, *J. MEMS*, 12(4), August 2003.

16. R.J. Smith, *Circuits, Devices, and Systems*, John Wiley & Sons Inc., New York, 1966.

17. C.T.A. Johnk, *Engineering Electromagnetic Fields and Waves*, John Wiley & Sons, New York, 1988.

18. F. Paschen, Über die zum Funkenübergang in luft, Wasserstoff and Kohlensäure bei verschiedenen Drücken erforderliche Potentialdifferenz, *Weid. Ann. der Physick*, 37, 69, 1889.

19. S.F. Bart, T.A. Lober, R.T. Howe, J.H. Lang, M.F. Schlecht, Design considerations for micromachined electric actuators, *Sensors Actuators*, 14, 269–292, 1988.

20. B. Bollee, Electrostatic motors, *Philips Tech. Rev.*, 30, 178–194, 1969.

21. I.J. Busch-Vishniac, The case for magnetically driven microactuators, *Sensors Actuators* A, A33, 207–220, 1992.

22. A. O'Keefe, D.A.G. Deacon, Cavity ring-down optical spectrometer for absorption measurements using pulsed laser sources, *Rev. Sci. Instrum.*, 59, 2544–2551, 1988.

23. P.F. van Kessel, L.J. Hornbeck. R.E. Meier, M.R. Douglass, A MEMS-based projection display, *Proc. IEEE* 86, 1687–1704, 1998.

24. D.M. Bloom, The grating light valve: revolutionizing display technology, *Proc. SPIE*, 3013, 165–171, February 1997.

25. A. Manz, N. Graber, H.M. Widmer, Miniaturized total chemical analysis systems: a novel concept for chemical sensing, *Sensors Actuators*, B1, 244–248, 1990.

26. K. Petersen, W. McMillan, G. Kovacs, A. Northrup, L. Christel, F. Pourahmadi, The promise of miniaturized clinical diagnostic systems, *IVD Technol.*, July 1998.

27. E. Oberg, Ed., *Machinery's Handbook*, 24th ed., Industrial Press, New York, 2000.

28. G. Boothroyd, W.A. Knight, *Fundamentals of Machining and Machine Tools*, Marcel Dekker, New York, 1989.

29. G.E. Moore, Cramming more components onto integrated circuits, *Electronics*, 38(8), April 19, 1965.

30. Y. Martin, C.C. Williams, H.K. Wickramasinghe, Atomic force microscope force mapping and profiling on a sub 100Å scale, *J. Appl. Phys.*, 61(9), 4723, 1987.

31. G. Benning, H. Rohrer, Scanning tunneling microscopy — from birth to adolescence, *Rev. Mod. Phys*, 59(3), Part 1, 615, 1987.

32. J.A. Stroscio, D.M. Eigler, Atomic and molecular manipulation with the scanning tunneling microscope, *Science*, 254, 1319–1326, 1991.

33. N. Taniguchi, Current Status in and future trends of ultraprecision machining and ultrafine materials processing, *Ann. CIRP*, 32(2), 573–582, 1983.
34. G.L. Benevides, D.P. Adams, P. Yang, Meso-scale machining capabilities and issues, Sandia National Laboratories report SAND2000-1217C, 2000.
35. A.H. Slocum, *Precision Machine Design*, Prentice Hall, Englewood Cliffs, NJ, 1992.
36. A.H. Slocum, Precision machine design: macromachine design philosophy and its applicability to the design of micromachines, *Proc. IEEE Micro Electro Mechanical Syst.* (MEMS '92). Travemunde, Germany, 1992, 37–42.
37. R. Donaldson, S. Patterson, Design and construction of a large-axis diamond turning machine, UCRL-89738, NTIS, 1983.
38. W.H. Press, B.P. Flannery, S.A. Teukolsky, W.T. Vetterling, *Numerical Recipes in C, The Art of Scientific Computing*, Cambridge University Press, 1990.
39. K.J. Bathe, E.L. Wilson, *Numerical Methods in Finite Element Analysis*, Prentice Hall, Inc., Englewood Cliffs, NJ, 1976.
40. O.C. Zienkiewicz, *The Finite Element Method*, McGraw-Hill Book Company, New York.
41. ANSYS proposed system of units for MEMS scale simulation: http://www.ansys.com/ansys/mems/mems_key_features/key_feature_units.htm.
42. W.G. Knauss, I. Chasiotis, Y. Huang, Mechanical measurements at the micron and nanometer scales, *Mechanics Mater.*, 35, 217–231, 2003.
43. J.W. Hutchinson, Plasticity at the micron scale, *Int. J. Solids Struct.*, 37, 225–238, 2000.
44. W.W. Van Arsdell, S.B. Brown, Subcritical crack growth in silicon MEMS, *J. MEMS*, 8(3), 319–327, September 1999.
45. S. Quilici, G. Cailletaud, FE simulation of macro-, meso- and micro-scales in polycrystalline plasticity, *Computational Mater. Sci.*, 16, 383–390, 1999.
46. W.R. Runyan, K.E. Bean, *Semiconductor Integrated Circuit Processing Technology*, Addison-Wesley, Reading, MA, 1990.

5 Design Realization Tools for MEMS

Just as MEMS fabrication has its roots in the microelectronics fabrication infra-structure, the MEMS design realization infrastructure has its roots in the micro-electronics infrastructure as well. However, the MEMS design realization require-ments are significantly different. MEMS design involves complex geometric, three-dimensional moving mechanical devices similar to macroworld machine design. The result is a MEMS design realization environment that leverages a significant portion from microelectronics while plotting a new path to meet the new demands.

5.1 LAYOUT

The design of a device that is to be fabricated via LIGA, surface micromachining, or bulk micromachining requires a mask to be made for the patterning step of the fabrication process. Figure 5.1 shows how the mask set, which is the interface for the design engineer's information (i.e., design), fits in the fabrication process flow. A mask is a two-dimensional design representation that will be patterned and etched into the working material. Bulk micromachining and LIGA products typically require a minimal number of masks, typically only one or two masks, to produce a high aspect ratio MEMS part. Surface micromachining can require as many as 14 masks to produce a complex MEMS design. Thus, the three MEMS fabrication technologies share a common need to interface design information with the mask-making infrastructure; however, surface micromachining is more complex due to the number of masks required. Surface micromachining will be emphasized in this chapter because the complexities of design in this MEMS fabrication technology is a superset of the issues involved with the others.

The infrastructure for mask making is an established industry primarily ser-vicing the microelectronics production complex. The two common data formats for the exchange of design layout information for use in the mask-making industry are GDSII stream format [1] and the Cal Tech Intermediate Format (CIF) [2]. GDSII is a binary format and CIF is an ASCII format; both have become de facto standards, but GDSII is much more prevalent.

Due to the geometric simplicity needed for microelectronics, the GDSII file formation is merely a sequence of closed polygons that may be an approximation

155

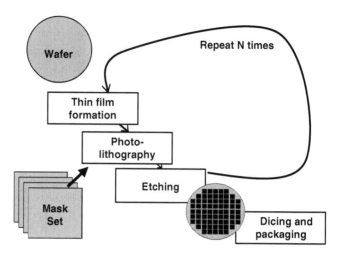

FIGURE 5.1 Surface micromachining process cycle.

of the true geometry. However, current implementations of the GDSII file format and the supporting translators enable a large number of line segments, which enable a faithful representation of the desired geometry. A closed polygon is a sequence of line segments that start and end at the same point (Figure 5.2). Obviously, the complete suite of two- and three-dimensional geometries (e.g., circles, ellipses, curves, cubes, spheres, cylinders) standard in the macroworld machine design infrastructure does not exist within the context of a GDSII file format. Figure 5.3 shows the mask data preparation path between the CAD layout tool and the GDSII file used by the mask supplier.

Current MEMS layout tools may have the ability to lay out complex geometric entities such as circles, ellipses, and curves; however, when the data are exported to a GDSII data file [1], the geometric entities are approximated by closed

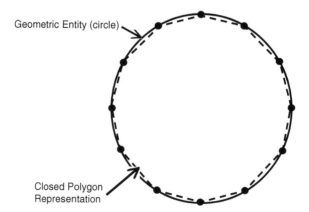

FIGURE 5.2 An example of a closed polygon approximation vs. the true geometric entity.

Design Realization Tools for MEMS 157

FIGURE 5.3 Mask data preparation flow.

polygons in a GDS file format. For example, the SUMMiT™ (Sandia's ultraplanar multilevel MEMS technology) design tool suite [3,4] utilizes the two-dimensional geometric layout capabilities of AutoCAD®, which has the full set of geometric entities to facilitate complex mechanical design. However, to transfer these data to the mask-making process requires that data pass through a translator to put the data in the GDSII format. In this example, the data are exported from AutoCAD [5] to a DXF™ file format (i.e., an AutoCAD proprietary file format) that retains the geometric entity information and then to the GDSII file format, which does not retain the geometric entity information. The translation from the DXF format to the GDSII format can be accomplished with a translator such as ASM3500 [6].

Geometric editing (e.g., changing the radius of a circle, radius of a fillet, etc.) of the design information can be easily done when the geometric entities are preserved. For example, the AutoCAD binary data file and DXF file utilized in the SUMMiT design tool suite retain the geometric entity information. Similar editing of design information utilizing a GDSII file format is severely inhibited because all curved geometry is merely approximated by a sequence of chords. Therefore, to edit a curved surface would involve the individual editing (i.e., repositioning) of a series of individual points that specify the curved surface.

Within the MEMS layout tool, layers corresponding to the masks that will be used in the fabrication processing are defined. Figure 5.4 shows the SUMMiT design environment [4] implemented in AutoCAD. Also shown in Figure 5.4 is the access to the drawing layers utilized for device layout as well as access to the standard components, design rule checking, and visualization tools discussed later in this chapter. Figure 5.5 shows the detailed layer definitions within the SUMMiT design tool suite; Table 5.1 lists the definitions of the mask names, layout layer names, and GDS layer numbers utilized in the SUMMiT design tool suite. Figure 5.6 illustrates the SUMMiT material layers deposited on the substrate. The mask name refers to the physical mask that will be utilized in fabrication, and the layer name refers to a drawing layer that will be used to define the design data used to make the physical mask. The additional layers shown in Figure 5.5 that are not listed in Table 5.1 are utilized for layout construction and notation purposes.

Table 5.1 has several instances in which two drawing layers correspond to one physical mask. For example, the MMPOLY2 mask is composed of information from the MMPOLY2 and MMPOLY2_CUT drawing layers. The two drawing layers are defined to provide ease in layout. For example, Figure 5.7 shows the layout for the seismic mass of an accelerometer in which only one drawing layer was used. Surface micromachine processes require etch release holes in layer

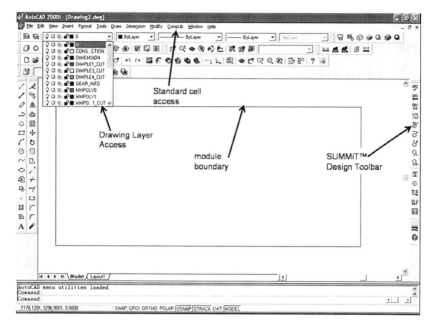

FIGURE 5.4 SUMMiT™ design environment within AutoCAD®.

surfaces to enable the release etchant to remove the sacrificial material below. Because GDSII is limited to closed polygons, the etch release holes are described by closed polygons. For the case shown in Figure 5.7, a polygon is formed by a continuous sequence of line segments that define the periphery of the mass as well as the individual etch release holes. This form of layout for a device is circuitous and difficult to modify if changes are required.

If two layers are used in the CAD layout tool and then combined, a much easier and more editable layout of the MEMS device can be performed. Figure 5.8 shows the layout of the same seismic mass shown in Figure 5.7, but two layers (i.e., *layer* and *layer_cut*) are utilized. These layers can be combined via a logical XOR operation (Figure 5.9) to form the mask description necessary for mask production. The XOR operation is typically performed by the mask production vendor. Proper use of this technique requires that a *master* layer be defined (denoted by "a" in Table 5.1). For example, MMPOLY2 is the master layer for the MMPOLY2 mask. The layer_cut is only valid when contained within the master layer, as shown in Figure 5.8. Figure 5.10 and Figure 5.11 show several approaches to the layout of some typical geometric shapes using the two layer concept.

5.2 SUMMIT TECHNOLOGY LAYOUT

This section will provide examples of the layout of devices and structures that can be manufactured in the SUMMiT technology [3]. Of the three categories of MEMS technologies (i.e., surface micromachining, bulk micromachining,

Design Realization Tools for MEMS 159

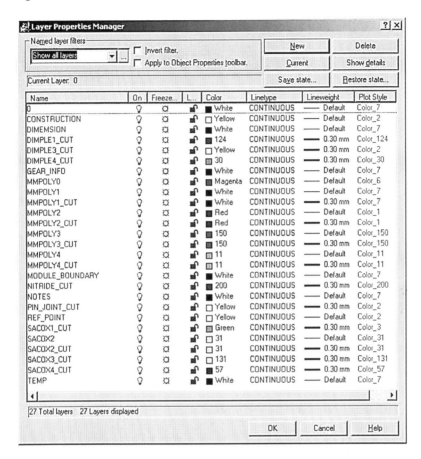

FIGURE 5.5 Layer definitions within the SUMMiT design tool suite.

LIGA), surface micromachining usually contains a superset of the issues involved in MEMS layout or design. The following subsections will discuss the layout details of five types of structures, which span the spectrum of device layout frequently encountered. Review of the layout of these structures will provide the basis for the discussion of manufacturing issues and design rules in subsequent sections. Although some specific details of these layouts may be unique to SUMMiT, the general techniques, issues, and concerns will be typical in comparable MEMS technologies. The layout of these devices will involve a combination of the 14 masks utilized in the SUMMiT V™ technology. Figure 5.12 shows the cross-hatch patterns for the various masks that will be utilized in the following discussions.

5.2.1 ANCHORING LAYERS

Anchoring or attaching mechanical layers to each other and to ground is the most basic function to be achieved in any surface micromachined device such as

TABLE 5.1
SUMMiT V Design Tool Mask and Layer Definitions

Mask			Drawing layer		
Name	Code	Field	Layer name	GDS	Color
NITRIDE_CUT	N1C	Dark	NITRIDE_CUT	21	Purple
MMPOLY0	P0	Light	MMPOLY0	22	Magenta
DIMPLE1_CUT	D1C	Dark	DIMPLE1_CUT	23	Dk blue
SACOX1_CUT	X1C	Dark	SACOX1_CUT	24	Green
MMPOLY1_CUT	P1C	Dark	MMPOLY1_CUT[a]	25	Black
			MMPOLY1	35	
PIN_JOINT_CUT	PJC	Dark	PIN_JOINT_CUT	26	Yellow
SACOX2	X1	Light	SACOX2	27	Tan
MMPOLY2	P2	Light	MMPOLY2[a]	28	Red
			MMPOLY2_CUT	38	
DIMPLE3_CUT	D3C	Dark	DIMPLE3_CUT	29	Yellow
SACOX3_CUT	X3C	Dark	SACOX3_CUT	30	Black
MMPOLY3	P3	Light	MMPOLY3[a]	31	Blue
			MMPOLY3_CUT	41	
DIMPLE4_CUT	D4C	Dark	DIMPLE4_CUT	34	Orange
SACOX4_CUT	X4C	Dark	SACOX4_CUT	42	Green
MMPOLY4	P4	Clear	MMPOLY4[a]	36	Peach
			MMPOLY4_CUT	46	

[a] Denotes the master layer.

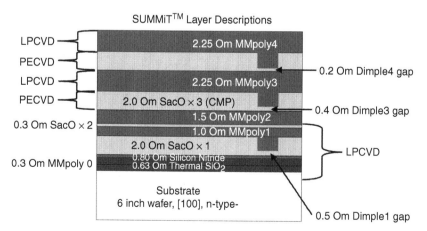

FIGURE 5.6 SUMMiT V (Sandia ultraplanar multilevel MEMS technology) technology material layer description.

Design Realization Tools for MEMS

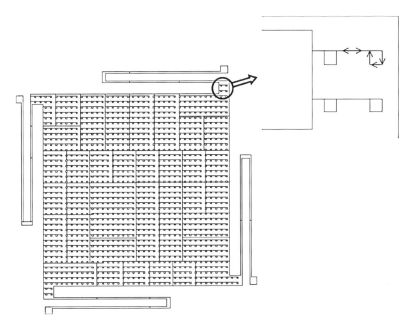

FIGURE 5.7 Layout of the seismic mass level of an accelerometer utilizing only one drawing layer.

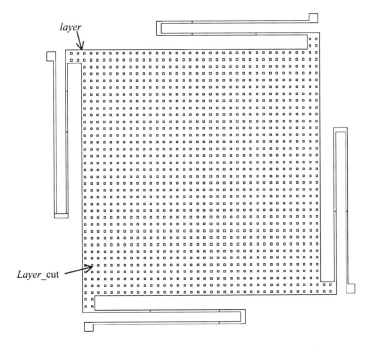

FIGURE 5.8 Layout of the seismic mass level of an accelerometer utilizing two layers.

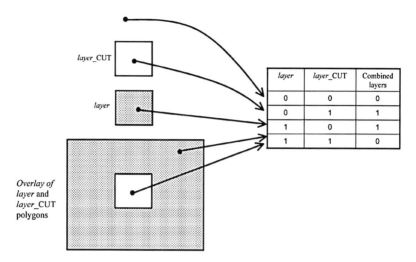

FIGURE 5.9 Use of a bit by bit XOR logical function to combine layers (*layer* and *layer*_CUT) to form a mask definition.

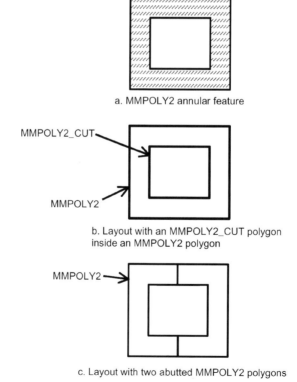

FIGURE 5.10 Alternative approaches to layout of an annular MMPOLY2 feature.

Design Realization Tools for MEMS   163

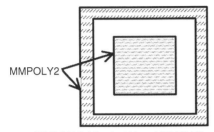

a. MMPOLY2 island within a MMPOLY2 annular feature.

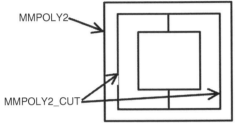

b. Layout with two MMPOLY2_CUT polygons within a MMPOLY2 polygon.

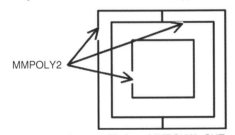

c. Layout with three MMPOLY2_CUT polygons.

FIGURE 5.11 Alternative approaches to the layout of an MMPOLY2 island inside an MMPOLY2 annular feature.

SUMMiT. Because surface micromachining is an alternating stack of two types of materials (i.e., structural and sacrificial), mechanical layers are attached by etching a hole or via in the sacrificial material; this enables the next mechanical layer material deposited to attach to the mechanical material below at the via. Figure 5.13 shows SEM images of various layers of SUMMiT anchored to each other. The attachment between the layers at the via produced by the SACOX_CUT can be seen between the adjacent layers. The term "SACOX_CUT" refers to a generic operation of opening a via in a sacrificial oxide layer to allow attachment of adjacent structural layers. SACOX#_CUT is the use of a SACOX_CUT on a specific SACOX# layer.

A particular layer cannot be directly anchored to ground in the SUMMiT technology because deep SACOX_CUTs are not allowed. The SACOX_CUT

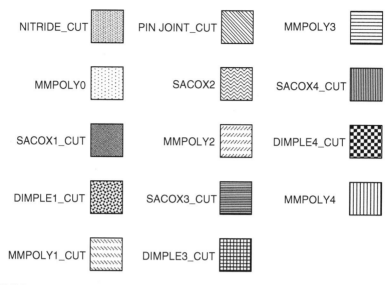

FIGURE 5.12 Cross-hatch patterns for the SUMMiT V masks.

enables the mechanical layers immediately above and below to be attached. For example, a SACOX3_CUT will enable MMPOLY3 and MMPOLY2 to be attached. Figure 5.14 shows a demonstration of the five layers of the SUMMiT process. Figure 5.15 show a cross-section visualization and the masks for a post that extends from MMPOLY0 to MMPOLY4. Notice that the SACOX_CUTs are all on top of each other and the size of the SACOX_CUT becomes bigger at each higher level. The size increase enables the mechanical material at the higher level to attach to the shoulder of the via to the mechanical material on the level immediately below. This is denoted as a *nested* anchor.

The nested anchor method can produce the smallest size post possible. However, a nested anchor will have encased silicon dioxide trapped inside the post as shown in the cross-section of Figure 5.15. The encased silicon dioxide is due to the inability of the etching processes to remove material completely at locations that have significant vertical topography. This artifact is known as a *stringer* and is discussed in Section 3.3 and Section 5.3.1.6. The residual stress of the entrapped silicon oxide in the post can cause slight deflections [7]; this may be a design consideration, depending upon the application.

A *staggered* anchor is a method of reducing the amount of encased silicon dioxide in an anchor; however, the required size of the post will increase. Figure 5.16 shows a cross-section visualization of a post utilizing a staggered anchor approach.

5.2.2 Rotational Hubs

Rotational hubs are structures that enable 360° rotation similar to a wheel and axle. The ability to implement this structure at the microscale with no assembly

Design Realization Tools for MEMS 165

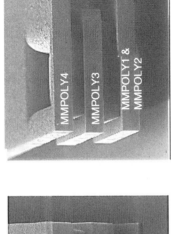

FIGURE 5.13 Scanning electron microscope images of nested SACOX_CUTs to anchor layers to each other and to ground. (Courtesy of Sandia National Laboratories.)

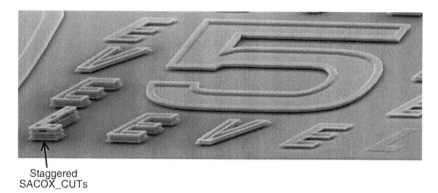

FIGURE 5.14 Example of the five mechanical levels of SUMMiT anchored to each other and to ground. The anchors utilized nested SACOX_CUTs except as noted in the figure where staggered SACOX_CUTs are used. (Courtesy of Sandia National Laboratories.)

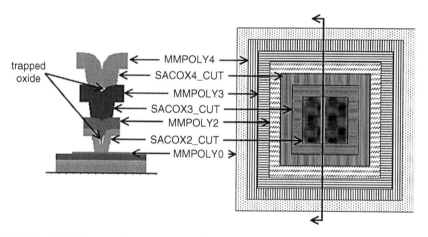

FIGURE 5.15 Masks and cross-section of a post composed of anchored layers utilizing nested SACOX_CUTs.

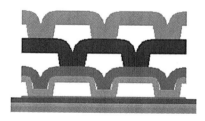

FIGURE 5.16 Cross-section of an anchored layer stack using staggered SACOX_CUTs.

Design Realization Tools for MEMS

necessary is an enabling feature for MEMS devices that require mechanisms. Two methods can be used to produce a rotational hub in the SUMMiT technology: a *cap and post hub* and a *low-clearance hub*, which are discussed next.

A cap and post hub can be implemented in any three-level surface micromachine technology, and it is the simplest hub design that can be utilized. Figure 5.17 shows the masks and a cross-section of a cap and post hub implemented in the SUMMiT technology. The central feature of this type of hub is a central post with a cap of sufficient diameter so that a rotating wheel will be constrained vertically. The figure's cross-section shows the rotating wheel composed of MMPOLY1 and MMPOLY2, which are laminated together. Functionally, the rotating wheel could be only one layer instead of two. An MMPOLY3 cap is supported by a post of MMPOLY1 and MMPOLY2.

The implementation of the cap and post structure is similar to the anchors discussed in the previous section. The clearance for the rotating wheel is defined by the ability of the lithography process to etch layers MMPOLY1 and MMPOLY2 at the rotating interface. The vertical clearance is defined by the thickness of the sacrificial oxide layer or the ability to produce structures such as dimples to constrain the vertical motion. Dimples are small "bumps" underneath surface micromachined layers that prevent broad area surface contact when the layers contact the substrate or each other. Dimples can also be used to minimize clearances.

The low-clearance hub is a feature that SUMMiT was especially designed to implement (Figure 5.18). This hub utilizes the ability to deposit and etch thin films of sacrificial material accurately (i.e., silicon dioxide) to control the clearance in the hub. Figure 5.18 shows the layout and cross-section of the low-clearance hub, and Figure 5.19 shows an FIB cross-section of a low-clearance hub and pin joint fabrication in SUMMiT. A pin joint is very similar to a hub, but is not attached to ground. A pin joint enables linkages between rotating members, as shown in Figure 5.19.

Figure 5.20 shows a cross-section of the SUMMiT fabrication sequence for the low-clearance hub at several key points in the process:

- Figure 5.20a shows the fabrication at the point at which SACOX1 has been deposited and patterned to produce dimples and anchor MMPOLY1. MMPOLY1 has been deposited and patterned with the PIN_JOINT_CUT mask. A combination of anisotropic and wet etching has been performed to form the features beneath MMPOLY1.
- Figure 5.20b shows the process after the SACOX2 layer has been deposited, patterned, and etched and the MMPOLY2 layer deposited. At this stage, SACOX2 can be seen to define the clearances in the internals of the low-clearance hub. The low-clearance hub lateral and vertical clearances in SUMMiT are 0.3 μm.
- Figure 5.20c shows the cross-section after the MMPOLY2 etch has been performed. This etch can etch the laminated MMPOLY1 and MMPOLY2 layers, thus providing an even outside surface for the

168 Micro Electro Mechanical System Design

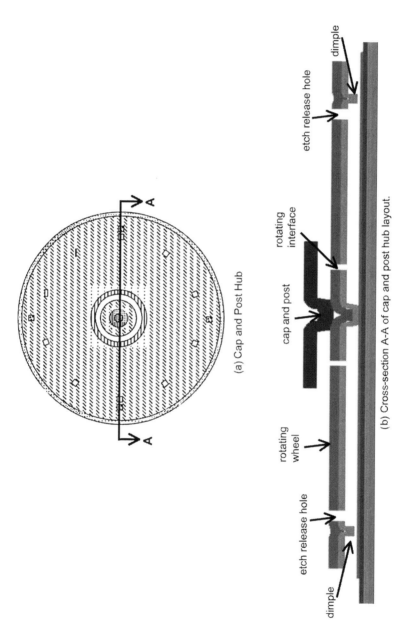

FIGURE 5.17 Cap and post hub layout and cross-section visualization.

Design Realization Tools for MEMS 169

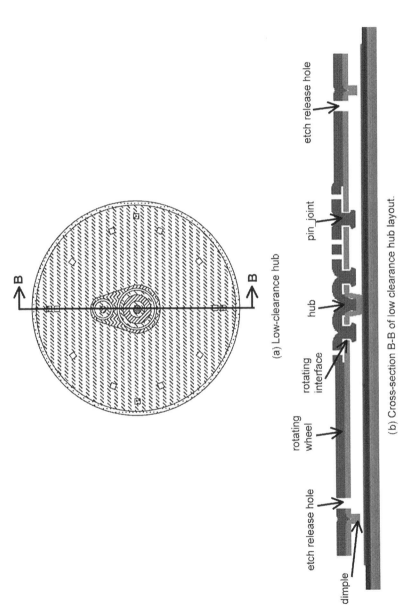

FIGURE 5.18 Low-clearance hub and pin joint layout and cross-section visualization.

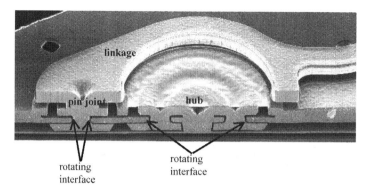

FIGURE 5.19 A focused ion beam (FIB) cross-section of a rotational hub and pin joint. (Courtesy of Sandia National Laboratories.)

rotating wheel and etch release holes through the rotating wheel disc. Note that the MMPOLY2 etch stops on the SACOX2 layer in the internal hub features.
- Figure 5.20d shows the cross-section of the released low-clearance hub and pin joint structure.

5.2.3 POLY1 BEAM WITH SUBSTRATE CONNECTION

The MMPOLY1 beam with a substrate connection is a simple structure illustrating the application of two features useful in design of a number of devices in the SUMMiT technology. The MMPOLY1 layer can be patterned in either of two ways in SUMMiT:

- The MMPOLY1 layer can be patterned directly using the MMPOLY1_cut mask and etch.
- The MMPOLY1 layer can also be patterned indirectly by using SACOX2 as a "hard" mask and etching with the MMPOLY2 etch.

In the previous section, Figure 5.20c showed that the MMPOLY2 etch would etch the MMPOLY1 and MMPOLY2 layers except when the MMPOLY1 layer is protected by SACOX2. In this case, the SACOX2 layer was used as a "hard" mask to stop the MMPOLY2 etch. For the MMPOLY1 beam shown in Figure 5.21, the SACOX2 mask is used to define the MMPOLY1 beam via the MMPOLY2 etch. The MMPOLY1 beam is attached to the substrate using a SACOX1_CUT as discussed in Section 5.2.1. If a connection is to be established to the substrate for electrical grounding purposes, the NITRIDE_CUT mask is used to define the etch of the NITRIDE layer.

5.2.4 DISCRETE HINGES

The concept of discrete hinges for MEMS applications was initially proposed by Pister [8]. Since that time, a number of different variations and types of hinges

Design Realization Tools for MEMS 171

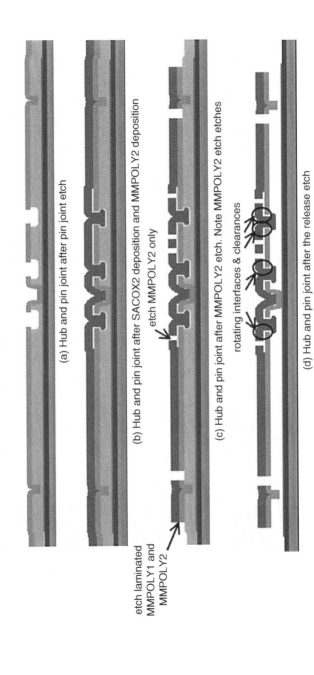

FIGURE 5.20 Low-clearance hub and pin joint cross-section visualization at various stages in the SUMMiT fabrication.

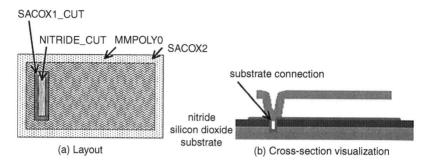

FIGURE 5.21 An MMPOLY1 beam with a substrate connection.

have been developed and applied [9]. Figure 5.22 shows SEM images of a "pop-up" mirror that utilizes two types of discrete hinges:

- *Staple and pin hinge.* This allows a plate to be attached to the substrate and rotate about an axis parallel to the substrate. Figure 5.23 shows the layout and cross-section for a staple and pin hinge implemented in SUMMiT. The cross-section shows two posts that go up to the MMPOLY3 level, which bridges between the posts to form the *staple*. The pin is a narrow piece of laminate MMPOLY1 and MMPOLY2 that connects out of the plane of the cross-section to the movable plate. As in a cap and post hub, the staple and pin hinge clearances are defined by the width of the MMPOLY2 etch.
- *Plate-to-plate hinge.* This couples two plates and allows them to rotate about an axis parallel to the plane of the plates and the hinge to deflect off the substrate, as shown in Figure 5.22. The design of the plate-to-plate hinge is quite complex. Figure 5.24 and Figure 5.25 show the layout and two cross-sections of the device. Figure 5.24 shows the individual masks required to fabricate the hinge in SUMMiT technology, as well as the composite mask, "stacked" together. To fabricate this device as well as any other device in SUMMiT requires the masks be aligned precisely to each other. The A-A cross-section shown in Figure 5.25c reveals the major parts of this hinge design. An MMPOLY2 pin is connected out of the plane of the cross-section to plate 2. The pin is trapped by the staple and floor and rotates within these objects. Figure 5.25b shows that the SACOX2 layer separates MMPOLY1 and MMPOLY2 and stops the MMPOLY2 etch so that an MMPOLY2 pin is formed. The staple is formed by utilizing SACOX3_CUTS to attach MMPOLY3 in two places to MMPOLY2. A subtle feature is the cutting of MMPOLY1 as defined by the MMPOLY1_CUT mask to separate the pin and the floor structures as shown in Figure 5.25d and Figure 5.25e. Without the separation of the floor and pin structures via the MMPOLY1_CUT, this hinge would rigidly attach plate 1 and plate 2 (i.e., not functional).

Design Realization Tools for MEMS   173

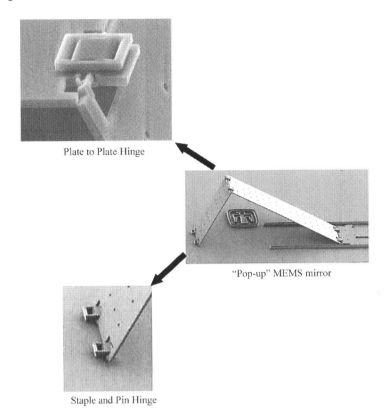

FIGURE 5.22 Discrete hinges utilized in a "pop-up" MEMS mirror design implemented in SUMMiT. (Courtesy of Sandia National Laboratories.)

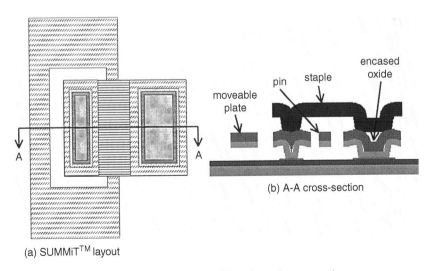

FIGURE 5.23 Staple and pin hinge SUMMiT masks and cross-section.

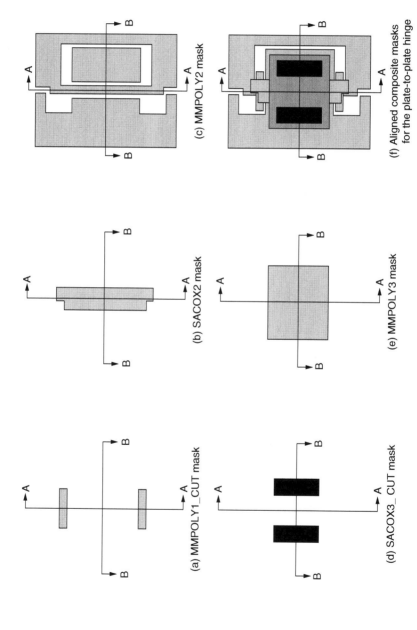

FIGURE 5.24 SUMMiT masks for the plate-to-plate hinge.

Design Realization Tools for MEMS 175

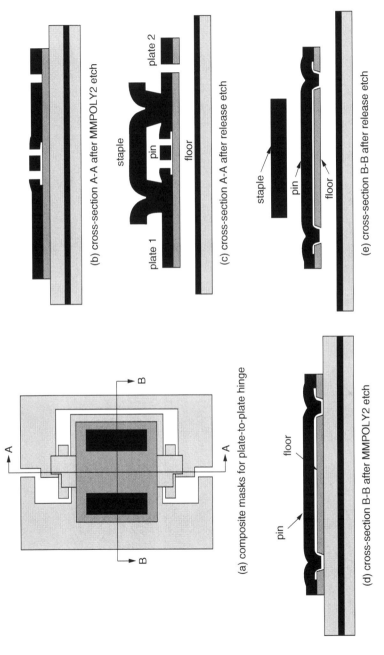

FIGURE 5.25 SUMMiT masks and cross-sections for the plate-to-plate hinge.

5.3 DESIGN RULES

The term *design rules* originally comes from the microelectronics industry. These rules are a formal communication between the fabrication engineer and the design engineer. For the microelectronics industry, design rules are the layout rules required to obtain optimum yield (functional devices vs. nonfunctional devices) in as small an area as possible without compromising circuit reliability.

Design rules for MEMS fabrication are also a formal communication between the fabrication and design engineers. MEMS fabrication processes are similar to microelectronics; however, due to the additional ability of motion MEMS devices (e.g., inertial sensors, mechanisms, pumps, valves, etc.) they are very different from microelectronics (i.e., electronic circuitry) and more varied in function and application. The varied function and application of MEMS devices make the assessment of yield from the perspective of design rules for a general-use MEMS fabrication process difficult to define. Design rules for MEMS fabrication processes are the layout rules required to produce MEMS devices with minimal defect with the smallest feature sizes possible. MEMS device functionality and reliability are generally very specific to the device design and cannot be totally encompassed by MEMS layout design rules alone.

The layouts for microelectronic and MEMS design are very complex and involve a number of mask layers. The mechanics of automated design rule checking for VLSI circuitry layout was established during the rise of the microelectronics industry [10–12]. Yarberry [4] discusses the implementation of automated design rule checking for a MEMS fabrication process, SUMMiT™.

Section 5.3.1 and Section 5.3.2 will discuss the manufacturing issues that are the basis for the specification of design rules and the implementation of design rules for a MEMS fabrication process.

5.3.1 MANUFACTURING ISSUES

5.3.1.1 Patterning Limits

The definition of MEMS device features is a function of the patterning and etch steps of the fabrication processes used. These processes may be utilized for patterning, such as photolithography and lift-off, and are discussed in Section 2.6. A number of isotropic and anisotropic etch processes (Section 2.5) may be used to etch the pattern into the MEMS material. The patterning limits will control the smallest feature that can be realized in a MEMS device. This patterning limit is frequently called the *feature size* or *CD* (critical dimension). The patterning limits of a fabrication technology are expressed as line width and space design rules for each layer in the fabrication technology. Figure 5.26 illustrates the expression of a line width and space design rule for a layer in the SUMMiT™ process. Figure 5.27 illustrates the result of attempting to pattern features in violation of the design rules for the process. Figure 5.28 shows that lithographically patterning has a "rounding" effect of the sharp-cornered features due to the patterning limitations.

Design Realization Tools for MEMS   177

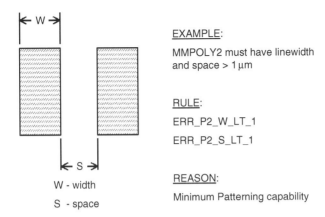

FIGURE 5.26 Example of a line width (*W*) and space (*S*) design rule.

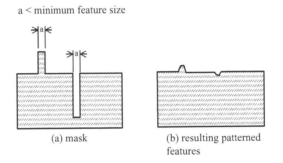

FIGURE 5.27 Expected result when line width and space design rule violated.

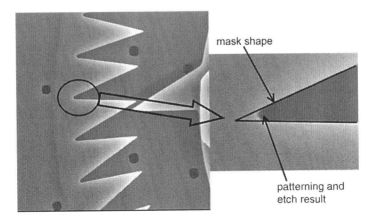

FIGURE 5.28 "Rounding" of a lithographically patterned angular feature. (Courtesy of Sandia National Laboratories.)

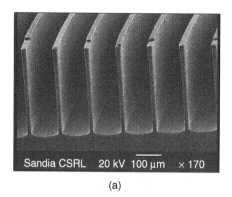
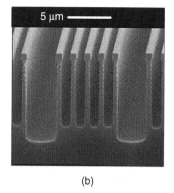

FIGURE 5.29 Pattern uniformity affects the ability to remove etching products and influences etch rate. (Courtesy of Sandia National Laboratories.)

5.3.1.2 Etch Pattern Uniformity

Etch pattern uniformity becomes an issue in many fabrication technologies due to optical effects of patterning large arrays of similar or varied structures, or etching an array of structures of varying size. Figure 5.29 illustrates the etching of a series of trenches in a bulk micromachining process. The ability to remove the etch products from the trench can influence the etch rate of the process. This can be an advantage or a disadvantage, depending upon the device to be produced.

An accelerometer produced in a surface micromachining process generally has large banks of comb fingers and electrodes. Lithographic optical issues of patterning a large arrays of repeating structures such as this may cause the patterning of the electrodes on the edge to become distorted *edge effects*. This can be accommodated by adding a few extra "dummy" electrodes to ensure that the functioning electrodes are fabricated without distortion.

5.3.1.3 Registration Errors

MEM fabrication technologies frequently require masks at different stages in the fabrication process to be aligned for the fabricated device to be produced as designed. For example, the SUMMiT utilizes 14 masks that need to aligned to each other. The alignment is accomplished by the aid of alignment targets (Figure 2.26), which are etched into each layer to enable the alignment. Positioning objects on a mask is accomplished to computer precision; however, aligning masks relative to each other has a finite precision, called *registration error*. Design rules can be expressed that address the registration error issue. Figure 5.30 illustrates an enclosure design rule that ensures that the SACOX1_CUT feature is inside the MMPOLY2 feature. One of the things that determines the amount of enclosure is determined by the registration errors of the mask alignment.

Design Realization Tools for MEMS

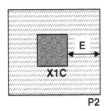

Rule:
MMPOLY2 must enclose SacOx1_Cut with an enclosure boundary > .5 µm

Design Rule Error:
ERR_P2_X1C_E_LT_0PT5

Reason:
Mask Registration error

Note: Layer named first must be outside

FIGURE 5.30 Example enclosure design rule to provide for finite mask alignment precision.

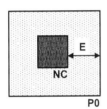

Rule:
MMPOLY0 must enclose NITRIDE_CUT with an enclosure boundary > .5 µm

Design Rule Error:
ERR_P0_NC_E_LT_0PT5

Reason:
Prevent exposure of underlying layers to subsequent etches.

Note: Layer named first must be outside

FIGURE 5.31 Example enclosure design rule to prevent exposure of underlying layers to subsequent etches.

5.3.1.4 Etch Compatibility

Alignment of masks within a registration tolerance is also necessary for etch compatibility reasons (Figure 5.36). Exposure of one material to an etch designed for another material or etching underlying materials can be an issue for fabrication technologies, such as surface micromachining, that involve a number of materials and mask levels. Figure 5.32a shows that an underlying layer of silicon dioxide was attacked by the release etch in a surface micromachine process because the masks were not enveloped sufficiently to allow for a finite registration error of the masks.

5.3.1.5 Stringers

Stringers are manufacturing artifacts produced when performing anisotropic etching on a surface that has topography. Figure 5.32c shows an example of a floating stringer that landed on the gear. Figure 5.33 shows the mask and a cross-section of how a floating and attached stringer can be formed in the SUMMiT™ process. Stringers are troubling when they are floating particles that can impede the

180 Micro Electro Mechanical System Design

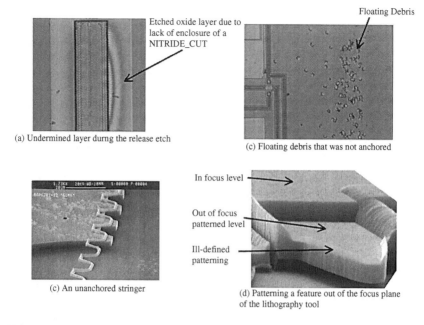

FIGURE 5.32 Examples of manufacturing defects that can be prevented by adherence to the appropriate design rule. (Courtesy of Sandia National Laboratories.)

operation of devices upon which they land. Fixed stringers are generally undesirable artifacts in a design. The stringer is formed because the anisotropic etch cannot remove all the material in a topographic discontinuity in the surface. The design rule that would have prevented the stringers shown in Figure 5.33 is "X1C WITHOUT P2" or "X3C WITHOUT P3." For example, "X1C WITHOUT P2" (read literally as SACOX1_CUT polygon without an enveloping MMPOLY2 polygon) would produce a stringer in the topographic discontinuity of the SACOX1 layer because of the inability of the anisotropic etch of MMPOLY2 to remove the material in the "corners."

5.3.1.6 Floaters

Floaters are just pieces of material that are not fastened to the substrate, another object attached to the substrate, or contained within another structure. Figure 5.34 shows a piece of MMPOLY2 that is not attached to the substrate with a SACOX1_CUT. An example of a design rule that would alert the designer to this problem is "P2 without X1C." Figure 5.32b shows an occurrence of unattached pieces of material that have floated away deposited elsewhere on the die.

5.3.1.7 Litho Depth of Focus

Patterning is generally performed with optical lithography. The depth of focus of a lithography tool that can pattern a small feature size is generally limited.

Design Realization Tools for MEMS

(a) Mask for a cut in the sacrificial material which is not enveloped by the structural material.

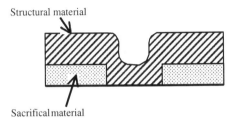

(b) Structural material covering an etch in the sacrificial material which produces topography.

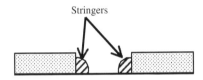

(c) Anisotropic etch to remove the structural material which leaves a stringer.

FIGURE 5.33 Example of stringer formation in a surface micromachine process.

Attempting to pattern an object outside the focus of the lithography tool will generally perform poorly. Figure 5.32d is an example of attempting to pattern MMPOLY2 within a cut in the sacrificial oxide cut.

5.3.1.8 Stiction (Dimples)

Dimples are small "bumps" on the bottom side of the mechanical layers of a surface micromachine process for the purpose of preventing broad area surface contact, which could cause layers to stick together or stick to the substrate. Dimples are placed at a certain frequency and prevent stiction of the layer. Design rules that would address this issue are usually informational or advisory because the designer may intentionally decide not to have dimples because the structure is sufficiently stiff to prevent stiction.

5.3.1.9 Etch Release Holes

Etch release holes are holes placed in a large area of a surface micromachined layer to allow the release etchant to remove the sacrificial material immediately below more readily. An example of a SUMMiT design rule that addresses this

(a) Mask for patterning the structural material which is not enveloping a cut in the sacrificial material that would anchor the material.

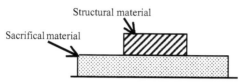

(b) Patterned and etched structural material which is not anchored by a cut in the sacrificial material.

(c) Structural material which is free to float away, *floater*, during the release etch.

FIGURE 5.34 Example of floater formation in a surface micromachine process.

issue is "MMPOLY2_CUT_SPACE GREATER THAN 38 µm." A violation of this design rule will result in a structure that is not fully released from the sacrificial layer.

5.3.1.10 Improper Anchor (Area of Anchor)

Anchoring one layer to another is a basic function required in a MEMS design. Frequently, a design rule specifying the minimum area of the anchor is utilized to ensure anchor strength.

5.3.2 Design Rule Checking

Layouts of devices are frequently large and complex and require an automatic check of the design rules to ensure compliance. Automated design rule checking was originated in the microelectronics industry [10–12]. Techniques were developed to check for design rule violations by automatically and efficiently scanning large microelectronic layouts containing many repeated cells. Microelectronic layouts are generally *Manhattan geometries*, which do not involve arbitrary angles or curves. However, MEMS layouts frequently include general geometric shapes that are non-Manhattan geometry. Most MEMS design tools suites incorporate design rule checking [3,4]. The design rules are frequently divided into two categories:

Design Realization Tools for MEMS

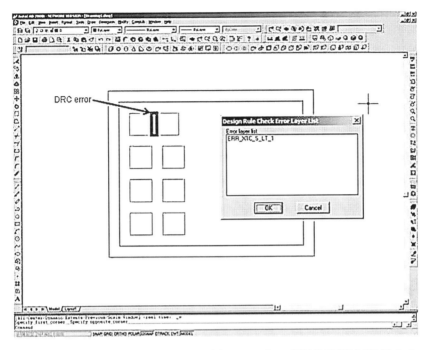

FIGURE 5.35 Automatic design rule checking operation. (SUMMiT™ MEMS design tools — courtesy of Sandia National Laboratories.)

- Design rule errors: a design rule violation that requires mandatory attention
- Design rule advisory: a design rule violation that requires the designer to evaluate the necessity for correction

Figure 5.35 shows the operation of the automatic design rule-checking capability within the SUMMiT™ design tool suite. The layout is scanned for design rule errors; these are noted and highlighted in the layout to assist the MEMS designer in locating and evaluating the design rule violation. The design rule-checking environment generally contains tools to navigate through the drawing to assess each violation.

5.4 STANDARD COMPONENTS

Standard components is another concept taken from the microelectronics world (i.e., standard cells) in which frequently used components are designed, fabricated, tested, and placed into a library for use in further design. Models of the components will have already been developed and available. Ideally, a designer would be able to implement a significant portion of any design by placing and coupling the components together and simulating the total system response uti-

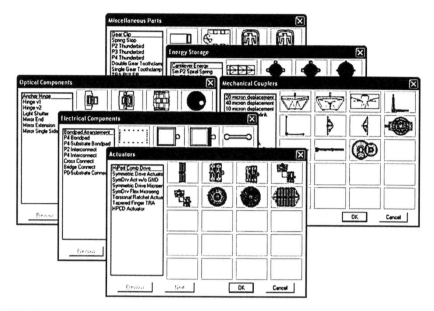

FIGURE 5.36 Standard components available in the SUMMiT™ design tool suite.

lizing existing models. This approach greatly reduces the risk and time inherent in new system design.

The validity of utilizing the standard cell concept for MEMS is challenged by the breadth of MEMS applications. However, many MEMS foundry technologies [3,13] utilize the standard cell concept to the degree practical. Figure 5.36 shows the standard components available in the SUMMiT™ design tool suite and Figure 5.37 is an example of a system made entirely from standard components. However, in general, it is difficult for standard components to span the breadth encountered in MEMS; therefore, the standard component library elements generally focus on the most frequently used components encountered in MEMS design, such as

- Actuators
- Electrical elements (bond pads, wiring elements, etc.)
- Mechanical coupling elements (springs, displacement multiplier, etc.)
- Optical elements (mirror plates, hinges, etc.)

5.5 MEMS VISUALIZATION

MEMS visualization tools are a significant help to the MEMS engineer. Two categories of tools have been developed and used to aid the visualization of geometry that MEMS processes produce:

- *Physics-based process modeling tools.* These tools can predicatively model the MEMS processes used to fabricate MEMS devices. They

Design Realization Tools for MEMS 185

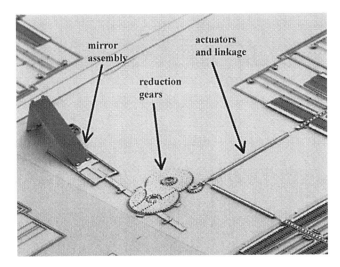

FIGURE 5.37 Example of a MEMS system (pop-up mirror and actuators) that can be made from standard components. (SUMMiT MEMS design tools — courtesy of Sandia National Laboratories.)

can provide detailed information on process parameters and visualization for MEMS devices. A number of tools have been developed to simulate anisotropic etching of silicon [14–21]. Modeling of thin film deposition and etch processes is significant in microelectronics and MEMS manufacture and the effort to create suitable simulators is ongoing [22–26]. This type of visualization, which is based upon detailed physical modeling of the processes, is computationally intense and provides useful information for the MEMS process engineer. However, it is frequently inappropriate for visualization for the MEMS design engineer.

- *Geometric emulation of MEMS processes.* This approach utilizes a geometric description of the result of each process step that is concatenated to build a geometric model of the complete MEMS device. The geometric emulation approach [27–31] has become a widely accepted approach for MEMS visualization applications. Koppelman first utilized geometric emulation in the OYSTER program [27], which was subsequently extended to the MemBuilder [28] and MEMulator™[34] modules incorporated into the MEMCAD program [32–34]. The geometric emulation approach is capable of modeling the spectrum of process steps (i.e., conformal depositions, planariziation, etch profiles, gap fills; Figure 5.38) encountered in MEMS processes [30,31] in a computationally efficient manner, thus providing the MEMS designer with meaningful information. Figure 5.39 illustrates how MEMS two- and three-dimensional visualization tools assist the MEMS designer from layout to the fabricated device. The three-dimensional visualiza-

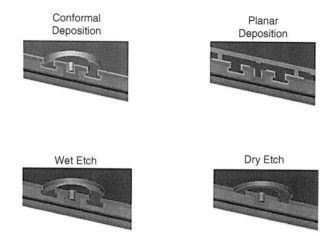

FIGURE 5.38 MEMS three-dimensional visualization tools can emulate a spectrum of process steps. (SUMMiT MEMS design tools — courtesy of Sandia National Laboratories.)

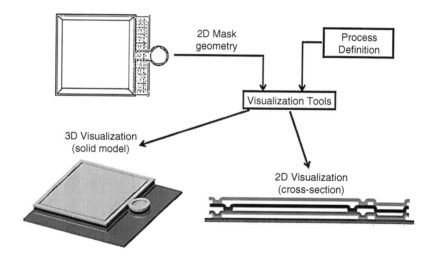

FIGURE 5.39 MEMS visualization tools enable a MEMS designer to access and verify the design before fabrication. (SUMMiT MEMS visualization tools — courtesy of Sandia National Laboratories.)

tion tools also provide a gateway to MEMS analysis through the solid model, which can be meshed with finite elements for detailed analysis (Figure 5.40).

5.6 MEMS ANALYSIS

MEMS analysis is the computations that enable the engineering of a MEMS device. The level of detail and effort expended on the analysis [35,36] depends

Design Realization Tools for MEMS

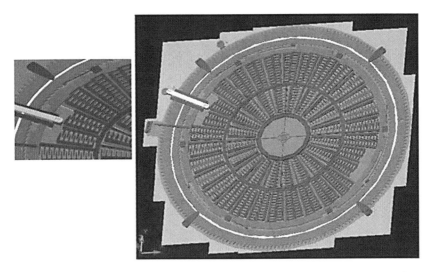

FIGURE 5.40 Three-dimensional solid model of a complex MEMS actuator, torsional ratcheting actuator. (Courtesy of Sandia National Laboratories.)

upon the stage of development and value of the MEMS device. Figure 5.41 shows a concept of the various levels of complexity of MEMS analysis. A *design synthesis model* is the first stage of analysis that is appropriate in the transition from concept to product. This type of model is generally low order, such as a lumped parameter model with a limited number of degrees of freedom (DOF). A design synthesis model may be developed by first-principle analysis of the MEMS device. These models may be incorporated into a model of the system into which the MEMS device will be incorporated. Frequently, the metric of evaluation of a MEMS device is how the system into which it is incorporated behaves.

At the other end of the scale of complexity is the *detailed design model*, which is frequently characterized by a large number of degrees of freedom utilizing finite element methods (FEMs) or boundary element methods (BEMs). The detailed design model may also involve multiple physics domains (i.e., structural, electrical, fluidic, etc.). A detailed design model can provide extremely fine detail on the operation of a MEMS device, but it may be at the expense of significant computation and model development. A system level model may require the information generated by a detailed design model; however, direct incorporation of a detailed design model into a system model is generally impractical. This may be accomplished by a *phenomenological or macromodel* [37,38], which may be developed by projecting the results of the detailed design model onto spaces spanned by a small number of DOF [38].

This book emphasizes the development of design synthesis models because the conceptual stage MEMS device design, in which modeling provides understanding of the physics and the trade-offs between design variables, is paramount. Chapter 7 will discuss development of design synthesis models utilizing Lagrange's equations to formulate the governing equations of the MEMS device.

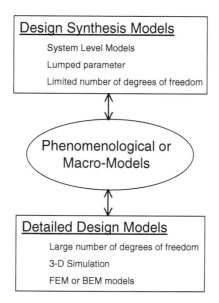

FIGURE 5.41 Hierarchy of MEMS analysis models. (After S.D. Senturia, *Sensors Actuators A*, 67, 1–7, 1998, and *Proc. IEEE*, 86(8), 1611–1626, 1998.)

A low-order nodal modeling technique [39–41] that can provide the governing equations and simulations will also be used.

The techniques for developing a detailed design model of MEMS devices have been an active research area for approximately 15 years. The early development of techniques for detailed design models of MEMS devices centered on the development of analysis methods for multiple physics domains encountered in MEMS devices. The first such domain MEMS problem that received significant attention was the structural–electrostatic problem [32,33], which resulted in the MEMCAD software that has been incorporated in commercial MEMS analysis packages [34]. Subsequently, many FEM and BEM analysis packages have linked multiple physics domains for MEMS analysis [34,43,44].

However, MEMS devices research in coupled domain MEMS analysis continues [38], with significant emphasis on the coupling of several domains such as fluidics, electrical, structural, and thermal. Simulation MEMS devices on the atomic scale [42] is also an active area of research. Atomic scale modeling becomes significant when the area of interest in a device becomes so small that continuum mechanics modeling with partial differential equations is no longer valid. An example of this is high-frequency radio frequency (RF) devices.

5.7 SUMMARY

MEMS design realization tools are essential to the ability of the MEMS design engineer to produce MEMS devices that meet specifications in an efficient manner. The MEMS design realization tools include the following capabilities:

Design Realization Tools for MEMS

- Layout tools (prototype or technology file and two-dimensional layout capability)
- Standard components
- Automated design rule checking
- MEMS visualization

The MEMS visualization tools that were discussed in Section 5.5 construct a cross-section or three-dimensional solid model (Figure 5.39) of the MEMS device given the device layout and a process description. This technique is called *art to part* (i.e., layout art to solid model of the part). Recent research [45–47] has developed the basic algorithms for the *part-to-art* problem (i.e., solid model of the part to the layout art). With further development and the use of the part-to-art algorithms, a new method of design realization for MEMS may be possible. Instead of the MEMS designer producing two-dimensional layouts to produce the three-dimensional MEMS device, he or she may in the future develop a three-dimensional solid model of the MEMS device; the part-to-art algorithm could produce the two-dimensional layouts of the masks to make the part in a particular technology. This would enable a MEMS designer to work in the same manner as a macroworld design engineer, who develops a three-dimensional solid model of a device to be designed.

MEMS analysis capability is essential to the ability of the MEMS designer to engineer a device. A range of complexity of MEMS analysis models exists:

- Design synthesis model: low number of DOF or lumped parameter models
- Phenomenological or macromodel: a model that can map complex phenomena or detailed model data to a low-order space
- Detailed design model: large number of DOF, FEM, or BEM models

In subsequent chapters, MEMS analysis — in particular methods for developing the design synthesis model — will be discussed in detail.

QUESTIONS

1. What information does a prototype file or technology file contain?
2. Why is the GDSII file important? Who uses it?
3. What is the difference in how a GDS file represents geometry and the binary layout file in the design environment?
4. What is the advantage of using two layers (i.e., layer and a layer_cut) and the XOR logical operation to create a mask layout? Why is a polygon drawn in layer_cut only valid within a poly drawn in layer? (Hint: see Figure 5.8 and Figure 5.9.)
5. Compare the advantages and disadvantages of the nested anchor vs. the staggered anchor.

6. Compare the advantages and disadvantages of the cap and post hub vs. the low-clearance hub.
7. What advantages do standard components have for MEMS device layout?
8. What is the difference between the philosophy for definition of MEMS and microelectronic design rules?
9. Consider the layout of a nested anchor vs. the staggered anchor. Assume that all SACOX_CUT for anchors should be a minimum of 2 µm square. The line width and space rules for all layers are 1 µm. The layer envelope rules for the MMPOLY layers enveloping the SACOX_CUT layers are >0.5 µm. What is the minimum size anchor for the nested anchor vs. the staggered anchor? Why would you choose one type of anchor over the other?
10. Layout a post in the SUMMiT™ technology that goes from MMPOLY0 to MMPOLY4 utilizing the staggered anchor concept.
11. What can be done with the staple and pin hinge (Figure 5.23) to reduce the clearances?
12. What are the types of design rule errors? What is their significance?

REFERENCES

1. *GDSII Stream Format Manual*, release 6.0, G.E. Calma, February 1987. See also http://www.cadence.com and http://www.cnf.cornell.edu/SPIEBook/spie9.htm).
2. A.M. Carver, L.A. Conway, *Introduction to VLSI Systems*, Reading, MA, Addison-Wesley, 1980.
3. E. Shepherd, Prototyping with SUMMiT™ technology, Sandia's ultraplanar multilevel MEMS technology, ASME International Mechanical Engineering Congress & Exposition, IMECE 2002, paper no. IMECE 2002-34258.
4. V.R. Yarberry, Meeting the MEMS design-to-analysis challenge: the SUMMiT™ V design tool environment, ASME International Mechanical Engineering Congress & Exposition, IMECE 2002, paper no. IMECE 2002-34205.
5. AutoCAD 2000i, User's Guide, Autodesk, Inc., San Rafael, CA, http://www.autodesk.com.
6. *ASM 3500 DXF to GDSII Bidirectional Translator*, Artwork Conversion Software, Inc., Santa Cruz, CA. http://www.artwork.com/gdsii/asm3500/index.htm.
7. M.S. Baker, M.P. de Boer, N.F. Smith, L.K. Warne, M.B. Sinclair, Integrated measurement-modeling approaches for evaluating residual stress using micromachined fixed-fixed beams, *J. Microelectromechanical Syst.*, 11(6), 743–753, December 2002.
8. K.S.J. Pister, M.W. Judy, S.R. Burgett, Fearing, Microfabricated hinges, *Sensors Actuators A*, 33, 249–256, 1992.
9. R. Yeh, E.J.J. Kruglick, K.S.J. Pister, Surface micromachined components for articulated microrobots, *J. Microelectromechanical Syst.*, 5(1), 10–17, March 1996.
10. C.M. Baker, C.J. Terman, Tools for verifying integrated circuit designs, *Lambda Mag. (VLSI Design)*, 4th quarter, 22–30, 1980.

Design Realization Tools for MEMS

11. T.G. Szymanski, C.J. Van Wyk, Space efficient algorithms for VLSI artwork analysis, *Proc. 20th Design Automation Conf.*, 734–739, June 1983.
12. H.S. Baird, Fast algorithms for LSI artwork analysis, *Proc. 14th Design Automation Conf.*, 303–311, 1977.
13. http://www.memscap.com/memsrus/crmumps.html.
14. R.A. Buser, N.F. de Rooij, ASEP: a CAD program for silicon anisotropic etching (micromechanical structure), *Sensors Actuators A*, A28(1), 71–78, 1991.
15. C.H. Sequin, Computer simulation of anisotropic etching, Proc. 6th Int. Conf. Solid-State Sensors Actuators (*Transducers '91*), San Francisco, CA, 28–28 June, 1991, 801–806.
16. T.J. Hubbard, E.K. Antonsson, Emergent faces in crystal etching, *J Microelectromech, Syst.*, 3, 19–28, 1994.
17. J. Fruhauf, K. Trautmann, J. Wittig, D. Zielke, A simulation tool for orientation dependent etching, *J. Micromech. Microeng.*, 3(3), 113–115, 1993.
18. D. Dietrich, J. Fruhauf, Computer simulation of the development of dish-shaped deepenings by orientation-dependent etching of (100) silicon, *Sensors Actuators A*, A39(3), 261–262, 1993.
19. U. Heim, A new approach for the determination of the shape of etched devices, *J. Micromech. Microeng.*, 3(3), 116–117, 1993.
20. D. Zielke, J. Fruhauf, Determination of rates for orientation-dependent etching, *Sensors Actuators A*, A48(a), 151–156, 1995.
21. H. Camon, A. Moktadir, M. Djafari-Rouhani, New trends in atomic scale simulation of wet chemical etching of silicon with KOH, *Mater. Sci. Eng. B*, B37(1–3), 142–145, 1996.
22. W.G. Oldham, A.R. Neureuther, J.L. Reynolds, S.N. Nandaonkar, C. Sung, A general simulator for VLSI lithography and etching processes, II: Application to deposition and etching, *IEEE Trans. Electron Devices*, ED-27(8), 1455–1459, 1980.
23. S. Yamamato, T. Kure, M. Ohgo, T. Matsuzama, S. Tachi, H. Sunami, A two-dimensional etching profile simulator: ESPRIT, *IEEE Trans. Computer-Aided Design*, CAD-6(3), 417–422, 1987.
24. J. Pelka, K.P. Muller, H. Mader, Simulation of dry etch processes by COMPOSITE, *IEEE Trans. Computer-Aided Design*, 7(2), 154–159, 1988.
25. C. Hedlund, C. Strandman, I.V. Katardjiev, Y. Backlund, S. Berg, H.O. Blom, Method for the determination of the angular dependence during dry etching, *J. Vac. Sci. Tech. B*, 14(5), 3239–3243, 1996.
26. M. Fujinaga, N. Kotani, 3-D topography simulator (3-D MULSS) based on a physical description of material topography, *IEEE Trans. Electron Devices*, 44, 226–238, 1997.
27. G.M. Koppelman, OYSTER, a three-dimensional structural simulator for microelectromechanical design, *Sensors Actuators*, 20(1/2), 179–185, 1989.
28. P.M. Osterberg, S.D. Senturia, Membuilder: an automated 3D solid-model construction program for microelectromechanical structures, *Proc. Transducers '95*, Stockholm, Sweden, June 1995, 2, 21–24.
29. N.R. Lo, K.S.J. Pister, 3D μV — a MEMS 3-D visualization package, *Proc. SPIE Micromachined Devices Components*, Austin, TX, Oct. 23–24, 1995, 290–295.
30. C.R. Jorgensen, V.R. Yarberry, A 3D geometry modeler for the SUMMiT V MEMS designer, *Proc. Modeling Simulation Microsyst. Conf.*, 606–609, 2001.

31. C.R. Jorgensen, V.R. Yarberry, A 2D visualization tool for SUMMiT V designs, *Proc. Modeling Simulation Microsyst. Conf.*, 594–597, 2001.
32. S.D. Senturia, R.M. Harris, B.P. Johnson, S. Kim, K. Nabors, M.A. Shulman, J.K. White, A computer-aided design system for microelectromechanical systems (MEMCAD), *J. Microelectromech. Syst.*, 1(1), 3–13, 1992.
33. J.R. Gilbert, P.M. Osterberg, R.M. Harris, D.O. Ouma, X. Cai, A. Pfajfer, J. White, S.D. Senturia, Implementation of MEMCAD system for electrostatic and mechanical analysis of complex structures from mask descriptions, *Proc. IEEE Micro Electro Mechanical Syst.*, Fort Lauderdale, FL, Feb. 7–10, 1993, 207–212.
34. Coventer: http://www.coventor.com.
35. S.D. Senturia, Simulation and design of microsystems: a 10-year perspective, *Sensors Actuators A*, 67, 1–7, 1998.
36. S.D. Senturia, CAD challenges for microsensors, microactuators, and microsystems, *Proc. IEEE*, 86(8), 1611–1626, 1998.
37. S.D. Senturia, N. Aluru, J. White, Simulating the behavior of MEMS devices: computational methods and needs, *IEEE Computational Sci. Eng.*, 30–43, Jan.–Mar., 1997.
38. R.M. Kirby, G.E. Karniadakis, O. Mikulchenko, K. Mayaram, An integrated simulator for coupled domain problems in MEMS, *J. Microelectromech. Syst.*, 10(3), 379–391, 2001.
39. SUGAR: http://bsac.berkeley.edu/cadtools/sugar/sugar/#overview.
40. J.V. Clark, D. Bindel, W. Kao, E. Zhu, A. Kuo, N. Zhou, J. Nie, J. Demmel, Z. Bai, S. Govindjee, K.S.J. Pister, M. Gu, A. Agogino, Addressing the needs of complex MEMS design, *Proc. MEMS 2002*, Las Vegas, January 20–24, 2002.
41. J.V. Clark, N. Zhou, K.S.J. Pister, Modified nodal analysis for MEMS with multienergy domains, *Int. Conf. Modeling Simulation Microsyst., Semiconductors, Sensors Actuators*, San Diego, CA, March 27–29, 2000, 31–34.
42. R.E. Rudd, The atomic limit of finite element modeling in MEMS: coupling of length scales, *Analog Integrated Circuits Signal Process.*, 29, 17–26, 2001.
43. MEMSCAP: http://www.memscap.com.
44. ANSYS: http://www.ansys.com.
45. V. Venkataraman, R. Sarma, S. Ananthasuresh, Part to art: basis for a systematic geometric design tool for surface micromachined MEMS, *Proc. ASME Design Eng. Tech. Conf.*, 1–14, 2000.
46. A. Perrin, V. Ananthakrishnan, F. Gao, R. Sarma, G.K. Ananthasuresh, Voxelbased hetrogeneous geometric modeling for surface micromachined MEMS, *Proc. Modeling Simulation Microsys. Conf.*, 136–139, 2001.
47. R. Schiek, R. Schmidt, A new, topology driven method for automatic mask generation from three-dimensional models, *NanoTech 2003*, San Francisco, CA, Feb. 2003.

6 Electromechanics

MEMS devices invariably involve engineering of multiphysics designs to attain a design objective. The two physical domains most frequently utilized in MEMS devices are structural and electrical dynamics. Regardless of the design objective, a structure invariably needs to be designed to support, contain, or possibly deflect to perform a function. An electrical system is needed to sense the mechanical motion of the structure. At the microscale, damping due to viscous losses of the device to the surrounding atmosphere greatly influences the dynamics of the system.

For example, a MEMS accelerometer requires the suspended seismic mass to have a preferred mode of vibration in the sensitive axis at a specific resonant frequency. This device would also have an electrical sense interface to transduce the motion of the seismic mass and, possibly, electrical force feedback to maintain the position of the accelerometer sense mass at a neutral position. The damping of the accelerometer seismic mass will greatly influence the dynamics of the system and needs to be considered in the design.

This chapter will present an overview of the important topics in structural mechanics, damping, and electrical circuit elements. Due to space limitations, an in-depth treatment of these topics is not possible; however, the topics relevant to the design of MEMS devices will be presented. References 1 through 4 provide a more complete background in structural mechanics.

Structural mechanics necessitates the development of the following concepts to obtain a basic understanding of the subject for purposes of MEMS design:

- Structural material models
- Models of the basic structural elements (bending, torsion, axial rods, columns)
- Combining the basic structural elements

Damping mechanisms for vertical and laterally moving MEMS devices will be presented. The basic electrical circuit elements and models for them will be presented along with methods for combining them to form a circuit. The set of equations that describe the electrical circuit elements will also be developed.

194 Micro Electro Mechanical System Design

6.1 STRUCTURAL MECHANICS

6.1.1 MATERIAL MODELS

The atomic structure of materials — broadly classified as *crystalline, polycrystalline,* and *amorphous* — is illustrated in Figure 2.2. A *crystalline* material has a large-scale, three-dimensional atomic structure in which the atoms occupy specific locations within a lattice (e.g., epitaxial silicon, diamond). The atomic packing may be in one of seven main crystal patterns with orientations measured via the *Miller indices* (discussed in Chapter 2).

A *polycrystalline* material consists of a matrix of grains, which are small crystals of material; the interface material between adjacent grains is called the grain boundary. Most metals, such as aluminum and gold as well as polycrystalline silicon, are examples of this material structure. A noncrystalline material that exhibits no large-scale structure is called *amorphous*. Silicon dioxide and other glasses are examples of this material structure.

The material type greatly influences fundamental structure and completeness of interatomic bonds. This basic material structure affects a number of material properties, such as the electrical and thermal conductivities, chemical reactivity, and mechanical strength. For example, the metallurgical processes of cold working and annealing greatly affect the material grains and grain boundary and the resulting material properties of strength, hardness, ductility.

The characteristics of a material that first come to mind in connection with the design of a structure are strength, elasticity, and ductility. These characteristics relate to the ability of the material to resist mechanical forces and how the material will fail. In order to establish a meaningful way to design with these considerations, it is necessary first to define some commonly used engineering terms.

Given a bar of material loaded with a uniform force distribution across the cross-sectional area, A, as shown in Figure 6.1, a quantity, *stress* σ, is defined as the total force, F, per unit cross-sectional area A (Equation 6.1). The applied load will deform the material, which will require the definition of a metric to describe the extent of deformation. The metric for localized deformation of a material, *strain*, is a dimensionless quantity defined as the change in length, δ, per length, L (Equation 6.2).

$$\sigma = F / A \qquad (6.1)$$

$$\varepsilon = \delta / L \qquad (6.2)$$

When an experiment is performed on the specimen of Figure 6.1 in which the load is increased in a controlled manner, stress vs. strain can be plotted (Figure 6.2). The material shown in this figure exhibits elastic strain. The material deforms under load as indicated by strain, but the deformation is not permanent. When the load is removed, the stress and strain return to zero.

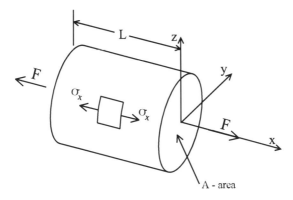

FIGURE 6.1 Loaded material specimen.

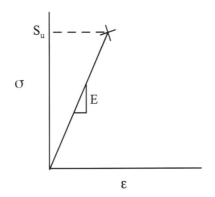

FIGURE 6.2 Elastic stress–strain relationship.

If the load on the material is increased further, the material will plastically deform or fail abruptly. Figure 6.2 shows a material that deforms elastically until it abruptly fails. The stress at failure is known as the *ultimate* strength of the material, S_u. This type of material failure is known as brittle. Figure 6.3 shows a material that deforms elastically until the material yields at a stress known as the *yield stress*, S_y. This is the *elastic limit* of the material. Increasing the stress (by increasing the load) beyond the S_y will induce *plastic strain*, which is a permanent deformation of the material. Unloading a material that has been stressed beyond S_y will cause a different path to be followed on the stress–strain curve upon unloading.

When the material is unloaded, a permanent deformation has been induced in the material as shown by a nonzero deformation existing at zero load. If the stress in the material (load on the specimen) is increased past the yield stress until the material eventually fails, the stress at failure is the ultimate strength of the material S_u. The shape of the stress–strain curve for different ductile materials stressed beyond the elastic limit can vary due to large changes in the material cross-section during plastic deformation. Some material will exhibit a distinct

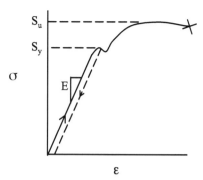

FIGURE 6.3 Plastic stress–strain relationship.

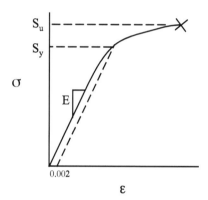

FIGURE 6.4 Plastic stress–strain relationship.

change in slope or distinct plastic deformation at the yield point, but others will be more subtle. When the yield point is not distinct (Figure 6.4), the yield point is generally defined as that stress, which induces 0.2% (0.002) plastic strain.

Most engineering applications will not intentionally stress a material past the yield strength. The system will be designed to operate within the elastic region of the material. The slope of the elastic region of the stress–strain curve is a widely used engineering property of a material known as *Young's modulus*, E, which has units of force per area and is a measure of material stiffness. Appendix E lists typical values of Young's modulus for a number of materials frequently used in MEMS devices.

A frequently used material model for operation within the elastic region of a material is *Hooke's law*, which states that the stress in a material is proportional to the strain that produced it. This is merely the mathematical relationship for the material operating within the elastic portion of the stress–strain curve:

$$\sigma = E\varepsilon \tag{6.3}$$

Electromechanics

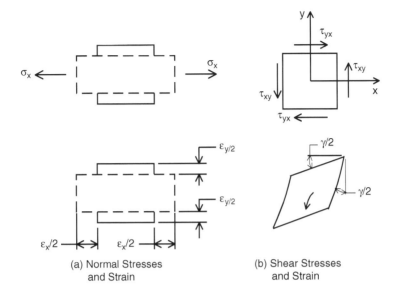

(a) Normal Stresses and Strain

(b) Shear Stresses and Strain

FIGURE 6.5 Planar unit element of material loaded with normal and shear stress.

The discussion thus far has centered on a material loaded normally to the cross-section of the bar, as shown in Figure 6.1. The load could be in tension or compression. Alternatively, a material could be loaded in shear (Figure 6.5). In this case, the load is in the plane of the loaded cross-section. Shear stress, τ, is defined as the load divided by the cross-sectional area, which is similar to the definition for normal stress, σ. However, shear strain, γ, is defined as the change in angle of a unit cube of the material shown in Figure 6.5. The development of shear stress and shear strain is similar to that presented for normal stress loading. Hooke's law for shear loading is shown in Equation 6.4. The constant of proportionality, G, is known as the modulus of rigidity or the shear modulus. E and G represent fundamental properties of a material, and they have units of force per area squared. E and G are measures of the stiffness or rigidity of a material for normal and shear loading, respectively.

$$\tau = G\gamma \tag{6.4}$$

It has also been observed that a material placed in tension also exhibits lateral strain in addition to axial strain. Poisson demonstrated that these two strains are proportional to each other within the elastic region modeled by Hooke's law. The proportionality constant is known as *Poisson's ratio*, ν (Equation 6.5). The Poisson ratio is dimensionless and typically has a value between 0 and 0.5. A solid with $\nu = 0.5$ does not undergo a change volume when strained uniaxially. For example, rubber is a material with $\nu = 0.5$. The common situation for most solid

198 Micro Electro Mechanical System Design

materials is for the volume to expand under uniaxial loading, which corresponds to $v < 0.5$.

$$v = -\frac{\text{lateral strain}}{\text{axial strain}} \qquad (6.5)$$

The three elastic constants, E, G, and v, are related to each other as shown in the following equation:

$$E = 2G(1+v) \qquad (6.6)$$

At this point, stress and strain for a one-dimensional situation have been discussed. Generalized Hooke's law for normal and shear stresses and strains in three dimensions for an isotropic material is shown in Equation 6.7. Isotropic material properties are not a function of spatial orientation. Figure 6.6 illustrates the six stresses (three normal stresses and three shear stresses) involved in the three-dimensional problem. This formulation is frequently appropriate for poly-crystalline and amorphous materials.

$$
\begin{Bmatrix} \varepsilon_x \\ \varepsilon_y \\ \varepsilon_z \\ \gamma_{xy} \\ \gamma_{yz} \\ \gamma_{zx} \end{Bmatrix} =
\begin{bmatrix}
\frac{1}{E} & -\frac{v}{E} & -\frac{v}{E} & 0 & 0 & 0 \\
-\frac{v}{E} & \frac{1}{E} & -\frac{v}{E} & 0 & 0 & 0 \\
-\frac{v}{E} & -\frac{v}{E} & \frac{1}{E} & 0 & 0 & 0 \\
0 & 0 & 0 & \frac{1}{G} & 0 & 0 \\
0 & 0 & 0 & 0 & \frac{1}{G} & 0 \\
0 & 0 & 0 & 0 & 0 & \frac{1}{G}
\end{bmatrix}
\begin{Bmatrix} \sigma_x \\ \sigma_y \\ \sigma_z \\ \tau_{xy} \\ \tau_{yz} \\ \tau_{zx} \end{Bmatrix} \qquad (6.7)
$$

However, for crystalline materials, the material properties will frequently be a function of the spatial orientation. An orthotropic material has three planes of material property symmetry. To describe this spatial material property, dependency for an orthotropic material requires nine independent material properties — an increase over the three independent material properties required for an isotropic material. There will be a Young's modulus for each axis (E_x, E_y, E_z); modulus of rigidity for the three shear planes (G_{xy}, G_{yz}, G_{xz}); and a Poisson ratio (v_{xy}, v_{yz}, v_{xz}) for each axis. Equation 6.8 shows the orthotropic stress–strain relations.

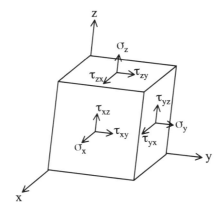

FIGURE 6.6 Unit cube with three-dimensional stresses.

$$\begin{Bmatrix} \varepsilon_x \\ \varepsilon_y \\ \varepsilon_z \\ \gamma_{xy} \\ \gamma_{yz} \\ \gamma_{zx} \end{Bmatrix} = \begin{bmatrix} \dfrac{1}{E_x} & -\dfrac{\nu_{xy}}{E_y} & -\dfrac{\nu_{xz}}{E_z} & 0 & 0 & 0 \\ -\dfrac{\nu_{xy}}{E_y} & \dfrac{1}{E_y} & -\dfrac{\nu_{yz}}{E_z} & 0 & 0 & 0 \\ -\dfrac{\nu_{xz}}{E_z} & -\dfrac{\nu_{yz}}{E_z} & \dfrac{1}{E_z} & 0 & 0 & 0 \\ 0 & 0 & 0 & \dfrac{1}{G_{xy}} & 0 & 0 \\ 0 & 0 & 0 & 0 & \dfrac{1}{G_{yz}} & 0 \\ 0 & 0 & 0 & 0 & 0 & \dfrac{1}{G_{zx}} \end{bmatrix} \begin{Bmatrix} \sigma_x \\ \sigma_y \\ \sigma_z \\ \tau_{xy} \\ \tau_{yz} \\ \tau_{zx} \end{Bmatrix} \quad (6.8)$$

An anisotropic material is the most general material that requires 21 material properties to model its behavior. Equation 6.9 shows the stress–strain relations that would model an anisotropic material. Isotropic and orthotropic material models are special cases of an anisotropic material model.

$$\begin{Bmatrix} \varepsilon_x \\ \varepsilon_y \\ \varepsilon_z \\ \gamma_{xy} \\ \gamma_{yz} \\ \gamma_{zx} \end{Bmatrix} = \begin{bmatrix} C_{11} & C_{12} & C_{13} & C_{14} & C_{15} & C_{16} \\ C_{12} & C_{22} & C_{23} & C_{24} & C_{25} & C_{26} \\ C_{13} & C_{23} & C_{33} & C_{34} & C_{35} & C_{36} \\ C_{14} & C_{24} & C_{34} & C_{44} & C_{45} & C_{46} \\ C_{15} & C_{25} & C_{35} & C_{45} & C_{55} & C_{56} \\ C_{16} & C_{26} & C_{36} & C_{46} & C_{56} & C_{66} \end{bmatrix} \begin{Bmatrix} \sigma_x \\ \sigma_y \\ \sigma_z \\ \tau_{xy} \\ \tau_{yz} \\ \tau_{zx} \end{Bmatrix} \quad (6.9)$$

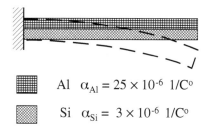

FIGURE 6.7 Aluminum–silicon cantilever beam.

6.1.2 Thermal Strains

When the temperature of an unconstrained elastic member is increased, the member expands in all directions. The normal strain produced in the material is called *thermal strain* and is proportional to the temperature increase, ΔT, shown in Equation 6.10. The proportionality constant, α, is a material property called the *coefficient of thermal expansion* [8]. Appendix E gives representative values for the coefficient of thermal expansion of a number of commonly used MEMS materials. If the elastic member is unconstrained, the temperature increase produces thermal strain; however, no stress is induced in the material.

$$\varepsilon_x = \varepsilon_y = \varepsilon_z = \alpha \Delta T \tag{6.10}$$

However, if a uniform rod is constrained at each end so that the material cannot expand when subjected to a temperature increase, a compressive stress (Equation 6.11) will be induced because of the constraint.

$$\sigma = \varepsilon E = \left(\alpha \Delta T\right) E \tag{6.11}$$

Thermal strains can also be developed in devices incorporating materials with different coefficients of thermal expansion. For example, a beam with aluminum deposited on silicon, as shown in Figure 6.7, will flex out of plane when exposed to a uniform temperature increase due to the different coefficients of thermal expansion of the materials.

In the discussion thus far, only the case in which a uniform temperature increase in a material produces a thermal strain has been considered. Another common and interesting situation is produced when temperature gradients due to nonuniform temperature distributions in the material exist. Temperature gradients can be due to thermal transients and nonuniform heat generation or heat deposition within the material. A *thermal stress* is developed due to a temperature gradient in a body.

Figure 6.8 illustrates the thermal stress induced in a MEMS die and device during transient heat transfer. The heat flux on the bottom surface of the substrate

Electromechanics

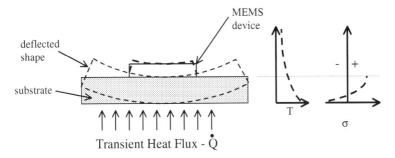

FIGURE 6.8 Thermal stress during a transient heat transfer.

produces a temperature gradient; the bottom surface is the hottest. The temperature gradient causes the material at different depths and temperature to expand differently, thus resulting in a thermal strain that causes out-of-plane deflections of the substrate.

The effect of thermally induced stresses on precision MEMS sensors will have a direct impact on performance and thus needs to be considered in the system design. Thermal stress is also utilized as a MEMS actuation mechanism (discussed in Chapter 8).

6.1.3 Axial Rod

The axial rod shown in Figure 6.1, which was discussed during the development of material models, is a fundamental structural element. A rod is an idealized one-dimensional structural element subjected only to axial forces, which can be tensile or compressive. Figure 6.9 is a schematic of the displacement of an element of material in the rod. In this schematic, the element of material, dx, has an applied force, F. Due to this loading, the material at position x undergoes a displacement

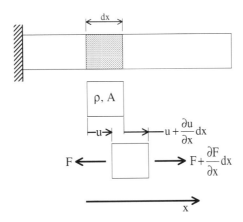

FIGURE 6.9 Loaded one-dimensional axial rod material element.

u, and the material at position $x + dx$ undergoes a displacement $u + \left(\dfrac{\partial u}{\partial x}\right) dx$. It can be seen that the element of material has changed length by the amount $\left(\dfrac{\partial u}{\partial x}\right) dx$. Therefore, the strain is $\left(\dfrac{\partial u}{\partial x}\right)$, which is a mathematical partial derivative expression for strain (i.e., change in length ∂u per length ∂x). Using Hooke's law, Equation 6.12 is obtained upon substituting for σ and ε.

$$\sigma = \frac{F}{A} = E\frac{\partial u}{\partial x} = E\varepsilon \qquad (6.12)$$

Differentiating F with respect to x yields Equation 6.13, which is an expression for the rate of change of force in the material.

$$\frac{\partial F}{\partial x} = AE\frac{\partial^2 u}{\partial x^2} \qquad (6.13)$$

Newton's law of motion is now applied for the element and the net elastic force in the material element is equated with the inertial force of the material element, where ρ is the density of the rod:

net elastic force = mass $\times$ acceleration

$$\frac{\partial F}{\partial x} dx = \left(\rho A dx\right)\frac{\partial^2 u}{\partial t^2} \qquad (6.14)$$

Substituting for $\dfrac{\partial F}{\partial x}$ yields the one-dimensional wave in Equation 6.15, which is a common governing equation for a number of physical phenomena in addition to axial vibration of a rod (e.g., string under tension, acoustics):

$$\frac{\partial^2 u}{\partial x^2} = \frac{1}{c^2}\frac{\partial^2 u}{\partial t^2} \qquad (6.15)$$

The quantity c is the *velocity of propagation* of displacement or stress in a rod. This metric includes information regarding the stiffness of the material as well as its density.

$$c = \sqrt{\frac{E}{\rho}} \qquad (6.16)$$

Electromechanics

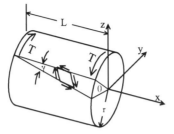

FIGURE 6.10 Torsion rod.

The stiffness coefficient, K, is another useful design metric that can be obtained for the axial rod. Using Equation 6.12 and recognizing that the strain is the change in length, δ, divided by the original length, L, yields the stiffness coefficient, K. The stiffness coefficient for the axial rod is the ratio of the force per unit deflection and has units of force divided by length. Whereas E and G were metrics for stiffness of a particular material, K is a measure of the stiffness of a particular structural element with a specified loading situation.

$$K = \frac{F}{\delta} = \frac{AE}{L} \qquad (6.17)$$

6.1.4 Torsion Rod

Another common structural element involved in structural design, torsion rods may be used as flexures to couple or suspend structures. The torsion rod is a length of material loaded with an applied torque that will produce an angular displacement, θ, and shear stress, τ, in the structural element (Figure 6.10). The theoretical development for torsion bars involves the following assumptions:

- The torsion bar is straight and of uniform circular cross-section (solid or concentrically hollow).
- The torsion bar is loaded by and opposite torques.
- The torsion bar is not stressed beyond it elastic limit.

The torsion bar twists as torque is applied where plane cross-sections remain plane and radii remain straight. Torsion produces a shear stress at any point in the cross-section, which is a maximum at the outer surface; the shear stress is proportional to the distance from the center. Equation 6.18 defines the shear stress due to torsion as a function of the distance from the center of the torsion bar, r. J is the area polar moment of inertia. The shear stress is a maximum at the outer surface, $r = R$ (Equation 6.19). This fact explains why many torsion drive shafts are annular tubes. The material near the shaft center carries little stress in pure torsion loading. J for circular cross-sections is given in Appendix G, Table A.G.1.

$$\tau = \frac{Tr}{J} \tag{6.18}$$

$$\tau_{max} = \frac{TR}{J} \tag{6.19}$$

The development of the governing equation for the torsion rod is similar to that for the axial rod. Once again, the governing equation is the one-dimensional wave equation (Equation 6.20) involving a speed of propagation, c, for torsional displacement and stress (Equation 6.21). A stiffness coefficient for a torsion rod that relates the applied torque to the angular displacement can also be developed (Equation 6.22). The stiffness coefficient, K, has units of torque per radian. These equations involve the coefficient of rigidity, G, and material density, ρ, as well as the length of the torsion rod, L.

$$\frac{\partial^2 \theta}{\partial x^2} = \frac{1}{c^2} \frac{\partial^2 \theta}{\partial t^2} \tag{6.20}$$

$$c = \sqrt{\frac{G}{\rho}} \tag{6.21}$$

$$T = K\theta \tag{6.22}$$

where,

$$K = \frac{JG}{L} \tag{6.23}$$

MEMS applications rarely encounter circular cross-sections. Rectangular and square cross-sections are more common due to the restrictions of MEMS fabrication techniques. When a torsion load is applied to a noncircular cross-section, some basic assumptions for torsion are violated. The cross-sections become warped; the greatest stress occurs at a point on the perimeter nearest the axis of twist and the corners of the rectangular and square cross-sections have no stress. The analysis of torsion for these cross-sections becomes complex and has been studied for years [4,5]. Table A.G.1 in Appendix G gives values for the torsional constant, J, that account for the effects of noncircular cross-sections; these can be used in calculations for torsional stiffness.

Electromechanics

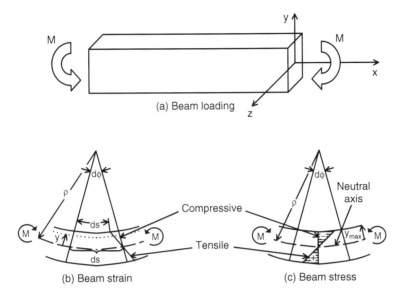

FIGURE 6.11 Beam bending.

6.1.5 BEAM BENDING

Lateral beam bending is a one-dimensional element in which the loading is perpendicular to the axis of the beam. The loading may be distributed along the length of the beam or concentrated at a specific location; it can also be a combination of these situations. A basic formulation of lateral beam bending is called the *Euler–Bernouli beam*. The assumptions involved in the development of the Euler–Bernouli beam model are:

- The material is homogenous and isotropic and obeys Hooke's law.
- The beam is initially straight with a constant cross-section.
- The beam is subjected to pure bending (i.e., no torsion or axial loads).
- Plane cross-sections (y–z plane) remain plane during bending.
- The beam has an axis of symmetry.

The lateral deflection of a Euler–Bernouli beam, shown in Figure 6.11, is due to a bending moment, M, that bends the beam in a curve with a radius of curvature, ρ. The radius of curvature is assumed to be related to the bending moment, as shown in Equation 6.24, where E is Young's modulus and I is the area moment of inertia about the axis of bending (z axis in Figure 6.11a). The EI product is the proportionality constant in Equation 6.24, which is frequently called the *beam cross-section stiffness*. The stiffness of the entire beam will involve other information, such as the beam length and the beam end conditions. The radius of curvature, ρ, is also shown in Equation 6.24 to be approximately equal to second derivative of y with respect to x. This approximation used in the

206 Micro Electro Mechanical System Design

formulation of the Euler–Bernouli beam will limit the resulting governing equation's validity to small deflections. The deflection of the beam has a "neutral axis" that is not stressed during bending deformation. Table A.G.2 and Table A.G.3 in Appendix G show the calculation of the area moment of inertia for the common beam cross-sections encountered in MEMS devices.

For the beam shown in Figure 6.11c, the material above the neutral axis is in compression and below the neutral axis is in tension. Equation 6.25 shows the mathematical expression for the stress distribution across the beam cross-section illustrated in Figure 6.11c. As shown in this figure, the stress is a maximum at the outer fibers of material (Equation 6.26). Depending on the direction of the bending moment at a particular cross-section, the stress is compressive on one side of the neutral axis and tensile on the other side, with the maximum tensile and compressive stresses occurring at the outer surfaces.

$$\frac{M}{EI} = \frac{1}{\rho} = \frac{d^2y/dx^2}{\left(1 + \left[dy/dx\right]^2\right)^{3/2}} \approx \frac{d^2y}{dx^2} \tag{6.24}$$

$$\sigma = -\frac{My}{I} \tag{6.25}$$

$$\sigma_{max} = -\frac{My_{max}}{I} \tag{6.26}$$

Figure 6.12 schematically illustrates a beam undergoing lateral bending due to a distributed load $q(x)$, which is a measure of applied force per length along the beam. The beam also has a mass per length, m. This figure also shows the forces and moments that will exist on a unit length, dx, of beam material, where M is a moment and V is a shear force.

The governing equation for a Euler–Bernouli beam can be developed from this schematic representation. Equation 6.27 shows an application of Newton's second law in the lateral direction for the unit length of material. Equation 6.28 can be obtained by summing moments with respect to the right face of the unit element of material. A limiting process results in the elimination of the dx^2 term.

$$m\frac{\partial^2 y}{\partial t^2}dx = -\frac{\partial V}{\partial x}dx + qdx$$

$$m\frac{\partial^2 y}{\partial t^2} = -\frac{\partial V}{\partial x} + q \tag{6.27}$$

Electromechanics

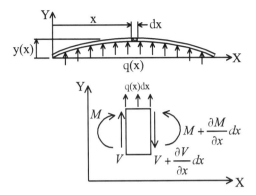

FIGURE 6.12 Element of material undergoing beam bending.

$$\frac{\partial M}{\partial x} dx - V dx - \frac{1}{2} q dx^2 = 0 \quad (6.28)$$

$$\frac{\partial M}{\partial x} = V$$

Combining Equation 6.24, Equation 6.27, and Equation 6.28 results in Equation 6.29, which is the governing partial differential equation for a Euler–Bernouli beam.

$$m \frac{\partial^2 y}{\partial t^2} + \frac{\partial^2}{\partial x^2}\left(EI \frac{\partial^2 y}{\partial x^2}\right) = q \quad (6.29)$$

This formulation can be further simplified for a beam with a uniform cross-section vs. length:

$$m \frac{\partial^2 y}{\partial t^2} + EI \frac{\partial^4 y}{\partial x^4} = q \quad (6.30)$$

Solution of the governing equation for a specific beam configuration necessitates the definition of the end conditions of the beam. Because the governing equation for a beam is higher order than the governing equation for torsion or an axial rod, the specification of the boundary condition at the ends of the beam requires two variables to be defined. A beam requires the specification of any two of the following: deflection (y), slope (θ), moment (M), or shear force (V). Table 6.1 shows examples of the various *classical* specifications of beam end conditions. The solutions of the Euler–Bernoulli beam equation have been studied extensively and solutions for a great variety of loading and boundary condition situations are readily available [2–4].

TABLE 6.1
End Conditions for a Beam

End condition	Deflection (y)	Slope (θ)	Moment (M)	Shear (V)
Fixed	$y = 0$	$\theta = 0$		
Free			$M = 0$	$V = 0$
Hinges	$Y = 0$		$M = 0$	

The stiffness coefficient, K, which is the proportionality constant between displacement and force, is a useful design parameter. For simple beam-bending situations, the stiffness coefficient, K, has a generic form shown in Equation 6.31, where EI is the beam cross-section stiffness, L is beam length, and C is a coefficient that depends on the beam-loading and boundary conditions. Appendix F gives the stiffness for some simple beam-bending situations as well as some more complicated flexures that may be used in MEMS design. Compare the stiffness coefficients listed there for some simple beam-bending situations such as a fixed–free beam (e.g., $C = 3$) with Equation 6.31.

$$K_{beam} = C \frac{EI}{L^3} \tag{6.31}$$

The Euler–Bernouli beam is a very useful formulation for beam bending and can be used extensively in MEMS design. Seely and Smith [1] have an in-depth discussion of the assumptions involved in the Euler–Bernouli beam. However, the designer needs to be aware of the limitations of this theory and have a feel for when an alternative theory is more appropriate. Euler–Bernouli beam theory is appropriate for small deflections of a beam that, due to the approximation Equation 6.24, is involved in development of the governing equation. As a result, the beam response predicted by this theory will become inaccurate when the deflections become large.

Shear deformation and rotary inertia effects can also affect the accuracy of Euler–Bernouli beam theory. Shear deformation effects become significant when the beam is short and mode of deformation is due to shear vs. bending. *Timoshenko* beam theory [7] is appropriate for situations with shear and rotary inertia effects; however, the governing equation is more complex. A beam with an initial curvature is another frequently encountered situation that is addressed in Roark [2].

6.1.6 FLAT PLATE BENDING

Flat plate bending (Figure 6.13) is a structural situation that involves bending of a two-dimensional flat plate due to transverse loading. Many of the same assumptions used in the development of the governing equations for one-dimensional beam bending are also used for flat plate bending; however, the analysis is more

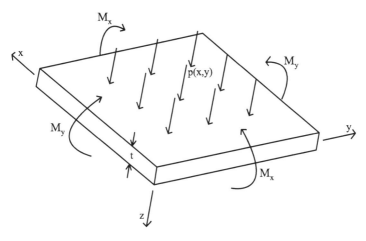

FIGURE 6.13 Flat plate bending schematic.

complex because of multiaxis bending. Soedel [6] and Timoshenko and Woinowsky-Krieger [7] provide the detailed governing equation development for the flat plate (Equation 6.32). The quantity, D, is known as the *plate flexural rigidity* (Equation 6.33). Beam bending and flat plate bending are fourth-order partial differential equations. The boundary conditions for a flat plate are conceptually the same as those for beam bending except that the boundary conditions are applied along the length of the plate edges.

$$\rho \frac{\partial^2 w}{\partial t^2} + D \left(\frac{\partial^2}{\partial x^2} + \frac{\partial^2}{\partial y^2} \right) \left(\frac{\partial^2 w}{\partial x^2} + \frac{\partial^2 w}{\partial y^2} \right) = p \quad (6.32)$$

$$D = \frac{Eh^3}{12(1-v^2)} \quad (6.33)$$

The analysis of flat plate bending is very important to the design of MEMS devices such as pressure sensors. Plate theory has been studied extensively and a number of theories have been developed to analyze plates of various configurations and loadings. For thick plates, transverse shear becomes important and for thin plates or membranes, the restoring force is primarily due to tension. Equation 6.32 can be used for the analysis of plates in which flexural stress dominates and the deflections are small relative to the thickness of the plate (e.g., $y < 0.4\ t$). Analytical solutions of flat plate bending [6,7] have been developed for a number of cases, but due to their complexity, they are not generally useful for design synthesis calculations. Figure 6.14 and Figure 6.15 show simplified formulas for pressure loaded circular and rectangular plates with fixed boundary conditions. These two cases are particularly relevant for MEMS devices.

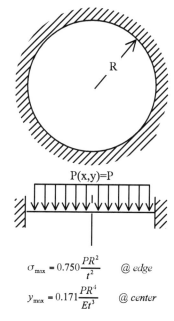

FIGURE 6.14 Pressure loaded, fixed-boundary circular plate deformation and stress.

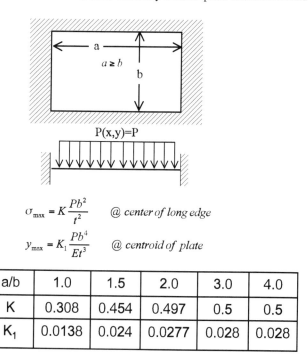

a/b	1.0	1.5	2.0	3.0	4.0
K	0.308	0.454	0.497	0.5	0.5
K_1	0.0138	0.024	0.0277	0.028	0.028

FIGURE 6.15 Pressure loaded, fixed-boundary rectangular plate deformation and stress.

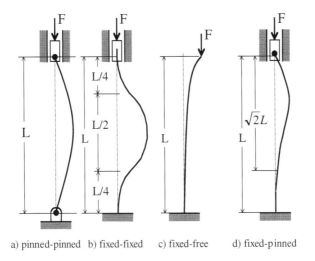

a) pinned-pinned   b) fixed-fixed   c) fixed-free   d) fixed-pinned

FIGURE 6.16 Axially loaded columns with different end conditions.

6.1.7 Columns

Axial loading of a structural member (Figure 6.16) is a frequently encountered situation. The axial load, F, can be tensile or compressive. A member in tension would behave as an axial rod (described in a previous section). However, a structural member under axial compression will have two modes of failure:

- *Material failure.* This occurs when a structural member behaves as an axial rod where the material will fail.
- *Instability failure.* If the structural member is sufficiently long and slender, the member will deflect laterally even though the loading is purely axial. At certain critical loads, this will cause a sudden and catastrophic collapse of the structure.

In structural mechanics, columns generally refer to axial compressive loading of a member and the analysis of this instability phenomenon.

Using a very similar approach to the development of the governing equations for a beam, but with axial loads included, the governing equation for a compressive axial loaded column can be developed (Equation 6.34). The resulting governing equation is the same as the Euler–Bernouli beam except for an additional term that accounts for the axial loading effects. The second and third terms on the left-hand side of the equation model the stiffness of the beam.

The overall stiffness of a beam or column is greatly influenced by the type of end condition. Various combinations of idealized end conditions for a column are illustrated in Figure 6.16. A pinned end condition constrains the lateral deflection of the beam, but the rotation is not constrained. The fixed end condition constrains both the deflection and rotation of the beam end, and conversely the free end condition constrains neither the deflection nor rotation. An idealized fixed

212   Micro Electro Mechanical System Design

end condition is difficult to realize in practice; therefore, the theoretical results developed using this idealization are sometimes modified (see note Table 6.2).

$$m\frac{\partial^2 y}{\partial t^2} + \frac{\partial^2}{\partial x^2}\left(EI\frac{\partial^2 y}{\partial x^2}\right) + F\frac{\partial^2 y}{\partial x^2} = q \tag{6.34}$$

A simple analysis of a compressive axially loaded column (Figure 6.16c) can be done with Equation 6.34 as the starting point. Assuming the beam in Figure 6.16a is statically loaded, only by the axial force, F, allows neglect of the first term on the left and right sides of Equation 6.34. This would result in the following equation after some simplification:

$$\frac{d^2 y}{dx^2} + \frac{F}{EI}y = 0 \tag{6.35}$$

This second-order differential equation for the column, which is pin supported at each end, would have the following simple solution:

$$A\sin\left(\sqrt{\frac{F}{EI}}L\right) = 0 \tag{6.36}$$

If the column is about the collapse, the constant, A, in the preceding equation cannot equal zero because the column will deflect laterally. Therefore, the sin term will need to equal zero. This will be true if $\sqrt{F/EI}\,L = \pi$. Solving for F will produce the critical load, F_{cr}, at which the column will collapse.

$$F_{cr} = \frac{\pi^2 EI}{L^2} \tag{6.37}$$

This analysis of a pinned-pinned column has resulted in the classical theory of column collapse known as *Euler column theory*. This basic analysis can be generalized for beam of various end conditions (Figure 6.16). Table 6.2 show the various fundamental column loading situations and the end condition constants, C, that can be used in the generalized form of the Euler column formula (Equation 6.38). This can be used to provide an estimate of the critical load for column collapse. However, the analysis and prediction of F_{cr} for a real-world situation is very difficult. Column instability is very sensitive to small perturbation in the end conditions and structural imperfections. Also, a fixed boundary condition is very difficult to achieve in practice. Therefore, use of a value of C of no greater than 1.2 is recommended to avoid predicting a value too high for F_{cr}.

TABLE 6.2
Column End Conditions and the End Condition Constraint Constant, C

Column end conditions	End condition constant, C
Fixed–free	1/4
Pinned-pinned	1
Fixed–pinned	2
Fixed–fixed	4

Note: Fixed end conditions are hard to achieve in practice.

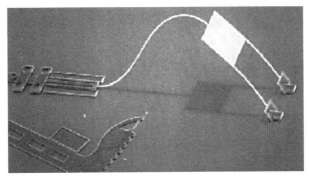

FIGURE 6.17 Mirror erected using elastic instability actuation. (Courtesy of E. Garcia, Sandia National Laboratories.)

$$F_{cr} = \frac{C\pi^2 EI}{L^2} \tag{6.38}$$

One of the difficulties in the analysis of column behavior is the imprecise demarcation between column behavior and an axial rod under compressive loading and the existence of non-axial-loading situations. This has resulted in a number of other approaches to improve column behavior prediction, such as Euler–Johnson column theory, and the secant formula; these are explained in detail in Shigley and Mitchell [3] and Rothbart [4]. Roark [2] discusses elastic instability for other structural situations such as plates and shells. Elastic instability can also be used to advantage. For example, a buckling beam has been used to produce out-of-plane motion for a MEMS mirror surface [9], as shown in Figure 6.17.

6.1.8 STIFFNESS COEFFICIENTS

The previous sections have developed the classical structural elements (axial rod, torsion rod, beam, plate, etc.) that can be used in combination to analyze more complex structures. A common metric for the analysis of structural elements

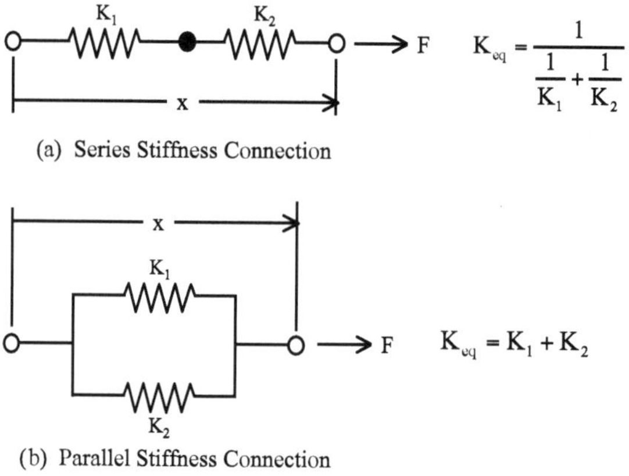

(a) Series Stiffness Connection

(b) Parallel Stiffness Connection

FIGURE 6.18 Combination of structural element stiffness coefficients.

separately or in combination is the stiffness coefficient, K, which has units of force or torque per deflection. To combine stiffness elements to access the stiffness of more complex structures, the methods of combining stiffnesses in series or parallel are used (Figure 6.18). A series spring combination has the same force or torque in both spring elements and the deflection is the total deflection of both springs. A parallel combination of spring elements has the force split between the springs and the deflection is the same for all the springs in parallel. Note: the mathematical relationship for combinations of mechanical stiffnesses vs. electrical elements is opposite. The use of simple models and the ability to combine them to make estimates of the system response are immensely valuable in design.

Example 6.1

Problem: Find the stiffness in the X direction of the crab-leg spring shown in Figure 6.19. Assume the spring is made of polysilicon with $E = 1.6 \times 10^5 \ \mu n/\mu m^2$.

Solution: The vertical beam has fixed–fixed end conditions and the stiffness for that element is given in Appendix F. The area moment of inertia, I, can be calculated using the relationships shown in Appendix G:

$$K_{beam} = \frac{12EI}{L^3} = \frac{12E}{L^3}\left(\frac{tw^3}{12}\right) = \frac{\left(1.6 \times 10^5\right)\left(2\right)\left(2^3\right)}{100^3} = 2.56 \ \frac{\mu n}{\mu m^2}$$

The stiffness coefficient of the horizontal rod element (Equation 6.17) at the top of the crab-leg suspension is:

$$K_{rod} = \frac{EA}{L} = \frac{Etw}{L} = \frac{\left(1.6 \times 10^5\right)\left(2\right)\left(2\right)}{100} = 6400 \ \frac{\mu n}{\mu m^2}$$

Electromechanics

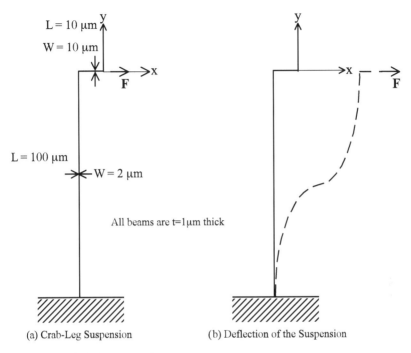

(a) Crab-Leg Suspension (b) Deflection of the Suspension

FIGURE 6.19 Crab-leg suspension.

The rod has a significantly higher stiffness than the beam element. The combination of these two stiffnesses is in series and can be calculated as shown next. Because K_{rod} is so much stiffer than K_{beam}, K_{total} is essentially K_{beam}. Stiffness is a measure of the force per deflection; therefore, an axial rod requires significantly more force to obtain the same deflection as that which can be obtained via beam bending.

$$K_{total} = \frac{1}{\frac{1}{K_{beam}} + \frac{1}{K_{rod}}} = \frac{1}{\frac{1}{2.56} + \frac{1}{6400}} \approx 2.56 \frac{\mu n}{\mu m^2}$$

Example 6.2

Problem: An object is supported by two beams with fixed–fixed boundary conditions and deflected in the lateral direction as shown in Figure 6.20. Find the total spring constant for this suspension.

Solution: The support beams in this situation are connected in parallel because they have the same deflection and the force split between them. Therefore, the total spring constant of this suspension is simply the sum of the two individual stiffnesses:

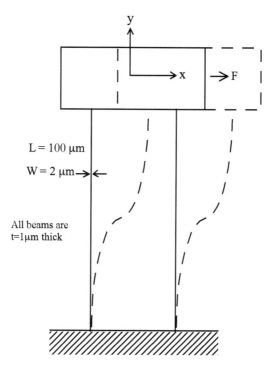

FIGURE 6.20 An object supported by two beams deflected in the lateral (x) direction.

$$K_{total} = K_{beam} + K_{beam} = 2.56 + 2.56 = 5.12 \frac{\mu n}{\mu m^2}$$

6.2 DAMPING

Damping refers to energy dissipation from a mechanical system. The effect of damping is to remove energy from the system; this can be accomplished by dissipation into heat energy. The exact mechanism of damping can take many forms, such as the heating of metal under repeated mechanical deformation; the radiation of sound when a structural object (a plate) is struck; or the aerodynamic drag on a moving automobile or airplane. MEMS devices also experience damping, which is mainly due to two sources:

- *Structural damping.* This form of damping or energy dissipation is due to an energy loss mechanism internal to the material. For large-amplitude, cyclic motion of a structural material, significant heat generation can become apparent. However, for low-amplitude oscillation of a structural member, such as may occur in a vibratory MEMS gyroscope, this form of damping is very small.

Electromechanics 217

- *Fluidic damping.* For MEMS devices, this form of damping is more significant and frequently needs to be considered in design. This damping is due to the cyclic motion of a structure with a gaseous and possibly liquid environment.

6.2.1 OSCILLATORY MECHANICAL SYSTEMS AND DAMPING

A dynamic mechanical system such as a spring-mass-damper can be mathematically described by a second-order differential equation (Equation 6.39). This equation has three force terms plus the external applied force, $f(t)$. The inertial force, F_m, is the mass–acceleration product related to the kinetic energy of the system. The stiffness force, F_k, is the stiffness–deflection product due to the elastic deformation. Elastic deformation produces the potential energy of the system. During oscillatory motion, energy oscillates between kinetic and potential energy. Energy is input to the system via the external applied force, $f(t)$. Energy is lost from the system via damping. The damping force, F_d, is the product of the damping coefficient and velocity, which is a linear viscous model of the damping. There are other damping models, such as Coulomb damping, but the linear viscous damping model is frequently used because it is readily analyzed mathematically.

$$m\ddot{x} + c\dot{x} + kx = f(t)$$
$$F_m + F_d + F_k = f(t)$$

(6.39)

The energy dissipated by damping, W_d, during oscillatory motion can be expressed by the path integral of the damping, F_d, over the displacement path, x, as shown in Equation 6.40. The force–displacement relationship expressed by Equation 6.40 can be illustrated by Figure 6.21, which shows a *hysteresis* loop. This loop is indicative of a dissipative system. The area of the *hysteresis* loop is proportional to the energy lost per cycle by the system. The exact force–displacement relationship may vary greatly due to the type of damping mechanism involved.

$$W_d = \oint F_d dx$$

(6.40)

The mass (m), stiffness (k), and damping (c) coefficients by themselves do not lend significant physical insight to the system response. However, a second-order system such as Equation 6.39 can be transformed into *modal coordinates* (Equation 6.41) by normalizing with respect to the mass and defining two variables known as the natural frequency, ω_n, and damping ratio, ζ (Equation 6.42 and Equation 6.43, respectively) [10,11]. An alternative metric for system damping is the quality factor, Q, which is related to the damping ratio by Equation 6.44.

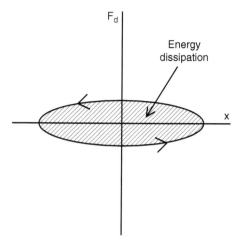

FIGURE 6.21 The force–displacement hysteresis loop, which illustrates the energy dissipated by damping.

$$\ddot{x} + 2\zeta\omega_n\dot{x} + \omega_n^2 x = f(t) \tag{6.41}$$

$$\zeta = \frac{c}{2\sqrt{km}} \tag{6.42}$$

$$\omega_n = \sqrt{\frac{k}{m}} \tag{6.43}$$

$$Q = \frac{1}{2\zeta} \tag{6.44}$$

The maximum system response, x, will be obtained if the system is excited with an oscillatory force at the natural frequency (e.g., $f \sin[w_n t]$). This condition is known as *resonance*. If the system has no damping (c and ζ are zero), the response at resonance is mathematically infinite; however, every system contains at least some small amount of damping, which will limit the response. At resonance, the kinetic and potential energy of the system are equal and the system energy is oscillating in form between kinetic and potential energy.

The damping ratio, ζ, is any positive real number. The damping ratio directly controls the nature of the system response as illustrated in Figure 6.22. Figure 6.22 shows the response for a spring–mass–damper system that is displaced and released. For values of the damping ratio $0 \leq \zeta < 1$, the system has an oscillatory

Electromechanics

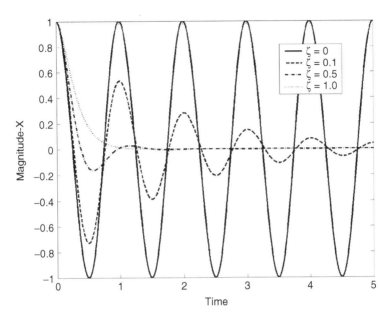

FIGURE 6.22 System oscillatory response for different values of damping ratio, ζ.

response. At $\zeta = 0$, the system is undamped with no dissipation; this allows the amplitude of oscillation never to decrease. As ζ increases toward 1, the damping increases, thus causing the oscillations to decay in amplitude; at $\zeta \geq 1$, the system response is not oscillatory. $\zeta = 1$ is called *critical damping* because it is the transition between oscillatory and decaying non-oscillatory motion.

The system damping also controls the amplitude of the response when excited at resonance. Figure 6.23 is a plot of the normalized response vs. normalized frequency for the second-order system (e.g., mass–damper–spring system). The system response is normalized to the system response at zero frequency (i.e., static response), x_0. The normalized system response is referred to as the amplification factor, M. The excitation frequency is normalized to the system natural frequency to define the frequency ratio, r. Resonance occurs when $r = \omega/\omega_n = 1$. As can be seen from Figure 6.23, the system damping controls the amplitude at resonance and the width of the normalized response curve at resonance. *The amplification factor, at resonance, M_{max}, is Q.*

The width of the normalized response curve at what is known as the half-power point is also directly correlated to the system damping. Because power is proportional to the amplitude squared, $X^2 \propto M^2$, the half-power point is defined as $M_{max}/\sqrt{2}$. If the system has very little damping, its response peak is very tall and narrow. For large amounts of damping, the peak is not much greater than unity and very broad. From this curve, it is evident that *Q is the amplification of system response at resonance.*

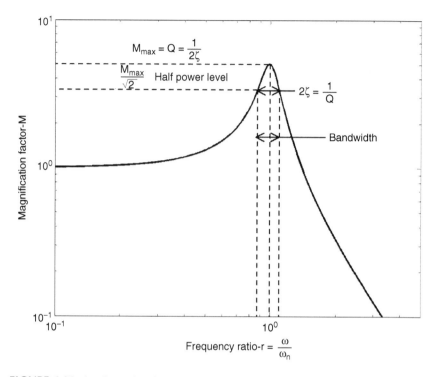

FIGURE 6.23 A schematic of a second-order system frequency response plot showing the effect of damping at the natural frequency.

6.2.2 Damping Mechanisms

Many MEMS devices involve an oscillating structure to implement a sensing methodology. In order to induce the oscillatory response of the MEMS device, the structure is electrostatically actuated in a controlled atmosphere and pressure environment. Zook et al. [12] studied the response of polysilicon resonant microbeams and showed that the system Q, vs. environmental pressure, P, had three response regimes (Figure 6.24). Regime I has the maximum Q attainable at very low pressure. The damping in regime I is extremely low and due primarily to the resonator material damping. In regime II, the pressure is still low; the gas molecules are not interacting with each other. Damping occurs by momentum transfer between gas molecules and the vibrating beam. Q is a function of pressure in regime II. At the higher pressures attained in regime III, the gas molecules interact and exert a viscous drag force on the vibrating beam.

The spaces between structures, walls, and substrate for MEMS devices can be as small as 1 or 2 μm. The development of damping models suitable for such applications may involve the utilization of fluid mechanics theory developed for macroscale applications. This will bring into question the appropriateness of continuum-based fluid theories for use in MEMS applications. Continuum theo-

Electromechanics

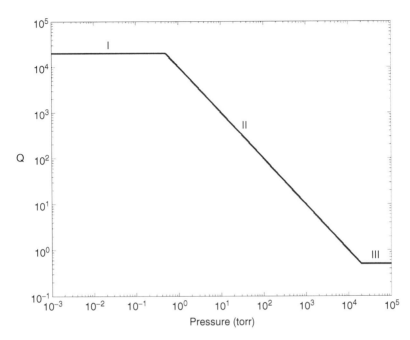

FIGURE 6.24 Quality factor, Q, vs. pressure regimes for a MEMS resonator. (Reprinted from J.D. Zook et al., *Sensors Actuators A*, 35, 51–59, 1992. With permission from Elsevier.)

ries assume that, in any analysis domain, a large number of atoms is present that use continuous partial differential equations to describe the relevant physics vs. the discrete dynamic interactions of individual molecules.

An appropriate metric to assess this situation is the mean free path, λ, of the gas surrounding the MEMS device; this is the distance that a molecule travels before it collides with another molecule. Collisions are the primary momentum transfer mechanism of gases. These collisions will give rise to energy dissipation on the surface of a moving MEMS device. If a molecule of diameter, d, travels within a distance, d, of another molecule of the same type, a collision will occur. Therefore, the collision cross-section of the molecule is πd^2. For molecules such as N_2 and O_2, $d \approx 3$ Å is a good estimate. The probability of collision, P, within a distance, D, of a volume containing a molecular density (molecules/volume), n, is given by:

$$P = D\pi d^2 n \qquad (6.45)$$

If the probability of a collision within a distance λ is 1, an estimate of the mean free path is approximately given by:

$$\lambda \approx \frac{1}{\pi d^2 n} \qquad (6.46)$$

A more statistically accurate estimate of the mean free path, λ, is given by:

$$\lambda \approx \frac{1}{\sqrt{2}\pi d^2 n} \qquad (6.47)$$

The molecular density, n, may be calculated from the ideal gas law:

$$n = \frac{p}{k_B T} \qquad (6.48)$$

Combining equations yields the mean free path for an ideal gas equation, where p is pressure; $k_B = 1.38 \times 10^{-23}$ J/K; and T is absolute temperature:

$$\lambda = \frac{k_B T}{\sqrt{2}\pi d^2 p} \qquad (6.49)$$

Example 6.3

Problem: (a) Plot the mean free path, λ, for N_2 at $T = 300°$K for pressure, p, varying from 50 mtorr to 1 atm pressure (760 torr). (b) What is λ at atmospheric pressure? (c) What is λ at 0.050 torr? (d) At what pressure is $\lambda = 2$ μm?

Solution: (a) Figure 6.25 is a plot of the mean free path vs. pressure obtained via evaluation of Equation 6.49; d was taken to be approximately 3Å for N_2. (b) At atmospheric pressure (760 torr), the mean free path in N_2 is 102 nm. (c) At a pressure of 50 mtorr, the mean free path in N_2 is 1554 μm. (d) The mean free path is $\lambda = 2$ μm in N_2 when the pressure is 38.8 torr.

6.2.3 VISCOUS DAMPING

Gas damping is the source of most MEMS damping and it is a strong function of viscosity. The absolute viscosity μ of a fluid or gas is a measure of the resistance to flow. The *absolute viscosity*, μ, is the ratio of the shear stress, τ, between flow layers to the velocity gradient, *dv/dy*, through the fluid channel (Equation 6.50). The flow field for this situation has the fluid fixed to the surface of both plates with a linear velocity profile between the plates as illustrated in Figure 6.26. The *kinematic viscosity*, v, is the absolute viscosity scaled by the density, ρ, of the fluid. This parameter is called kinematic because the units are length^2 per time.

Electromechanics

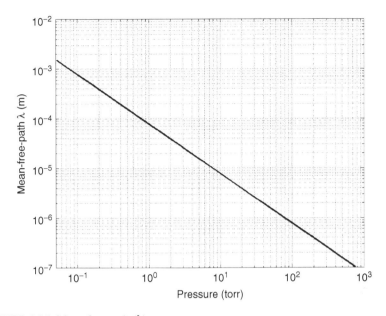

FIGURE 6.25 Mean free path (λ) vs. pressure.

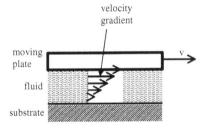

FIGURE 6.26 Linear velocity gradient between two plates as a result of the shear stress generated in the fluid due to viscosity.

$$\mu = \frac{\tau}{dv/dy} \tag{6.50}$$

$$\upsilon = \mu/\rho \tag{6.51}$$

The nature of viscosity of a liquid and a gas are different. Viscosity in liquid results from cohesion between adjacent molecules, but viscosity in a gas results from intermolecular collisions within the gas. Because the molecular collision rate depends on temperature in a gas, viscosity increases as temperature is increased; however, it is unaffected by pressure change in most macroscale engineering problems.

Boltzmann [13] calculated the viscosity coefficient (Equation 6.52), utilizing the concept of these intermolecular collisions, where $\bar{v}$ is the average velocity of the gas molecules. In an ideal gas, the density mean free path product, $\rho\lambda$, is constant (Equation 6.47); therefore, viscosity is constant with respect to pressure.

$$\mu = 0.3502\rho\bar{v}\lambda \qquad \text{for} \quad d > \lambda \tag{6.52}$$

However, when the distance between the plates become comparable to the mean free path, $d \approx \lambda$, *the viscosity becomes sensitive to pressure variation.* In this situation, molecular collisions rarely occur within the gas layer; momentum is directly transferred between the gas molecules and the MEMS surface, yielding a pressure-dependent viscosity. For example, Andrews et al. [17] showed that the effective viscosity of a 2-μm thick nitrogen film remained approximately constant at 1.56×10^{-5} kg/m/s near atmospheric pressure (760 torr) and 20°C, but dropped as pressure was decreased (e.g., 7.4×10^{-6} kg/m/s at 130 torr decreased to 7.4×10^{-5} kg/m/s at 13 torr).

At extremely low pressure and very narrow gaps, the viscosity becomes proportional to pressure and plate separation.

6.2.4 Damping Models

There are two frequently encountered configurations of moving MEMS surfaces interacting with a viscous gas flow that produce energy dissipation, which give rise to damping (Figure 6.27). *Squeeze film damping* occurs when the relative motion between two surfaces is perpendicular to the plane of the surfaces. This situation requires that gas between the plates be pushed (pumped) in and out of the gap between the plates. The other configuration is *slide film damping*, in which two surfaces are moving parallel to each other, shearing the fluid between the surfaces. In a real MEMS application, either or both of these configurations can occur. Models of the damping produced in these configurations are necessary for dynamic design of the MEMS device and estimation of the Brownian noise produced by a sensor where these forms of damping appear.

6.2.4.1 Squeeze Film Damping Model

Squeeze film damping occurs when the gap between the two closely spaced parallel surfaces changes, as shown in Figure 6.27. This type of damping occurs frequently in MEMS devices. For example, a pressure sensor diaphragm that deflects relative to the fixed base is an example of an instance in which squeeze film damping would apply. The motion of the diaphragm will force gas in or out of the gap. References 14 through 17 model the damping caused by squeezing the thin films of gas using the compressible Reynolds gas-film equations; this means that the gap is assumed to be greater than the mean free path of the gas.

Electromechanics

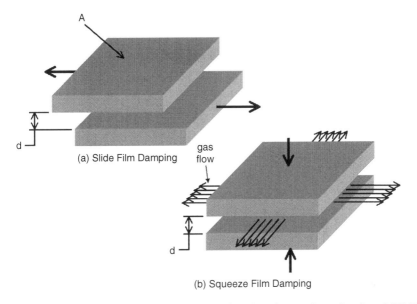

FIGURE 6.27 Squeeze film damping and slide film damping configuration for a MEMS device.

The *squeeze number*, σ, shown in the next equation represents the compressibility of the squeeze film, where μ is the viscosity; L and W are the plate length and width, respectively; P is the gas pressure; d is the fluid gap; and ω is the excitation frequency. For large plates and thin gaps, the squeeze number becomes large; the gas is trapped between the plates due to viscous effects at the plate edges. Conversely, a small plate with a large gap will produce a small squeeze number, which indicates that lateral motion of the gas is easily accomplished. The squeeze number is also proportional to the excitation frequency, ω, which indicates that the gas will have greater difficulty escaping the plate gap at higher excitation frequency.

$$\sigma = \frac{12\mu L^2}{Pd^2}\omega \tag{6.53}$$

The damping coefficient for squeeze film damping is

$$c = \frac{L^2 \beta P}{\omega d} \frac{64\sigma}{\pi^6} \left(\frac{1 + \left(\frac{1}{\beta}\right)^2}{\left[1 + \left(\frac{1}{\beta}\right)^2\right]^2 + \frac{\sigma^2}{\pi^4}} \right) \tag{6.54}$$

where

$$\beta = \frac{W}{L} \tag{6.55}$$

Equation 6.42 and Equation 6.44 can be used to put the damping coefficient in terms of damping ratio or quality factor, which may be more useful metrics in some cases. Equation 6.54 can accurately describe damping between parallel plates when the plate dimensions (L, W) are much greater than the gas gap (d), and the gas gap (d) is greater than the mean free path (λ) of the gas.

Squeeze film damping can be large and affect the dynamics and noise floor of sensors; it can be reduced by vacuum packaging or damping holes. However, vacuum packaging can significantly increase the cost and manufacturing complexity and any leakage in the vacuum package will be a source of long-term drift in the sensor. Damping holes can be incorporated into the device design to allow the gas to escape from between the plate and substrate. The etch release holes in a surface micromachine device can also be used as a damping hole. The damping factor of a perforated plate with damping holes of size a, arrayed with a pitch, p, can be calculated [18] by modeling the plate as N smaller plates of length L_p. The effective area of the idealized small plate, $L_p{}^2$, is merely the pitch area, p^2, minus the hole area, a^2 (Equation 6.56). The total damping factor will be N times the damping factor of the smaller plate.

$$L_p = \sqrt{p^2 - a^2} \tag{6.56}$$

For the case of narrow plate widths, edge effects represent a significant portion of the total damping. *Hagen–Poiseulle* flow [19] may provide a more accurate estimate of the damping coefficient for narrow widths such as comb fingers in an electrostatic actuator. The damping coefficient of a narrow width gap is given by:

$$c \approx 7.2\mu \left(\frac{L}{d} \right)^3 \tag{6.57}$$

6.2.4.2 Slide Film Damping Model

Laterally driven MEMS devices that move parallel to the substrate are frequently encountered in many applications. Two models for viscous slide film damping are useful in modeling the damping of laterally driven MEMS devices: *Couette* damping and *Stokes* damping (Figure 6.28). The differences between these two models are the gas velocity fields used to develop the damping model. The analysis and assumptions involved in the development of the Stokes and Couette dampers are discussed in detail in Cho et al. [20,21].

The *Couette damper* assumes that the plate instantaneously develops a fully established linear velocity profile in the fluid medium (Figure 6.28a). The velocity

Electromechanics

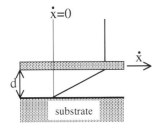
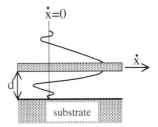

a) Couette Damper Velocity Profile   b) Stokes Damper Velocity Profile

FIGURE 6.28 Couette and Stokes damping velocity profiles for an infinite plate.

gradient at the top surface of the plate is assumed to be zero; this means that the ambient fluid above the plate oscillates with the plate motion, producing a negligible amount of viscous damping.

The *Stokes damper* is based upon the steady state solution of the one-dimensional Navier–Stokes and continuity equations that result in a one-dimensional diffusion equation to model the situation depicted in Figure 6.28b. This results in the plate motion propagating into the fluid with rapidly diminishing amplitude and a phase shift in the fluid motion.

The quality factor generated by the Couette damper, Q_{cd}, is given in Equation 6.58 for a plate of area, fluid gap, and viscosity of A, d, μ, respectively, attached to a spring mass system of k and m.

$$Q_{cd} = \frac{\mu A}{d}\sqrt{mk} \tag{6.58}$$

The conversion factors between Couette damping quality factor underneath the plate, Q_{cd}, and the Stokes damping quality factor underneath the plate, Q_{sd}, and the ambient fluid above the plate, $Q_{s\infty}$, are:

$$\frac{Q_{cd}}{Q_{s\infty}} = \beta d$$

$$\frac{Q_{cd}}{Q_{sd}} = \beta d \left(\frac{\sinh 2\beta i + \sin 2\beta i}{\cosh 2\beta o + \cos 2\beta o} \right) \tag{6.59}$$

where

$$\beta = \sqrt{\frac{\omega}{2\upsilon}} \tag{6.60}$$

and ω, υ are the frequency of oscillation and kinematic viscousity, respectively.

TABLE 6.3
Electrical Quantities and Units

Electrical quantity	Symbol	Units	Type
Charge	q	Coulomb (C)	Scalar
Electric potential	e or V	Volt (V)	Scalar
		$V = (N - m)/C$	
Electric field intensity	**E**	V/m or N/C	Vector
Magnetic flux density	**B**	Tesla (T)	Vector
		$T = Wb/m2 = (V - s)/m^2$	
Current	i	Ampere (A)	Scalar
		$A = C/s$	
Flux linkage	λ	Weber (Wb)	Scalar
		$Wb = V - s$	
Capacitance	C	Farad (F)	Scalar
		$F = C/V$	
Inductance	L	Henry (H)	Scalar
		$H = Wb/A = (V - s)/A$	
Resistance	R	Ohm (Ω)	Scalar
		$\Omega = V/A$	

6.3 ELECTRICAL SYSTEM DYNAMICS

The objective of this presentation of electromechanics is to provide the basis for analysis of actuator and sensor systems, which frequently involve a coupling of the mechanics of a structure (i.e., membrane, seismic mass, flexural supports) and an electrical system that senses the motion or provides a force for actuation. This section will provide an overview of the terminology and key physics necessary for the analysis of an electromechanical system. Table 6.3 is a list of electrical quantities, symbols, and units discussed in this section.

The mechanics of electrical systems are described by Maxwell's equations, which can be written in differential or integral forms and incorporate a number of assumptions for specific applications. An in-depth treatment of Maxwell's equations is a very complex subject and beyond the scope of this book. However, a number of references can provide a variety of views on this subject. Johnk [22] provides an in-depth theoretical treatment of Maxwell's equations. Moon [23] provides a link between the theoretical study of Maxwell's equations and applications that utilize electromechanics. Smith [24] provides an introductory, very practical application of electrical systems and electromechanics to engineering solutions. Crandall et al. [25] present a detailed explanation of the analysis of electromechanical systems that is very useful for the analysis of transducers such as those utilized in MEMS applications.

Electromechanics

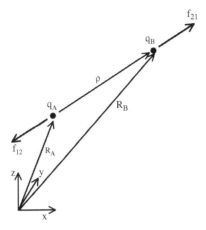

FIGURE 6.29 Coulomb's law modeling the forces between two charged particles.

6.3.1 ELECTRIC AND MAGNETIC FIELDS

A fundamental electrical quantity is *electrical charge*, q. Charge, q, is carried on particles such as electrons or ions and its unit of measure is the *coulomb*. When charge flows along a path such as a metallic wire, one may speak of a continuous flow of charge called current, i. *Current* is the rate of flow of charge (Equation 6.61) measured in units of *amperes* (i.e., coulomb per second). By convention, positive current flow results from a flow of positive charge in a positive direction.

$$i = \frac{dq}{dt} \quad (6.61)$$

Coulomb's law (Equation 6.62) describes the relationship of the force between electrically charged particles (Figure 6.29). The two charged particles, q_1 and q_2, are separated by a distance, ρ, along the unit vector, $\boldsymbol{u}_\rho$. The *permittivity* of the medium between the particles is ε. For free space (i.e., vacuum), the permittivity is $\varepsilon_0 = 8.85 \times 10^{-12}$ coulombs per newton-meter squared (farads per meter).

Coulomb's law can be applied to many particles by a sequential application of Equation 6.62 to every pair-wise combination and vectorially summing the forces. This law is similar in form to the relationship for gravitational forces, which also produce force over distance. The forces act along a line between the particles, and the magnitude varies inversely with the square of the distance between particles. However, Coulomb's law can be attractive or repulsive depending upon the sign of the charges, q_1 and q_2 (i.e., like charges repel, unlike charge attract).

$$-f_{12} = f_{21} = \frac{1}{4\pi\varepsilon}\frac{q_1 q_2}{\rho^2}\mathbf{u}_\rho \tag{6.62}$$

The forces produced by the electrostatic charges described by Coulomb's law are a vector field, $\mathbf{f(R)}$. These forces are a conservative field with the following meaning:

- The forces are a function of position only (i.e., $\mathbf{f[R]}$).
- The line integral of the electrostatic force field from point $\mathbf{R}_A$ to point $\mathbf{R}_B$ (Equation 6.63) is a function of the end points only and is independent of the path taken.

$$\int_{\mathbf{R}_A}^{\mathbf{R}_B} \mathbf{f} \cdot d\mathbf{R} \tag{6.63}$$

Thus, the electrostatic forces on a reference particle at location R_0 among an array of other charges at fixed locations are a function of reference particle position only. This is a conservative field and the potential energy per unit charge, $e(\mathbf{R})$, on the reference particle is shown in Equation 6.64. R_0 is the datum (reference point) for the integral of Equation 6.65. The potential energy per unit charge, $e(\mathbf{R})$, is also known as the *electrical potential,* which has units of a volt or, alternatively, a newton-meter per coulomb.

The change in electrical potential, Δe or *electrical potential difference,* due to moving a charge from point $\mathbf{R}_A$ to point $\mathbf{R}_B$ with respect to the array of other charges is simply the difference in the electric potential at point A and point B (Equation 6.65) because the electrostatic force field is conservative. The choice of the datum, $\mathbf{R}_0$, for the electric potential, $\mathbf{e(R)}$, does not influence the value of the electrical potential difference, Δe.

$$e(\mathbf{R}) = -\int_{\mathbf{R_0}}^{\mathbf{R}} \frac{\mathbf{f} \cdot d\mathbf{R}}{q} \tag{6.64}$$

$$\Delta e = e(\mathbf{R}_B) - e(\mathbf{R}_A) \tag{6.65}$$

The potential energy, W, of a charge, q, in an electric field is given by Equation 6.66 and has units of newton-meters. The datum chosen for the electrical potential difference, $e(\mathbf{R})$, does not change the potential energy.

$$W = q\Delta e \tag{6.66}$$

Electromechanics   **231**

If the charge is moving along a path from R_A to R_B within a time interval, dt, the potential energy is time varying. The increase in potential energy of the charge per unit time is power, P, which has a unit of a watt (volt-ampere).

$$P = i\Delta e \qquad (6.67)$$

Maxwell's equations are a set of equations [22] that can be expressed in differential or integral form involving a number of variables; the *magnetic field density*, **B**, and *electric field intensity*, **E**, predominate. Mathematically, **B** and **E** are vector fields that have magnitude, direction, and the properties previously described for the forces produced by charges in Coulomb's law (i.e., a function of position, their line integral is not path dependent).

The *electric field intensity*, **E**, which is frequently referred to as *electric field*, is the negative of the gradient of the electric potential e in Equation 6.68. The gradient, ∇, is a spatial rate of change operation on a scalar that produces a vector. In the one-dimensional case, the gradient reduces to a simple differentiation. Therefore, the *electric field*, **E**, is the spatial rate of change of *electric potential*, e.

$$\mathbf{E} = -\nabla e \qquad (6.68)$$

The *electric field intensity*, **E**, is also the direction and magnitude of the force on a positive charge in the electric field (Equation 6.69), which can be obtained from Equation 6.64 by differentiation. As illustrated in Figure 6.30, the force produced on the charge by the electric field is in the same direction as the electric field. The electric field, **E**, has units of newton per coulomb or volts per meter, which is consistent with Equation 6.68 and Equation 6.69 definitions.

$$\mathbf{f} = q\mathbf{E} \qquad (6.69)$$

Traditionally, magnetic fields were first described in terms of *lines of force* or *flux*. This is due to the familiar experiment that allows these magnetic lines of force to be visualized by iron filling sprinkled on a surface in the presence of a magnetic field produced by a permanent magnet or electromagnet. *Magnetic flux*, ϕ, is a scalar quantity (i.e., described only by magnitude) with units of webers. However, the vector field *magnetic flux density*, **B**, utilized in Maxwell's equations is generally considered the primary metric to describe a magnetic field. The magnetic flux density, **B**, has units of tesla or weber per meter squared and is related to the magnetic flux, ϕ, by Equation 6.71, which is the integration of the magnetic flux density, **B**, over an area, **A**:

$$\phi = \int \mathbf{B} \cdot d\mathbf{A} \qquad (6.70)$$

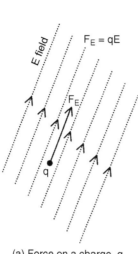
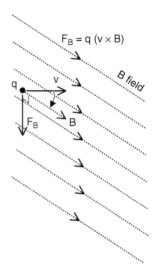

(a) Force on a charge, q, in an Electric Field, E.

(b) Force on a charge, q, with velocity, v, in an Magnetic Field, B.

FIGURE 6.30 Forces acting upon a charge in an electric field and magnetic field.

A magnetic field is due to the movement of electric charge or current. Examples of this magnetic effect are the spinning of electrons around an atom. When these atoms are aligned, this effect is magnified and produces a *permanent magnet* or *lodestone*. The intensity of the magnetic effect is measured by the magnetic flux density, **B**, which is a vector defining the magnitude and direction of the force on a charge, q, moving with a velocity, **v**. Equation 6.71 is the mathematical expression of the magnetic field force, which is always perpendicular to the magnetic flux density, **B**, due to the cross-product (Figure 6.30):

$$\mathbf{f} = q\mathbf{v} \times \mathbf{B} \tag{6.71}$$

The *Lorentz force law* equation is simply the vector sum of the forces on a charged particle, q, due to the electric field, **E**, and magnetic field, **B**. These forces are illustrated in Figure 6.30. The force due to the electric field, **E**, is a function of position of the charge and time; however, the force due to the magnetic field, **B**, is a function of position and velocity of the charge and time.

$$\mathbf{f} = q\left(\mathbf{E} + \mathbf{v} \times \mathbf{B}\right) \tag{6.72}$$

Faraday's law states that voltage (electric potential e) can be induced by motion of a conductor in a magnetic field density or by a time-varying magnetic field density (Equation 6.73). Figure 6.31 shows a schematic of the line segment,

Electromechanics

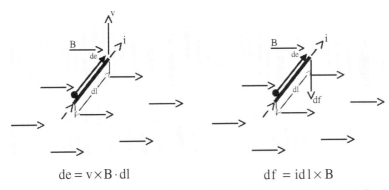

FIGURE 6.31 Faraday's law illustrating a time-varying magnetic field, $\partial \mathbf{B}/\partial t$, and motion of a line segment, dl, in a magnetic field, $\mathbf{B}$.

dl, area and line integrals involved in Equation 6.73. The first term of the equation expresses the voltage induced by a time-varying magnetic field density, which occurs within the closed conducting path over which the surface integral is taken. The second term is the voltage induced in a moving conduction within a magnetic field. The mechanics of Faraday's law explain the physics involved in common devices such as electric generators, electric motors, and inductors.

$$e(t) = \int_l \mathbf{E} \cdot dl = -\int_S \frac{\partial \mathbf{B}}{\partial t} \cdot d\mathbf{s} + \int_l \left(\mathbf{v} \times \mathbf{B} \right) \cdot dl \qquad (6.73)$$

Equation 6.74 states that a small segment of conducting wire of length dl moving with a velocity, $\mathbf{v}$, in a magnetic field will produce an electric potential, de, across the wire segment induced by the motion of the conductor in a magnetic field:

$$de = \mathbf{v} \times \mathbf{B} \cdot dl \qquad (6.74)$$

The force generated on the conductor segment, dl, can be obtained from the second term of the Lorentz force law (Equation 6.72) by utilizing the relationship $i\, dl = q\, \mathbf{v}$. Equation 6.74 shows the *right-hand rule* relationship among velocity, magnetic field, and electric potential, which is a result of the vector cross-product. Current–magnetic field–force has a similar relationship due to the vector cross-product in Equation 6.75.

$$d\mathbf{f} = i d\mathbf{l} \times \mathbf{B} \qquad (6.75)$$

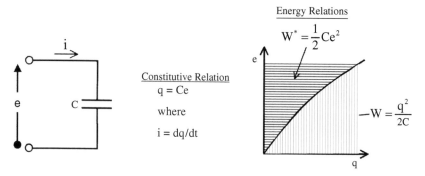

FIGURE 6.32 Parallel plate capacitor schematic and relationships.

6.3.2 ELECTRICAL CIRCUITS — PASSIVE ELEMENTS

The three passive circuit elements to be discussed are a capacitor, inductor, and resistor. These elements are considered passive because they do not contain an energy source. The capacitor and inductor can store energy and subsequently return that energy to the circuit. The resistor dissipates energy.

6.3.2.1 Capacitor

A capacitor consists of two conductors separated by a dielectric material (Figure 6.32). The dielectric material is an insulator that will keep electric charges apart and prevent the charge on the two conductors from equalizing. The capacitor is said to be uncharged when the amount of charge is the same on both conductors; this will result in zero voltage across the capacitor. When the capacitor is charged, one conductor will have acquired q units of charge and the other will have lost q units of charge, which will produce a voltage across the terminals of the capacitor.

The constitutive equation, which describes the relationship between the electrical variables (e, q) of a capacitor, is shown in Equation 6.76. Capacitance, C, is the proportionality variable between the electric potential across the capacitor terminals and the charge, q, on the capacitor. The unit of capacitance is the farad (coulomb per volt). Current flowing into or out of the capacitor will change the amount of charge on the capacitor. The capacitance, C, may be a constant for a fixed circuit element or a function of a variable such as conductor spacing or the dielectric material property change due to an environmental variable such as temperature, pressure, or humidity. The change in capacitance, C, can be used as a sensing mechanism in a transducer.

The electrical energy, W, stored on a capacitor is the work done in charging the capacitor. The energy stored in a capacitor is stored in the electric field. The electrical energy, W, may be calculated by integrating the area under the electrical potential, e, curve as a function of charge, q (Equation 6.77).

Electromechanics

$$q = Ce \tag{6.76}$$

$$W(q) = \int_0^q edq \tag{6.77}$$

Alternatively, a complementary energy function, W^*, may be calculated by integrating the area under the charge q curve as a function of electrical potential, e:

$$W^*(e) = \int_0^q qde \tag{6.78}$$

Utilizing the constitutive equation for a capacitor (Equation 6.76), the energy and complementary energy functions can be found:

$$W(q) = \frac{q^2}{2C} \qquad W^*(e) = \frac{Ce^2}{2} \tag{6.79}$$

The energy function, W, and complementary energy function, W^*, have a functional duality that can be seen by Equation 6.81, which differentiates W and W^* with respect to their independent variable and utilizing the capacitor constitutive equation, Equation 6.76. W and W^* describe the same constitutive relation with different independent variables. The relationship (Equation 6.81) between W and W^* can readily be graphically obtained from Figure 6.32 and is called a *Legendre transformation*. Dual functions of this nature are utilized frequently in thermodynamics where enthalpy and internal energy are dual. These energy and complementary energy functions will be employed in the development of the governing equations of electromechanical systems, where the use of different independent variables can be used to advantage.

$$\frac{dW}{dq} = e \qquad \frac{dW^*}{de} = q \tag{6.80}$$

$$W^*(e) = eq - W(q) \tag{6.81}$$

6.3.2.2 Inductor

An inductor consists of a conducting coil of one or more loops, Figure 6.33. When current flows through an inductor, a magnetic field is produced that surrounds the coils; this is called *flux linkage*, λ. Assuming that the magnetic flux density, **B**, is uniform inside the coil, the flux linkage is the product of the number

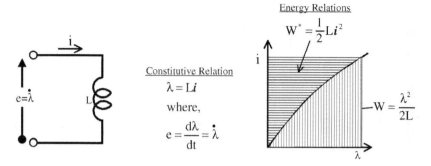

FIGURE 6.33 Inductor schematic and relationships.

of coils, N; the internal area of the coil, A; and the magnetic field density produced by the coil (Equation 6.82).

The relationship between electric potential, e, and flux linkage, λ, for an inductor (Equation 6.83) is developed from the first term of Faraday's law (Equation 6.73), which involves a time-varying magnetic field. The constitutive relation that relates flux linkage and current for an inductor is shown in Equation 6.84. The proportionality constant, L, is the inductance, which has a unit of the henry (weber per ampere or volt-second per ampere).

$$\lambda = NA\mathbf{B} \tag{6.82}$$

$$e = \frac{d\lambda}{dt} \tag{6.83}$$

$$\lambda = Li \tag{6.84}$$

The energy in an inductor is stored in the magnetic field. The energy and complementary energy for an inductor are given in Equation 6.85 and Table 6.4. These energies are related by the Legendre transformation (Equation 6.86).

$$W(\lambda) = \frac{\lambda^2}{2L} \quad W^*(i) = \frac{Li^2}{2} \tag{6.85}$$

$$W^*(i) = \lambda i - W(\lambda) \tag{6.86}$$

6.3.2.3 Resistor

The resistor is a dissipative circuit element that transfers electrical energy to thermal energy, which cannot be recovered (Figure 6.34). The constitutive equa-

Electromechanics

TABLE 6.4
Passive Circuit Element, Constitutive Equation, Energy, and Complementary Energy

Circuit element	Constitutive equation	W	W*
Resistor — R	$e = iR$	$\dfrac{1}{2}Ri^2$	$\dfrac{1}{2}\dfrac{e^2}{R}$
Capacitor — C	$q = Ce$ $\left(i = C\dfrac{de}{dt}\right)$	$\dfrac{1}{2}\dfrac{q^2}{C}$	$\dfrac{1}{2}Ce^2$
Inductor — L	$\lambda = Li$ $\left(e = L\dfrac{di}{dt}\right)$	$\dfrac{1}{2}\dfrac{\lambda^2}{L}$	$\dfrac{1}{2}Li^2$

Note: $e = \dot{\lambda}$ and $i = \dot{q}$.

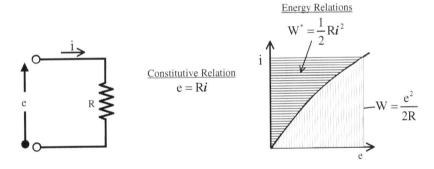

FIGURE 6.34 Resistor schematic and relationships.

tion for a resistor is *Ohm's law*, which states that the flow of current, i, though a resistor, R, requires electric potential, e. The resistance, R, has a unit of the ohm that is 1 V/A. The energy and complementary energy functions for the resistor are given in Equation 6.88 and Table 6.4.

$$e = Ri \tag{6.87}$$

$$W(i) = \frac{Ri^2}{2} \quad W^*(e) = \frac{e^2}{2R} \tag{6.88}$$

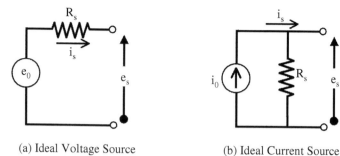

FIGURE 6.35 Real voltage and current sources and their voltage vs. current characteristic.

6.3.2.4 Energy Sources

Two idealizations of electrical energy sources are:

- The *ideal voltage source*, which will produce a prescribed voltage, $e(t)$, regardless of the current requirements
- The *ideal current source*, which will produce a prescribed current, $i(t)$, regardless of the voltage requirements

However, ideal sources cannot be realized due to the infinite power required by their definition. An example is an ideal voltage source with a resistor, R, across the terminals. As the resistance is reduced to zero, the current requirements due to Ohm's law will become infinite, which implies infinite power. A real voltage and current source can be modeled by inclusion of a resistance, R_S, as shown in Figure 6.35.

6.3.2.5 Circuit Interconnection

Combining the circuit elements discussed previously (resistor, capacitor, inductor, voltage and current sources) is required to produce a working device or transducer. This section will discuss the method of *Kirchhoff's law* to combine these circuit elements. The development of an analytical model of a transducer often will require the combination of electrical as well as mechanical components. The combination of electrical and mechanical components can be accomplished using Kirchhoff's laws and the physics of the electromechanical coupling. Chapter 7 will discuss methods for coupling electromechanical transducers utilizing the methods of analytical mechanics (*Lagrange's equations*). The discussion in this chapter of the energy stored or dissipated in components is a prelude to that.

Two forms of Kirchhoff's law can be used to combine circuit elements to form an equation describing the system:

- *Kirchhoff's current law.* The algebraic sum of the currents into a node at any instant is zero. Figure 6.36 is a circuit with a junction or node shown in which current $(i = \dot{q})$ can flow in or out.

Electromechanics

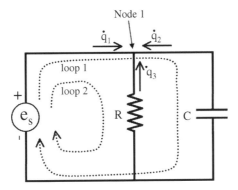

FIGURE 6.36 A voltage source, resistor, capacitor circuit.

$$\sum i = 0 \tag{6.89}$$

- *Kirchhoff's voltage law.* The algebraic sum of the voltages around any closed path of the circuit is zero. Figure 6.36 is a circuit with two circuit paths or loops annotated.

$$\sum e = 0 \tag{6.90}$$

Example 6.4

Problem: (a) Apply Kirchoff's current law to find the system equations for the circuit shown in Figure 6.36. (b) Apply Kirchoff's voltage law to find the system equations for the circuit shown in Figure 6.36. (c) Compare the two sets of equations.
Solution: (a) Application of Kirchhoff's current law yields:

$$\sum i = 0$$
$$\dot{q}_1 + \dot{q}_2 + \dot{q}_3 = 0 \tag{6.91}$$

Using the constitutive relations for the capacitor and resistor, Equation 6.76 and Equation 6.87, respectively yields the following relations to the source voltage, e_s, and other currents and charges in the system:

$$\begin{aligned} q_2 = e_s C &\Rightarrow \dot{q}_2 = \dot{e}_s C \\ e_s = R\dot{q}_3 &\Rightarrow \dot{q}_3 = \frac{e_s}{R} \end{aligned} \tag{6.92}$$

240   Micro Electro Mechanical System Design

Substituting Equation 6.92 into Equation 6.91 yields the following system equation in which e_s, C, and R are specified variables and q_1 is the unknown variable to be solved for:

$$\dot{q}_1 + \dot{e}_s C + \frac{e_s}{R} = 0 \tag{6.93}$$

(b) Application of Kirchhoff's voltage law around the two loops shown in Figure 6.36 yields the following two equations:

$$\sum e = 0$$

$$\text{Loop 1:} \quad e_s - \frac{q_2}{C} = 0 \tag{6.94}$$

$$\text{Loop 2:} \quad e_s - R\dot{q}_3 = 0$$

This method results in two equations and two unknowns (q_2, q_3) in terms of the specified variables e_s, C, and R.

(c) Kirchhoff's current law resulted in one first-order differential equation in terms of q_1. However, Kirchhoff's voltage law results in two equations: a first-order differential equation in terms of q_3 and an algebraic equation in terms of q_2. The reason for this discrepancy is that only one independent node is in this circuit, but two independent loops are. For more complex circuits, a judicious choice of approach (i.e., voltage law vs. current law) can be important to result in a smaller number of equations to describe the system.

QUESTIONS

1. A structural member has a thickness, width, and length of t, w, and L, respectively. The area moment of inertia about the bending axis of interest is $I = wt^3/12$.
   - a. List the stiffness equations for torsion, axial tension, and bending.
   - b. If t is doubled in thickness, what are the percent changes in the three stiffnesses?
   - c. Which of the stiffnesses is the most sensitive to a change in length?
2. Given a 500-μm long polysilicon beam with a 2-μm square cross-section and fixed–fixed boundary conditions. Assume polysilicon has a failure stress of 1600 μn/μm². What is the compressive load that will cause the beam to fail (consider both material failure and column buckling failure)? What is the tensile load that will cause the beam to fail?

Electromechanics   **241**

3. Given a 100-µm square plate made of 2.5-µm thick polysilicon suspended by a 1-N/M spring. This device is in an air environment at 1 atm, 20°C, with $\upsilon = 0.15$ cm^2/s. The air gap is 2 µm and the plate will oscillate at 1000 Hz. Calculate the quality factors, Q, for normal and lateral motion relative to the substrate. Which type of motion is more heavily damped?

REFERENCES

1. F.B. Seely, J.O. Smith, *Advanced Mechanics of Materials*, 2nd ed., John Wiley & Sons, Inc., New York, 1952.
2. R.J. Roark, *Formulas for Stress and Strain*, McGraw-Hill Company, New York, 1965.
3. J.E. Shigley, L.D. Mitchell, *Mechanical Engineering Design*, McGraw-Hill Book Company, New York, 1983.
4. H.A. Rothbart (Ed.), *Mechanical Design Handbook*, McGraw-Hill, New York, 1996.
5. R. Plunkett (Ed.), *Mechanical Impedance Methods for Mechanical Vibrations*, American Society of Mechanical Engineers, New York, 1958.
6. W. Soedel, *Vibrations of Shells and Plates*, Marcel Dekker, Inc., New York, 1981.
7. S.P. Timoshenko, S. Woinowsky–Krieger, *Theory of Plates and Shells*, McGraw-Hill, New York, 1959.
8. B.A. Boley, J.H. Weiner, *Theory of Thermal Stresses*, John Wiley & Sons, New York, 1960.
9. E.J. Garcia, Micro-flex mirror and instability actuation technique, *11th IEEE Int. Workshop Micro Electro Mechanical Syst.*, Janaury 25–29, 1998 in Heldelberg, Germany.
10. W.T. Thomson, *Theory of Vibration with Applications*, Prentice Hall, Inc., Englewood Cliffs, NJ, 1981.
11. F.S. Tse, I.E. Morse, R.T. Hinkle, *Mechanical Vibrations, Theory and Applications*, Allyn and Bacon, Inc., Boston, 1978.
12. J.D. Zook, D.W. Burns, H. Guckel, J.J. Sniegowski, R.L. Engelstad, Z. Feng, Characteristics of polysilicon resonant microbeams, *Sensors Actuators A*, 35, 51–59, 1992.
13. S. Dushman, *Scientific Foundations of Vacuum Technique*, John Wiley & Sons, New York, 1962.
14. J.J. Blech, On isothermal squeeze films, *J. Lubrication Technol.*, 105, 615–620, October, 1983.
15. W. Griffin, H. Richardson, S. Yamanami, A study of squeeze film damping, *ASME J. Basic Eng.*, 451–456, June 1966.
16. M. Andrews, I. Harris, G. Turner, A comparison of squeeze-film theory with measurements on a microstructure, *Sensors Actuators A*, 36, 79–87, 1993.
17. M.K. Andrews, G.C. Turner, P.D. Harris, I.M. Harris, A resonant pressure sensor based on a squeezed film of gas, *Sensors Actuators A*, 36, 219–226, 1993.
18. N. Yazdi, K. Najafi, An all-silison single-wafer micro-g accelerometer with a combined surface and bulk micromachining process, *JMEMS*, 9(4), 544–550, Dec. 2000.

242   Micro Electro Mechanical System Design

19. W. Kuehnel, Modeling of the mechanical behavior of a differential capacitor acceleration sensor, *Sensors and Actuators A*, 48, 101–108, 1995.
20. Y.H. Cho, B.M. Kwak, A.P. Pisano, R.T. Howe, Slide film damping in laterally driven microstructures, *Sensors Actuators A*, 40, 31–39, 1994.
21. Y.H. Cho, A.P. Pisano, R.T. Howe, Viscous damping model for laterally oscillating microstructures, *J. MEMS*, 3(2), 81–87, June 1994.
22. C.T.A. Johnk, *Engineering Electromagnetic Fields and Waves*, John Wiley & Sons, New York, 1988.
23. F.C. Moon, *Magneto-Solid Mechanics*, John Wiley & Sons, New York, 1984.
24. R.J. Smith, *Circuits, Devices and Systems*, John Wiley & Sons, New York, 1968.
25. S.H. Crandall, D.C. Karnopp, E.F. Kurtz, D.C. Pridemore-Brown, *Dynamics of Mechanical and Electromechanical Systems*, Original Edition 1968, McGraw-Hill, reprint edition, 1982, Krieger Publishing.

7 Modeling and Design

This chapter will discuss a number of topics relating to modeling and simulation of MEMS devices. As a design progresses from concept to a fully mature product, the modeling and simulation needs change. In the early stages of transitioning a design from a concept in the designer's imagination to an initial design layout of the device, the modeling and simulation requirements are different from those for perfecting the operation of an existing MEMS device design.

The *concept* to *first design* phase is *design synthesis*, which requires information of what design parameters are important, how they interact with each other, and the sensitivity of the device performance metric to the design variables. This type of information can most easily be obtained from an analytical model or low-order lumped parameter models of the device. In these types of models, the design variables of interest are frequently explicit and a large number of calculations to determine variable relationships and sensitivities can be easily performed.

The detailed *design analysis* of existing MEMS design to further optimize performance may require high-order numerical analysis calculations. For example, the analysis of a MEMS support anchor to determine dissipation losses to the substrate, which is important for a resonant device, will require a detailed solid model of the anchor–substrate region. This type of modeling is not feasible with design synthesis approaches and requires a large-order solid model of detailed effects such as material damping and hysteresis.

7.1 DESIGN SYNTHESIS MODELING

The challenge of design synthesis modeling is to capture as succinctly as possible the fundamental physics of a device. The method most frequently taught in undergraduate engineering courses is the use of Newtonian mechanics, free-body diagrams to develop the governing equations of a device, and lumped parameter circuit models. As anyone who has survived these courses knows, there is plenty of room for error. The main roadblock is maintaining correct signs for variables and dealing with multiple interacting bodies and systems.

Analytical mechanics [1–6] is an alternative approach for the development of the equations of motion for a system based on the minimization of the energy functions of the system. Hamilton's and Lagrange's equations are two alternative formulations of this approach. Lagrange's equations will be presented here as well as MATLAB™-based functions [7] (Sections H.1 and H.2 in Appendix H)

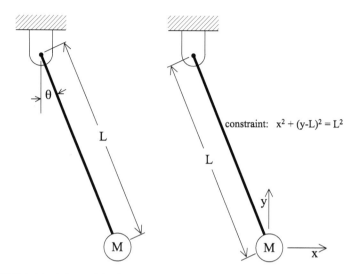

FIGURE 7.1 Pendulum with alternative choices of generalized coordinates.

to aid in the required computations. These functions require MATLAB and the Symbolic Math Toolbox.

7.2 LAGRANGE'S EQUATIONS

Lagrange's equations are the differential equations of motions of a device expressed in terms of *generalized coordinates*, which are a set of coordinates that completely describes the system dynamics. Generalized coordinates are not unique for a system and can be defined in several ways. They can be quantities such as angles, linear displacements, voltages, or electric charges, depending on the physical system that is to be modeled. Generalized coordinates will be denoted by q_k ($k = 1,2,...n$), where n is the number of generalized coordinates. The minimum number of independent generalized coordinates required to describe the dynamics of a system fully is referred to as the *system degree of freedom (DOF)*, N.

Figure 7.1 shows a diagram of a simple pendulum that has 1 DOF, with alternative expressions for the generalized coordinates that could be used to describe the system. The most direct and simple choice for the system generalized coordinate is θ. X could be used for the generalized coordinate, but the resulting equation would be not as meaningful. X and Y can be used as generalized coordinates, but they are not independent and a method for dealing with the constraint equation relating them ($r^2 = X^2 + Y^2$) needs to be developed.

A basic version of Lagrange's equations can be derived from the principle that the sum of the kinetic and potential energies of a conservative system is constant:

$$d(T + U) = 0 \tag{7.1}$$

Modeling and Design

TABLE 7.1
Mechanical and Electrical Energy Functions

	Mechanical		Electrical	
Generalized coordinate q_i	x	θ	Q	e or λ
Kinetic energy T	$\frac{1}{2}M\dot{x}^2$	$\frac{1}{2}M\dot{x}^2$	$\frac{1}{2}L\dot{Q}^2$	$\frac{1}{2}Ce^2 = \frac{1}{2}C\dot{\lambda}^2$
Potential energy U	$\frac{1}{2}Kx^2$	$\frac{1}{2}K_\theta\theta^2$	$\frac{1}{2}\frac{1}{C}Q^2$	$\frac{1}{2}\frac{1}{L}\lambda^2$
Raleigh dissipation function D	$\frac{1}{2}C\dot{x}^2$	$\frac{1}{2}C_\theta\dot{\theta}^2$	$\frac{1}{2}R\dot{Q}^2$	$\frac{1}{2}\frac{1}{R}e^2 = \frac{1}{2}\frac{1}{R}\dot{\lambda}^2$
Nonpotential force work W	$F \cdot \delta x$ Applied force	$T \cdot \delta\theta$ Applied torque	$e \cdot \delta Q$ Voltage source	$i \cdot \delta\lambda$ Current source

Note: $e = \dot{\lambda}$ and $i = \dot{q}$.

The kinetic energy, T, is a function of the generalized coordinates, q_i, and the generalized velocities, $\dot{q}_i$. The potential energy, U, is a function only of the generalized coordinates, q_i. Table 7.1 provides a definition of the kinetic and potential energy functions for lumped parameter mechanical and electrical systems.

Taking the differential of T and U, the invariance of the sum of the kinetic and potential energies becomes

$$d\left(T+U\right) = \sum_{i=1}^{N}\left[\frac{d}{dt}\left(\frac{\partial T}{\partial \dot{q}_i}\right) - \frac{\partial T}{\partial q_i} + \frac{\partial U}{\partial q_i}\right]dq_i \qquad (7.2)$$

If the N generalized coordinates are independent of one another, dq_i is arbitrary and the preceding equation will be satisfied only if the term in brackets equals 0. Equation 7.3 is Lagrange's equation for a conservative system with no *nonpotential* forces or constraints expressed with the generalized coordinates. Friction forces and externally applied forces are examples of nonpotential forces.

$$\frac{d}{dt}\left(\frac{\partial T}{\partial \dot{q}_i}\right) - \frac{\partial T}{\partial q_i} + \frac{\partial U}{\partial q_i} = 0 \qquad i = 1, 2, \ldots, N \qquad (7.3)$$

246   Micro Electro Mechanical System Design

Nonpotential forces can be included in the formulation of Lagrange's equations by the addition of the work of the nonpotential forces into the system total energy expression of Equation 7.1. This results in Equation 7.4, where Q_i is a generalized force or moment acting over an infinitesimal displacement of the generalized coordinate, q_i:

$$d\left(T+U\right) = dW = \sum_{i=1}^{N} Q_i dq_i \qquad (7.4)$$

Mathematically proceeding as before to find an expression of Lagrange's equations that includes the nonpotential forces results in:

$$\frac{d}{dt}\left(\frac{\partial T}{\partial \dot{q}_i}\right) - \frac{\partial T}{\partial q_i} + \frac{\partial U}{\partial q_i} = Q_i \qquad i = 1, 2, \ldots, N \qquad (7.5)$$

7.2.1 Lagrange's Equations with Nonpotential Forces

Viscous damping is a nonconservative force frequently arising in system modeling of devices that deserves special attention. If the damping forces are proportional to the generalized velocities, it is possible to devise a function, the *Raleigh dissipation function*, D (Equation 7.6), where the proportionality constant, c_{rs}, is the viscous damping coefficient.

$$D = \frac{1}{2} \sum_{r=1}^{n} \sum_{s=1}^{n} c_{rs} \dot{q}_r \dot{q}_s \qquad (7.6)$$

From this definition of D, it is obvious that the viscous damping force is merely a constant viscous damping coefficient multiplied by the generalized velocity:

$$Q_i = -\frac{\partial D}{\partial q_i} \qquad i = 1, 2, \ldots, n \qquad (7.7)$$

If the nonpotential forces, Q_i, in Equation 7.5 are separated into externally applied nonpotential forces and dissipative forces, Lagrange's equations can be rewritten as

$$\frac{d}{dt}\left(\frac{\partial T}{\partial \dot{q}_i}\right) - \frac{\partial T}{\partial q_i} + \frac{\partial U}{\partial q_i} + \frac{\partial D}{\partial q_i} = Q_i \qquad i = 1, 2, \ldots, n \qquad (7.8)$$

Modeling and Design

At this point, a form of Lagrange's equations, Equation 7.8, can be used to model a device for which a kinetic energy function (T); a potential energy function (U); a Raleigh dissipation function (D); and the nonpotential applied forces, Q_i, can be written in terms of the generalized coordinates and velocities ($q_i, \dot{q}_i$). This is a frequently used form of Lagrange's equations.

7.2.2 LAGRANGE'S EQUATIONS WITH EQUATIONS OF CONSTRAINT

One additional modification of Lagrange's equations may be useful in situations when defining the number of generalized coordinates, n, to be equal to the system degree of freedom, N, cannot be easily accomplished, or it is deemed advantageous to deal with an excess of generalized coordinates ($n > N > M$). To utilize this approach, M number of constraint equations (Equation 7.9) need to be defined and Lagrange's equations modified to account for the additional variables necessary to satisfy the constraints.

$$g_1\left(q_1, q_2, \ldots, q_N\right) = 0 \qquad 1 = 1, 2, \ldots, M \tag{7.9}$$

To have a uniquely solvable set of equations defining the system, the number of generalized coordinates, constraint equations, and system degrees of freedom will have the following relationship:

$$n - M = N \tag{7.10}$$

If $N + M$ generalized coordinates are identified for use with the Lagrange equation formulation of a system with N degrees of freedom, this means that M equations of constraint must be involved in the formulation so that all of the $N + M$ variables can be uniquely found. The result of having constraints between the generalized coordinates of the equations modeling any physical system will result in M constraint forces, Q_i^{con}, which are needed to enforce the constraint conditions (Equation 7.9) upon the system.

$$\frac{d}{dt}\left(\frac{\partial T}{\partial \dot{q}_i}\right) - \frac{\partial T}{\partial q_i} + \frac{\partial U}{\partial q_i} + \frac{\partial D}{\partial \dot{q}_i} = Q_i + Q_i^{con} \qquad i = 1, 2, \ldots, n \tag{7.11}$$

This additional constraint force term in Lagrange's equations is developed via the Lagrange multiplier method, which results in new variables, *Lagrange multipliers*, λ_l $1 = 1, \ldots, M$.

$$Q_i^{con} = \sum_{l=1}^{M} \lambda_l \frac{\partial g_l}{\partial q_i} \tag{7.12}$$

The new formulation of Lagrange's equations contains all the capabilities previously developed plus the ability to handle constraints among the generalized coordinates. The resulting system of equations that model the system dynamics are the n Lagrange equations, Equation 7.13, and the m constraint equations, Equation 7.14. The variables to be solved to determine system dynamics are the n generalized coordinates, q_i, and the m Lagrange multipliers, λ_l.

$$\frac{d}{dt}\left(\frac{\partial T}{\partial \dot{q}_i}\right) - \frac{\partial T}{\partial q_i} + \frac{\partial U}{\partial q_i} + \frac{\partial D}{\partial \dot{q}_i} = Q_i + \sum_{l=1}^{m} \frac{\partial g_l}{\partial q_i} \qquad i = 1, 2, \ldots, n \qquad (7.13)$$

$$g_l(q_1, q_2, \ldots, q_n) = 0 \qquad l = 1, 2, \ldots, M \qquad (7.14)$$

7.2.3 Use of Lagrange's Equations to Obtain Lumped Parameter Governing Equations of Systems

This section will provide several examples using Lagrange's equations to obtain the governing equations. This approach is quite useful in obtaining design synthesis models. The examples include electrical and mechanical systems by themselves as well as coupled electromechanical systems involving electrostatics or electromagnetics.

Example 7.1: One Degree of Freedom Mechanical and Electrical System Modeling

Problem: (a) Figure 7.2 shows an idealization of a one DOF mechanical system. This description could describe MEMS devices such as a mechanical accelerometers or resonators. Find the equations of motion for this system using Lagrange's equations. (b) Figure 7.3 shows a schematic of an RLC circuit. Find the equations of motion for this system using Lagrange's equations.

Solution: (a) The obvious generalized coordinate that will describe this one DOF system is the displacement of the mass, M, which is designated as x. The mechanical elements in this model are the mass, M, stiffness of the suspension,

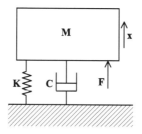

FIGURE 7.2 Schematic of a one degree of freedom mechanical system.

Modeling and Design

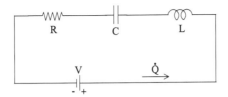

FIGURE 7.3 Schematic of a one degree of freedom RLC circuit.

K, damping, C, and an applied force, F, which could be generated by electrostatics, inertia effects, or electromagnetics depending on the specifics of the application.

For this simple system, the equations of motion could be easily generated by application of Newton's laws, but to illustrate the use of Lagrange's equations, the following function can be defined. (Table 7.1 provides the energy functions for a translational mechanical system.)

Kinetic energy: $T = \frac{1}{2} M \dot{x}^2$

Potential energy: $U = \frac{1}{2} K x^2$

Raleigh dissipation function: $D = \frac{1}{2} C \dot{x}^2$

Virtual work function: $W = F \, \delta x$

Applying Lagrange's equations (Equation 7.13) will result in the classic second-order oscillator differential equation, shown in Equation 7.15. Section H.3 in Appendix H shows the MATLAB commands that will call the function LagEqn.m to perform the computations to obtain the equations of motion. The file xcel1.m defines the symbolic variables, generalized coordinates, and energy functions (T, U, D, W) that will obtain the lumped parameter equations of motion for this system. (Note: in the context of the data for the function LagEqn.m, if x is the generalized coordinate, Dx is $\dot{x}$). Because this is a one DOF system and one generalized coordinate, x, is defined, no constraint equations are involved in this application of Lagrange's equations.

$$M\ddot{x} + C\dot{x} + Kx - F = 0 \qquad (7.15)$$

(b) Lagrange's equations can also be used to obtain the equations of motion of an electrical system as well as combinations of mechanical and electrical systems. To obtain the equations of motion of the simple RLC (resistance, inductance, capacitance) circuit shown in Figure 7.3, the generalized coordinate and energy functions need to be defined. A good choice for the generalized coordinate in this case is charge, Q; the energy functions are defined in Table 7.1 and shown below:

Kinetic energy: $T = \dfrac{1}{2}L\dot{Q}^2$

Potential energy: $U = \dfrac{1}{2}CQ^2$

Raleigh dissipation function: $R = \dfrac{1}{2}R\dot{Q}^2$

Virtual work function: $W = V\,\delta Q$

Once again, there are no equations of constraint because this is a one degree of freedom system and one generalized coordinate (Q) is defined. This electrical analogy of the mechanical system also produces a governing equation that is a second-order oscillator.

$$L\ddot{Q} + R\dot{Q} + CQ - V = 0 \qquad (7.16)$$

The governing equations of either of these systems could have been just as easily obtained by application of Newton's laws for the mechanical system or Kirchoff's laws for the electrical system. However, Lagrange's equation excels in situations with multiple degrees of freedom, combined physical domains, or constraints. In subsequent examples, it will be shown that, in general, it is easier to obtain the energy functions necessary for application of Lagrange's equations than for other approaches.

Example 7.2: Multibody Mechanical Resonator

Problem: Figure 7.4 shows a schematic of a two-body mechanical oscillator. A device such as this could be used in applications for MEMS actuation, inertial sensing, or mechanical filters. Figure 7.5 shows a surface micromachine (SUM-MiT™ [Sandia ultraplanar multilevel MEMS technology]) realization of a two-body MEMS device designed as a secondary mass drive for use in inertial sensing applications [8]. The mass, M_x, has bidirectional parallel-plate actuators. Because the mass, M_x, will move very little due to parallel-plate actuation and the method of operation [8], ignore damping on M_x, but include damping effects on mass, M_y. Find the equations of motion for this system using Lagrange's equations.

Modeling and Design

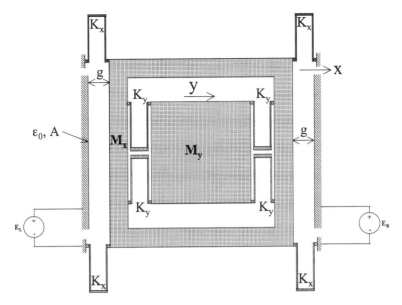

FIGURE 7.4 Two-body mechanical oscillator.

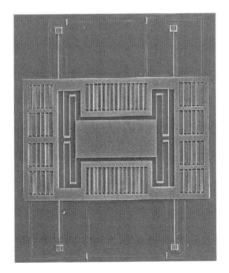

FIGURE 7.5 SEM of a two-body mechanical oscillator. (Courtesy of Sandia National Laboratories.)

Solution: This example involves the motion of two masses supported and coupled by a number of springs with stiffness specified by K_x and K_y. The generalized coordinates for this two DOF problem are specified by x and y. Once again, the energy functions for this application need to be specified. The specification of the energy functions requires only that the total potential kinetic energy

252   Micro Electro Mechanical System Design

of the system be defined, rather than the explicit definition of the interconnection forces required for use of Newton laws. For example, the definition of the potential energy, U, of the four springs with spring constant, K_y, that connect M_x and M_y requires recognition of the fact that the potential energy for these elements is a function of the difference in displacement of each end of the spring. The difference in displacement may be expressed as $(x - y)$ or $(y - x)$ because this term is squared and has no effect on the potential energy. Also, the total potential energy of the four K_x or K_y springs is represented by the appropriate multiplier in the potential energy term.

Kinetic energy:
$$T = \frac{1}{2} M_x \dot{x}^2 + \frac{1}{2} M_y \dot{y}^2 \cdots$$
$$+ \frac{1}{2} \frac{\varepsilon_0 A E_L^2}{(g + x)} + \frac{1}{2} \frac{\varepsilon_0 A E_R^2}{(g - x)}$$

Potential energy: $\quad U = \frac{1}{2} 4 K_x x^2 + \frac{1}{2} 4 K_y (x - y)^2$

Raleigh dissipation function: $\quad D = \frac{1}{2} C_y \dot{y}^2$

Virtual work function: $W = 0$

Application of Lagrange's equations to these functions results in the following equations of motion for this device. These equations can be solved to find the two natural frequencies of the system. The form of these equations of motion can be seen to be of the form of the classic *dynamic* or *vibration absorber* [9] frequently presented in introductory controls, dynamics, and vibrations courses.

$$M_x \ddot{x} + 4 \left(K_x + K_y \right) x - 4 K_y y = \frac{1}{2} \frac{\varepsilon_0 A E_L^2}{(g + x)^2} - \frac{1}{2} \frac{\varepsilon_0 A E_R^2}{(g - x)^2} \qquad (7.17)$$

$$M_y \ddot{y} + C_y \dot{y} - 4 K_y x + 4 K_y y = 0$$

Example 7.3: Constrained Motion of a Linkage

Problem: Figure 7.6 shows a schematic of the constrained motion of a linkage with an applied force. For this analysis, only the static deflection of the link due to the applied force (i.e., kinetic energy = T = 0) is considered. One end of the linkage bar is constrained to move in the horizontal direction. The other end is constrained to move in the vertical direction and is connected to a spring, K.

Modeling and Design

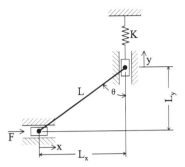

FIGURE 7.6 Constrained motion linkage.

Assume that the constrained motion of the link ends are frictionless and damping and that gravity effects are ignored. The linkage is rigid and of length L. This system has only one degree of freedom; however, the specification of a generalized coordinate for that one degree for freedom is challenging. The options include the rotation angle of the linkage or the linear displacement of either end of the rigid link.

Solution: Sometimes it is desirable to develop a set of governing equations that contain extra degrees of freedom because they are important for the system design, easy to measure, or make the definition of the energy functions easier. For this situation, the deflections at either end of the link, x and y, will be defined as the generalized coordinates. The energy functions for this example are:

$$T = 0$$
$$U = \frac{1}{2}Ky^2$$
$$R = 0 \qquad (7.18)$$
$$W = \begin{bmatrix} F\delta x \\ 0\delta y \end{bmatrix}$$

Because the system has one DOF and two generalized coordinates are specified, one constraint equation is to be defined. The constraint equation, Equation 7.19, is developed from the fact that the link is rigid and this must be accounted for as the link moves to various positions defined by the generalized coordinates, x and y. The constraint equation is an expression of the Pythagorean theorem because the link moves in the positive directions of the generalized coordinates.

$$G = (L_x - x)^2 + (L_y + y)^2 - L^2 = 0 \qquad (7.19)$$

254

Micro Electro Mechanical System Design

Applying Lagrange's equations for constrained motion (Equation 7.13) results in the governing equation for the link, Equation 7.20. The system is fully described by Equation 7.19 and Equation 7.20 and the variables x, y, and λ_1. The solution for the system deflections will involve the solution of constraint equation and the two governing equations for the variables x, y, and λ_1. The Lagrange multiplier used in the formulation is λ_1. The terms in the governing equation that involve λ_1 can be physically interpreted as the forces of constraint required to enforce the constraint.

$$F - \lambda_1 2\left(L_x - x\right) = 0$$
$$Ky - \lambda_1 2\left(L_y + y\right) = 0 \tag{7.20}$$

Equation 7.20 can be solved for the Lagrange multiplier, λ_1, to yield a relationship with the generalized coordinates, x or y.

$$\lambda_1 = \frac{F}{2\left(L_x - x\right)} = \frac{Ky}{2\left(L_y + y\right)} \tag{7.21}$$

This equation can be used in conjunction with the link length constraint equation, Equation 7.19, to form an equation in terms of either generalized coordinate. For example, Equation 7.22 is the equation that relates y to the given constants of K, L, L_y, and F. This is a nonlinear equation that could be solved numerically for y. A solution for the generalized coordinate, x, can be similarly obtained.

$$\left(Ky\right)^2 L^2 = \left(\left(Ky\right)^2 + F^2\right)\left(L_y + y\right)^2 \tag{7.22}$$

Equation 7.23 is a relationship between x and y and another possibility for a generalized coordinate, θ, which can be obtained from Equation 7.21. Equation 7.23 provides a relationship between the applied force, F, and the spring force, Ky. This shows the force multiplication advantage of this mechanism at various angles.

$$\frac{F}{Ky} = \frac{\left(L_x - x\right)}{\left(L_y + y\right)} = \tan\left(\theta\right) \tag{7.23}$$

Example 7.4: Parallel RLC Circuit

Problem: Find the governing equations for the RLC circuit shown in Figure 7.7 using Lagrange's equations.

Modeling and Design

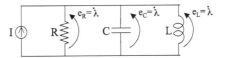

FIGURE 7.7 Parallel RLC circuit with a current source.

Solution: The circuit in Figure 7.7 has an ideal current source and a resistor, capacitor, and inductor in parallel. Using Kirchhoff's voltage law, the voltage across each circuit element will be the same. Also, noting the relationship between voltage and flux linkage, Equation 6.84, the energy functions for each of the circuit elements can be written in terms of flux linkage. This system has one degree of freedom and the generalized coordinate will be flux linkage λ. There are no equations of constraint. The energy functions for this system are defined as:

Kinetic energy: $T = \dfrac{1}{2} C \dot{\lambda}^2$

Potential energy: $U = \dfrac{1}{2} L \lambda^2$

Raleigh dissipation function: $D = \dfrac{1}{2} R \dot{\lambda}^2$

Virtual work function: $W = I \delta \lambda$

Applying Lagrange's equations produces a governing equation that is a second-order oscillator in terms of the flux linkage generalized coordinate. This is another analogy to the translational mechanical system, Equation 7.15, and the RLC circuit with circuit elements in series with charge as a generalized coordinate, Equation 7.16.

$$C\ddot{\lambda} + \frac{1}{R}\dot{\lambda} + \frac{1}{L}\lambda - I = 0 \qquad (7.24)$$

Example 7.5: Solenoid

Problem: Use Lagrange's equations to find the governing equations for the electromechanical device, a solenoid, shown in Figure 7.8.

Solution: A solenoid is a linear electromechanical device that produces linear motion of the mass, M, when an AC voltage, E, is applied to the circuit. The mass is a ferromagnetic material that will change the inductance as it moves into

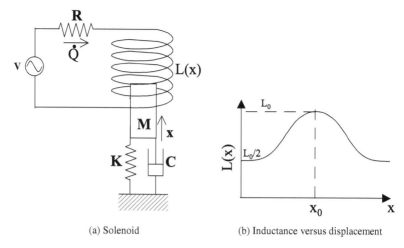

FIGURE 7.8 Schematic of a solenoid device.

the coil of the inductor. Thus, the inductance, $L(x)$, is a function of the displacement of x. The inductance is a minimum when the mass is at either edge of the coil and a maximum when the mass is fully inserted. Assume that the variation of inductance vs. displacement of the mass into the coil is defined by Equation 7.25. This definition of $L(x)$ will have a maximum, $L(x) = L_0$ at $x = x_0$, and a lower value when the mass is at either edge of the coil, $x = 0 = x_0$.

This system has two degrees of freedom, which can be described by the generalized coordinates of motion of the mass, x, and charge, Q, in the electrical circuit. An alternative choice for the electrical circuit generalized coordinate could have been flux linkage, but an equation of constraint (Kirchhoff's voltage law to define the voltage across the resistor) would have been necessary. The energy functions for this system are

$$L(x) = \frac{L_0}{1 + \left(\dfrac{x}{x_0} - 1\right)^2} \tag{7.25}$$

Kinetic energy: $T = \dfrac{1}{2}L\dot{Q}^2 + \dfrac{1}{2}M\dot{x}^2$

Potential energy: $U = \dfrac{1}{2}Kx^2$

Raleigh dissipation function: $D = \dfrac{1}{2}R\dot{Q}^2 + \dfrac{1}{2}C\dot{x}^2$

Modeling and Design

Virtual work function: $W = \begin{Bmatrix} V\delta Q \\ 0 \end{Bmatrix}$

Using Lagrange's equations and the energy preceding functions yields the governing equations for the solenoid system. The solenoid is described by a pair of coupled second-order differential equations.

$$\frac{L_0}{1+\left(\dfrac{x}{x_0}\right)^2}\ddot{Q} + R\dot{Q} - \frac{2L_0x}{\left(1+\left(\dfrac{x}{x_0}\right)^2\right)^2 x_0^2}\dot{Q}\dot{x} = E(t)$$

$$M\ddot{x} + C\dot{x} + Kx - \frac{L_0x}{\left(1+\left(\dfrac{x}{x_0}\right)^2\right)^2 x_0^2}\dot{Q}^2 = 0$$

(7.26)

Using the preceding definition of $L(x)$, the equation can be put in terms of the inductor, L, and its derivative, $\dfrac{dL}{dx}$. Equation 7.27 shows that the force applied to the mass is a function of the change in inductance and the current, $\dot{Q}$, supplied by the circuit.

$$L\ddot{Q} + R\dot{Q} + \frac{\partial L}{\partial x}\dot{Q}\dot{x} = V(t)$$

$$M\ddot{x} + C\dot{x} + Kx = \frac{1}{2}\frac{\partial L}{\partial x}\dot{Q}^2$$

(7.27)

7.2.4 ANALYTICAL MECHANICS METHODS FOR CONTINUOUS SYSTEMS

The Lagrange equation methods discussed thus far have developed *lumped* parameter governing equations for devices. A lumped parameter model is a type of representation in which the electrical or mechanical parameter values are considered to be represented by a quantity concentrated at a location in space. The parameters for a *continuous system* can vary with location. For example, a continuous model of a beam will have its displacement (y) and parameters (E, I) varying with position within the domain of the beam.

Analytical mechanics or energy methods can still be used to analyze these situations. The energy functions must now be integrated over the domain to

TABLE 7.2
Elastic Strain Energy for a Beam, Axial Rod, and Torsion Rod

Type	U	U*
Beam bending	$\int \dfrac{EI}{2}\left(\dfrac{\partial^2 y}{\partial x^2}\right)^2 dx$	$\int \dfrac{M^2}{2EI}dx$
Axial rod	$\int \dfrac{EA}{2}\left(\dfrac{\partial u}{\partial x}\right)^2 dx$	$\int \dfrac{F^2}{2EA}dx$
Torsion rod	$\int \dfrac{GJ}{2}\left(\dfrac{\partial \theta}{\partial x}\right)^2 dx$	$\int \dfrac{T^2}{2GJ}dx$

include the parameter variations in space. Table 7.2 provides the elastic strain energy integrals for a beam, axial rod, and torsion rod. Table 7.2 shows the energy, U, and complementary energy, U^*, formulations discussed in Chapter 6.

A valuable technique in analytical mechanics that operates on the complementary energy function to obtain displacement of an elastic system is *Castigliano's theorem*. This theorem states that if an external force, F, acts on a member, the deflection in the direction the force, δ_F, at the point of application is equal to the partial derivative of the strain energy with respect to the force (Equation 7.28). Because the derivative is with respect to force, the complementary energy, U^* is used. This method was used to calculate the stiffness of various configurations of MEMS suspensions (Appendix F). This type of compact solution for elastic members is very valuable in design calculations. The following example illustrates the use of Castigliano's theorem.

$$\delta_F = \frac{\partial U^*}{\partial F} \tag{7.28}$$

Example 7.6: Application of Castigliano's Theorem

Problem: The simply supported beam shown in Figure 7.9 has a constant cross-section (E and I are constant). A force, F, is applied at the center of the beam. (a) Find the deflection, δ_F, at the point of force application. (b) Find the stiffness of the beam at the point of force application.

Solution: (a) The moment is a function of the distance, x, along the beam as shown in the following equations. Starting at the point $x = 0$ and moving to the right, the moment is the reaction force at the left support, $F/2$ times the distance

Modeling and Design

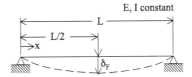

FIGURE 7.9 Simply supported beam with center load.

x. At $x = L/2$, the applied force at the center of the beam is encountered and must now be included in the moment equation. Thus, moment is a function of distance along the beam for two sections of the beam.

$$M(x) = \frac{F}{2} x \qquad 0 \leq x \leq \frac{L}{2} \qquad (7.29)$$

$$M(x) = \frac{F}{2} x - F\left(x - \frac{L}{2}\right) \qquad \frac{L}{2} \leq x \leq L \qquad (7.30)$$

The strain energy of the beam is calculated using the complementary beam strain energy formulation. The strain energy integral is the sum of two parts due to the preceding moment equations.

$$U^* = \frac{1}{2EI} \int_0^{L/2} \left(\frac{Fx}{2}\right)^2 dx + \frac{1}{2EI} \int_{L/2}^L \left(\frac{FL}{2} - \frac{Fx}{2}\right)^2 dx \qquad (7.31)$$

$$U^* = \frac{1}{96} \frac{F^2 L^3}{EI} \qquad (7.32)$$

Applying Castigliano's theorem, Equation 7.28, yields the deflection of the beam at the position of the center load.

$$\delta_F = \frac{\partial U^*}{\partial F} = \frac{1}{48} \frac{FL^3}{EI} \qquad (7.33)$$

(b) The stiffness of the beam at the center load is simply the ratio of force to deflection.

$$K = \frac{F}{\delta_F} = 48 \frac{EI}{L^3} \qquad (7.34)$$

260   Micro Electro Mechanical System Design

The Raleigh–Ritz method approximates the solution of a continuous system through the use of a summation of assumed solutions. If the system variable of interest, $u(x,t)$, is a function of space and time, assume the solution can be approximated by a summation of functions that are products of separate functions of space, $\Phi_i(x)$ and time $a_i(t)$ as shown in Equation 7.35. Equation 7.36 shows this summation written in matrix notation. The function $\Phi_i(x)$ is generally referred to as a *basis* or *shape* function. The basis function, $\Phi_i(x)$, which approximates the solution, u, within the domain of the system, must satisfy two requirements:

- $\Phi_i(x)$ must satisfy the geometric boundary conditions of the system.
- $\Phi_i(x)$ must be differentiable at least to the order appearing in the energy functions.

For most situations, a number of possible basis functions can satisfy these requirements. The number of functions that will approximate the solution to a desired level of accuracy depends on the function selected as well as the loading applied in the domain of the system. The Raleigh–Ritz method has transformed the problem from an exact continuous representation of the system to an approximate discrete system representation in terms of the coefficient, a_i. The use of the Raleigh–Ritz method is best illustrated in the following example.

$$u\left(x,t\right) = \sum_{i=1}^{n} \Phi_i\left(x\right)a_i\left(t\right) \tag{7.35}$$

$$u\left(x,t\right) = \begin{bmatrix} \Phi_1 & \Phi_2 & \cdots & \Phi_n \end{bmatrix} \begin{Bmatrix} a_1 \\ a_2 \\ \vdots \\ a_n \end{Bmatrix} = \Phi a \tag{7.36}$$

Example 7.7: Raleigh–Ritz Solution for a Beam

Problem: Find the deflection at the center of a fixed–fixed beam with a distributed load, w, across the length of the beam.

Solution: Choose the basis functions as shown next. This series of basis functions has zero deflection and slope at the beam ends ($x = 0$ and $x = L$), which satisfy the geometric boundary conditions. They can be twice differentiated as required by the beam strain energy function (see Table 7.2).

$$\Phi_i\left(x\right) = \cos\left(2\pi i \frac{x}{L}\right) - 1 \tag{7.37}$$

Modeling and Design

The beam displacement is approximated by the following series:

$$y(x,t) = \sum_{i=1}^{n} \Phi_i(x) a_i(t) \tag{7.38}$$

Lagrange's equations can be used to find the governing equations. Because this is a static problem, the kinetic energy and Raleigh dissipation function are zero. The Raleigh–Ritz approach has discretized the problem, and the generalized coordinates for this problem are the basis function coefficients, a_i. The energy functions for this situation are:

Potential energy:

$$U = \int_0^L \frac{EI}{2} \left(\frac{\partial^2 y}{\partial x^2} \right)^2 dx$$

$$= \frac{EI}{2} \sum_{i=1}^{n} \left(\int_0^L \left(\frac{\partial^2 \Phi_i}{\partial x^2} \right)^2 dx \; a_i \right)$$

Virtual work function: $W = w\Phi_i \delta a_i$

MATLAB and the function LagEqn were used to assist in the calculations. The MATLAB file used for this example is shown in Appendix H. Lagrange's equation yields the following series of equations in terms of the generalized coordinates, a_i.

$$8(i\pi)^4 \frac{EI}{L^3} a_i + Lw = 0 \quad i = 1, n \tag{7.39}$$

This system of equations can be solved for the generalized coordinates, a_i.

$$a_i = \frac{1}{8(i\pi)^4} \frac{L^4}{EI} w \tag{7.40}$$

Given the basis function, Equation 7.37, and the generalized coordinates, a_i, the displacement of the beam at any position x is:

$$y(x) = \sum_{i=1}^{n} \left(\cos\left(2\pi i \frac{x}{L} \right) - 1 \right) a_i \tag{7.41}$$

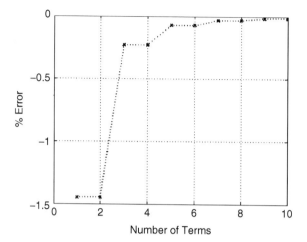

FIGURE 7.10 Solution error for beam deflection at ($x = L/2$) vs. the number of Raleigh–Ritz terms.

A comparison error of the preceding solution vs. the number of terms included can be seen in Figure 7.10. This is a plot of the error of the displacement at the center of the beam ($x = L/2$) vs. the number of terms, n, included in the Raleigh–Ritz solution. The exact solution for this configuration is shown in Equation 7.42. The Raleigh–Ritz solution improves only on every second term because the basis function is zero at $x = L/2$. As shown in Figure 7.10, the Raleigh–Ritz solution always converges from below (i.e., exact solution displacement > Raleigh–Ritz solution) because of the minimization of the system potential energy involved in the formulation.

The concept of approximating the solution of a continuous system with a basis or shape function used in the Raleigh–Ritz method is an essential concept utilized in the finite element method for numerical modeling of systems.

$$y_{exact} = \frac{1}{384} \frac{L^4}{EI} w \qquad (7.42)$$

7.3 NUMERICAL MODELING

The three major categories of numerical approaches to solving engineering mechanics problems are finite difference (FDM); finite element (FEM); and boundary element (BEM) methods. All three of these methods utilize the following steps:

- They replace differential or partial differential equations with approximate algebraic relationships.

Modeling and Design   **263**

- They define a computational grid or mesh that describes the geometry of the solution domain. The boundary conditions and responses are calculated at these grid positions.
- They solve the set of simultaneous algebraic equations to determine the response at the grid points.

Figure 7.11 shows examples of the FDM, FEM, and BEM meshes. The intersection of the grid lines is called nodes. The FDM approach [10] approximates the governing equation within the domain of the mesh with a finite difference approximation at each node. The FDM mesh size is chosen to achieve an accurate solution. Generally, the finer the mesh is, the more accurate the solution is. An FDM mesh is generally uniform due to the nature of the finite difference approximations.

The BEM [11] approach transforms the governing differential equations into equivalent integral equations using calculus relationships (e.g., Gauss–Green theorem) that contain no volume integrals involving the unknown response. This transformation utilizes fundamental known solutions (Green's functions).

The FEM [12] approach utilizes a subdomain or "element" in which the governing differential equation is transformed into an equivalent integral equation. The integral equation is discretized by an interpolation function that interpolates the response inside the element to the nodes on the periphery of the element. The elements are assembled to form the entire domain by ensuring that the responses of adjacent elements at their nodes are equivalent.

Each of these approaches to numerical modeling has advantages and disadvantages. Issues such as the ability of model boundary conditions of complex shapes; nonlinearities; infinite boundary conditions; calculation complexity; and user interaction to set up the problem will tend to make one approach more frequently used in one application over another. Table 7.3 is a comparison of the different approaches. MEMS analysis frequently requires the analysis of different physical phenomena such as structure, electrostatics, and fluidics. The FEM approach is most frequently used, except that electrostatics that encounter infinite boundary conditions are sometime best modeled with the BEM approach. The BEM and FEM methods can be coupled to model these multiphysics problems. Due to space limitation, the details of the FDM and BEM methods will not be discussed in detail. However, an example of the use of a simple, easy-to-use FEM software that has been developed for MEMS analysis will be discussed.

SUGAR is an open source simulation tool for MEMS [13] that has been under development at the Berkeley Sensor and Actuator Center (BSAC) at the University of California for several years [14,15].The SUGAR software is built upon MATLAB, which is frequently used for engineering analysis of various types. SUGAR was developed for MEMS with the same philosophy that has made simulation program with integrated circuit emphasis (SPICE) an essential tool for microelectronic designers. SUGAR is not an acronym but was named because of its duality with SPICE. SUGAR provides an environment for the development of low-order models (*design synthesis*) of a MEMS device, thus enabling the designer to explore the design space quickly.

264 Micro Electro Mechanical System Design

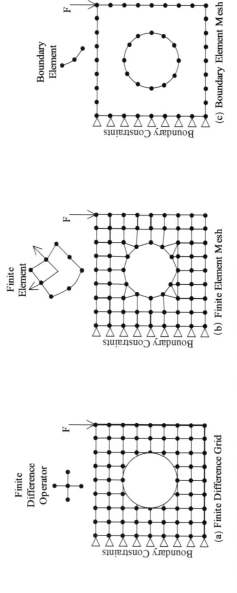

FIGURE 7.11 Finite difference, finite element, and boundary element meshes.

Modeling and Design

TABLE 7.3
Comparison of the FDM, FEM, and BEM Approaches to Numerical Modeling

Method	Advantage	Disadvantage
FDM	Mature, extremely general	Requires fine-structured grid
	Extensively used in fluid mechanics, aerothermal simulations	Nonuniform grid difficult
	Sparse matrices	Modeling geometrically nonuniform boundaries and boundary conditions difficult
FEM	Mature, extremely general	Requires detailed definition (meshing) of the analysis domain volume
	Extensively used in linear and nonlinear structural and thermal analysis	Cannot handle infinite domain problem well
	Sparse and frequently symmetric matrices	Large amounts of data must be stored as compared to BEM or FDM
	Handles 2-D and 3-D geometrically complex analysis well	Resolution of response gradients requires mesh refinement
BEM	Not as mature at FDM or FEM	Difficulty modeling nonlinear problems
	Handles infinite domain problems	Unsymmetric matrices
	Used in structural thermal, acoustics, electrostatic and electromagnetic analysis	Nonsparse matrices
	Requires only surface grids	Significant numerical integration of complex functions
	Unknowns are located only on the boundaries, simplifying mesh generation and storage requirements	Not easily applied to all types of problems

SUGAR describes the device in a SPICE *netlist* type of format with the capability to model beams, electrostatic gaps, circuit elements, and other elements modeled by small, coupled systems of differential equations. It is capable of static and dynamics analyses. SUGAR has been used to model a number of prototype MEMS devices [16]. A *design analysis model* to examine the subtle, second-order effects of the MEMS design is often developed as well to complement the design synthesis model. These higher order models are generally developed using one of a multitude of commercial packages developed for multiphysics MEMS devices [17–19].

Example 7.8: Leveraged Bending Modeling of a Beam with SUGAR

Problem: Figure 7.12 shows a schematic and SUGAR model of fixed–fixed leveraged beam bending. This beam is 600 μm long and 10 μm wide with no residual stress. The model consists of two beam-electrostatic gap elements and

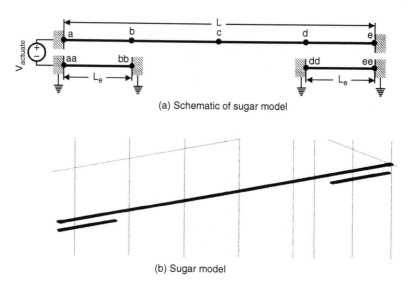

FIGURE 7.12 Schematic and SUGAR model of fixed–fixed leveraged beam bending.

two elastic beam elements. The polysilicon beams have a Young's modulus of 160 GPa and are actuated by applying voltage, $V_{actuate}$, between the beam and the electrodes positioned at each end of the beam. The electrode length is L_e. The gap between the electrodes and the beam is 2.0 μm in the undeflected position.

Solution: The displacement of node c must be able to be positioned at any deflection in the range of 0 to –2 μm. If an electrode is positioned beneath node c, the beam can only be deflected a fraction of the gap before the elastic-electrostatic system would become unstable and collapse. The electrostatic instability phenomenon experienced in parallel-plate electrostatic actuation is discussed in detail in Section 8.1.1.1. *Leveraged bending* [20] is a technique for amplifying the displacement of an electrostatically actuated elastic member before the system becomes unstable and collapses. A greater range of deflection at the expense of higher voltage can be obtained by actuating the beam at the outer edges of the beam as shown in Figure 7.12.

Calculate the voltage vs. displacement at node c for the leveraged bending beam for an electrode lengths of L_e = 80, 100, 200 μm.

Section H.8 through Section H.10 in Appendix H show the three files used to implement this SUGAR analysis. Detailed definition of the SUGAR program and functions are contained in Bindel and Garmire [21], which can be obtained from the Web site listed in Reference 13.

- summit.m (Section H.8, Appendix H): technology definition file that defines the process used to fabricate the device
- Lev_bend.net (Section H.9, Appendix H): a SPICE-like netlist that defines the model consisting of beam elements, beam-gap elements, anchors, electrical grounds, and parameters

Modeling and Design

- Lev_bend.m (Section H.10, Appendix H): MATLAB file that loads the netlist, defines the parameters, loops through a voltages to be applied, and calculates the beam deflections

The SUGAR netlist is the starting point for most analyses. The first line calls summit.net, which defines the fabrication technology parameters such as material properties, layer names, and thicknesses. This particular file has been specialized for the SUMMiT surface micromachine technology. The SUGAR Web site proves a few other frequently used technology files. The next statement calls stdlib.net, which is a SUGAR-specific file that defines the functions for SUGAR. The *param* statement defines the parameters, which can be changed from outside this netlist (i.e., these parameters can be defined by the MATLAB file, which runs SUGAR and controls the program flow). The beams, beam-gap, anchors, electrical grounds, and voltage source parameters are defined in the Lev_bend.net file.

Lev_bend.m is the MATLAB file that executes SUGAR. This file defines the problem parameters and performs a loop from 0 to 200 V applied to the beam. The solution is checked for convergence. Nonconvergence indicates that the system has become unstable at a particular voltage and the solution loop should be terminated.

Figure 7.13 is a node c displacement vs. voltage for three different electrode lengths, L_e. The deflection of a system with $L_e = 200$ μm becomes electrostatically unstable at approximately 35 V and 0.65 μm displacement, which does not achieve the stated objective. As the electrode length is shortened, the stable deflection range and the required voltage increase.

7.4 DESIGN UNCERTAINTY

Many MEMS devices must attain specified design metrics accurately to achieve proper performance. For example, inertial sensors such as an accelerometer may require the natural frequency of the suspension that defines the mechanical sensitivity to be within a specified range or have a specified value as accurately as possible. A gyroscope may require the drive mode natural frequency to be higher and within a specified ratio of the sense mode natural frequency [22]. These are examples of *absolute* and *relative design metrics*, respectively. For many MEMS fabrication technologies, it is easier to attain a relative design metric (e.g., matched devices such as transistors in microelectronics) than an absolute design metric. Many MEMS processes are starting to obtain process uncertainty information [23,24]; however, the MEMS designer may need to design specific test structures (Section 3.5) to quantify process uncertainty relevant to a particular design. The ability of a design to meet a design metric is a function of the *tolerances of the fabrication process* and the *device design*.

MEMS fabrication techniques utilize a combination of deposition, patterning, and etching techniques (Chapter 2), which can make devices with feature sizes of ~1 μm; however, these process do not have a high relative tolerance as compared to traditional machining processes (see Section 4.3). *Relative tolerance*

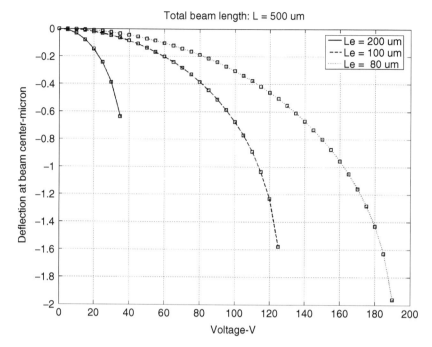

FIGURE 7.13 Node c displacement vs. voltage for three electrode lengths, L_e.

is defined as the feature size divided by part size; this provides a measure of the precision so that a fabrication process can produce a part of any given size. This means that the MEMS design will need to consider and accommodate these fabrication uncertainties to a much larger degree than macroscale design will.

Example 7.9

Problem: Consider beam bending where the stiffness, K, has the following relationship to the beam parameters of width, length, thickness, and Young's modulus (Section 6.1.5):

$$K \propto \frac{EI}{L^3} = \frac{Ewt^3}{12L^3}$$

where
E = Young's Modulus (160 GPa)
I = area moment of inertia ($wt^3/12$)
w = beam width (2 μm)
t = beam thickness (2 μm)
L = beam length (102 μm)

Modeling and Design

For the nominal parameters, the beam stiffness $K = 0.2$ N/m. Assume that the RIE etch can narrow the line width up to a maximum of 0.2 μm from the nominal line width drawn on the mask used to photolithographically pattern the material.

Solution: (a) If the RIE fabrication variation is applied to the width, w, of a spring, what is the effect on the spring constant?

$$K = \frac{E(w + \Delta w)t^3}{12L^3} = .22 \text{ N/m}, \text{ a stiffness variation of } 10\%.$$

(b) What is the effect on stiffness of this fabrication variation applied to the spring length?

$$K = \frac{Ewt^3}{12\left(L + \Delta L\right)^3} = .201 \text{ N/m}, \text{ a stiffness variation of } 0.5\%.$$

A variation in an aspect of fabrication, such as discussed earlier, will produce an effect on several parameters (e.g., length, width) that will affect the design metric (K in this case) unequally. In most real design situations, multiple uncertainties are present, such as line width variations, thickness variations, and material property variations, that will affect multiple parameters in a design (spring width, length, thickness, material) to produce various effects in a design metric, K.

Heuristically, the preceding example would illustrate that increasing the width of the beam to reduce the percentage of uncertainty in the width and increasing the beam length to maintain the desired spring constant is a viable approach.

The estimation of the effects of uncertainty is a stochastic problem that may be approximated using the partial derivative rule, Equation 7.43. This method of uncertainty prediction is very useful when an analytical model of the design metric, f, which can be used to calculate the partial derivative of the design metric with respect to the design parameter, p_i, is available. If the standard deviation of the design parameters is available, Equation 7.43 will allow the prediction of the standard deviation of the design metric.

$$\sigma_f^2 = \left(\frac{\partial f}{\partial p_1}\right)^2 \sigma_1^2 + \left(\frac{\partial f}{\partial p_2}\right)^2 \sigma_2^2 + \cdots + \left(\frac{\partial f}{\partial p_i}\right)^2 \sigma_i^2 \tag{7.43}$$

where
f = design metric
p_i = design parameter
σ_f = standard deviation of f
σ_i = standard deviation of p_i

An alternative approach to obtain the effect on the design metric due to parameter variations is to perform a direct statistical study. This could be accomplished with a detailed numerical model of the design in which a selection of design parameter perturbations is used to propagate and obtain their effect on the design metrics of interest. This may also be done by building the design with a number of design parameter perturbations. These approaches are generally computationally intensive and very costly in time and money; however, they are used when analytical models of sufficient complexity or accuracy cannot be obtained and uncertainty quantification is deemed necessary.

The implementation of these approaches to uncertainty quantification will require sampling of the design parameter perturbations that will be propagated through the model or built for testing. A number of sampling techniques that will ensure that a spectrum of variations is available for assessment exist [26,27]. Sampling technique examples include pure random, Monte Carlo, and Latin hypercube methods of sampling. Pure random sampling consists of choosing the samples without regard for how they are distributed. The Monte Carlo technique will consider the distribution of the input parameters to select the samples. The Latin hypercube uses a stratification method to ensure full coverage of the range of the input variables, which is often a problem with Monte Carlo sampling. Although the quantification of uncertainty of MEMS designs has been addressed in a few instances [25], much more needs to be done.

QUESTIONS

1. Use Lagrange's equations to find the equations of motion for the pendulum shown in Figure 7.1.
    a. Choose a set of generalized coordinates for this problem.
    b. Define the energy functions and constraint equations (if necessary) required for use with Lagrange's equations.
    c. Apply Lagrange's equations to find the governing equations.
2. Given the electrostatically actuated lever system shown in Figure 7.14, assume the parallel plate electrode remains parallel to the fixed electrode but translates vertically.

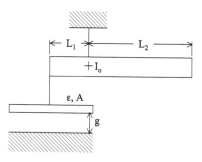

FIGURE 7.14 Electrostatically actuated lever system.

Modeling and Design

    a. How many degrees of freedom does this system have?

    b. Choose a set a generalized coordinates for this problem.

    c. Define the energy functions.

    d. Find governing equations for this system.

    e. Does this system have an electrostatic instability (i.e., pull-in)? If so, what is the pull-in voltage? What is the displacement in terms of the generalized coordinate that you defined?

3. Use the Raleigh–Ritz method to find the deflection of the simply supported beam shown in Figure 7.9.

    a. Define an appropriate set of basis functions.

    b. Compare the Raleigh–Ritz solution to the solution obtained for the deflection in Example 7.6.

    c. Plot the number of terms used in the Raleigh–Ritz solution vs. the solution accuracy.

4. Use Appendix F to design a double-folded beam with a transverse stiffness of $K_x = 1$ N/m. For the process that will be used to fabricate this beam, the minimum line width is 1 μm, and the layer thickness is 2 μm. The fabricated line width will have a standard deviation of $\sigma_w = 0.1$.

    a. What is the standard deviation of the stiffness for your design?

    b. What would you do to minimize the stiffness variability?

    c. What are some practical constraints that may limit your ability to minimize the stiffness variability?

5. Determine the K_x stiffness of the crab-leg flexure shown in Appendix F, using Castigliano's theorem. Use only bending strain energy in your formulation.

6. Explore the literature to find a MEMS sensor such as an accelerometer, gyroscope, pressure sensor, or magnetometer that you can analyze using SUGAR. Perform the analysis required to determine the operational dynamics of the system (e.g., natural frequencies, deflections, transfer function) needed for design.

REFERENCES

1. C. Lanczos, *The Variational Principles of Mechanics*, University of Toronto Press, 1970.
2. L. Meirovitch, *Methods of Analytical Dynamics*, McGraw-Hill Book Company, New York, 1970.
3. L. Meirovitch, *Elements of Vibration Analysis*, McGraw-Hill Book Company, New York, 1975.
4. M.R. Spiegel, *Theoretical Mechanics*, McGraw-Hill Book Company, New York, 1967.
5. D.A. Wells, *Lagrangian Dynamics*, McGraw-Hill Book Company, New York, 1967.

6. S.H. Crandall, D.C. Karnopp, E.F. Kurtz, D.C. Pridmore-Brown, *Dynamics of Mechanical and Electromechanical Systems*, Krieger Publishing Company, 1982. Original edition published by McGraw-Hill, Inc, 1968.
7. The Mathworks Inc., http://www.mathworks.com/.
8. C.W. Dyck, J.J. Allen, R.J. Huber, Parallel-plate electrostatic dual-mass resonator, *Proc. SPIE*, 3876, 198–209, 1999.
9. W.T. Thomson, *Theory of Vibration with Applications*, Prentice Hall, Inc., Englewood Cliffs, NJ, 1981.
10. R.G. Stanton, *Numerical Methods for Science and Engineering*, Prentice Hall, Englewood Cliffs, NJ, 1961.
11. P.K. Kythe, An *Introduction to Boundary Element Methods*, CRC Press, Boca Raton, FL.
12. O.C. Zienkiewicz, *The Finite Element Method in Engineering Science*, McGraw-Hill, New York, 1971.
13. http://www-bsac.eecs.berkeley.edu/cadtools/sugar/sugar/.
14. J.V. Clark, N. Zhou, K.S.J. Pister, Modified nodal analysis for MEMS with multienergy domains, *Proc. Int. Conf. Modeling Simulation Microsyst., Semiconductors, Sensors Actuators*, MSM 2000, 31–34, 2000.
15. J.V. Clark, D. Bindel, W. Kao, E. Zhu, A. Kuo, N. Zhou, J. Nie, J. Demmel, Z. Bai, S. Govindjee, K.S.J. Pister, M. Gu, A. Agogino, Addressing the needs of complex MEMS design, *MEMS 2002*, Las Vegas, Nevada, January 20–24, 2002.
16. S.X.P. Su, H.S. Yang, Analytical modeling and FEM simulations of single-stage microleverage mechanism, *Intl. J. Mechanical Sci.*, 44, 2217–2238, 2002.
17. MEMSCAP: http://www.memscap.com.
18. ANSYS: http://www.ansys.com.
19. CoventorWare: http://www.coventor.com/.
20. E.S. Hung, S.D. Senturia, Extending the travel range of analog-tuned electrostatic actuators, *JMEMS*, 8(4), 497–505, Dec. 1999.
21. D. Bindel, D. Garmire, *SUGAR 3.0: A MEMS Simulation Program (User's Guide)*, March 2002.
22. W.A. Clark, Micromachined vibratory rate gyroscopes, Ph.D. dissertation, University of California, Berkeley, Fall 1997.
23. D.A. Koester, R. Mahadevan, B. Hardy, K.W. Markus, *MUMPs™ Design Handbook*, Revision 5.0, Cronos Integrated Microsystems.
24. S. Livmary, H. Stewart, L. Irwin, J. McBrayer, J. Sniegowski, S. Montague, J. Smith, M. de Boer, J. Jakubczak, Reproducibility data on SUMMiT, *Proc. SPIE*, 3874, 102–112, September 1999.
25. J.W. Wittwer, T. Gomm, L.L. Howell, Surface micromachined force gauges: uncertainty and reliability, *J. Micromech. Microeng.*, 12, 13–20, 2002.
26. M.D. McKay, R.J. Beckman, W.J. Conover, A comparison of three methods for selecting values of input variables in the analysis of output from a computer code, *Technometrics*, 21(2), 239–245, 1979.
27. S.F. Wojtkiewicz, M.S. Eldred, R.V. Field, A. Urbina, J.R. Red-Horse, Uncertainty quantification in large computational engineering models, American Institute of Aeronautics and Astro., AIAA-2001-1455.

8 MEMS Sensors and Actuators

Sensors and actuators are forms of transducers, which are devices that transform energy from one form to another. Table 8.1 lists a number of transduction schemes, mostly centered about mechanical and electrical transduction methods. A sensor is an input transducer that senses an input form of energy (i.e., mechanical, thermal, chemical, etc.) and, in most cases, transforms it to electrical energy, which facilitates integration into systems. Alternatively, an actuator is an output transducer that, in most cases, transforms one form of energy into a mechanical output.

8.1 MEMS ACTUATORS

8.1.1 ELECTROSTATIC ACTUATION

Electrostatic actuation is produced by the electric field of a capacitor. Figure 8.1 illustrates the two basic configurations of a capacitor for electrostatic actuation of a MEMS device: the parallel plate and the interdigitated comb capacitor configurations. The interdigitated comb capacitor is dominated by the fringe electrostatic field, and the parallel plate capacitor is dominated by the direct electrostatic field.

8.1.1.1 Parallel Plate Capacitor

The capacitance of a fixed parallel plate capacitor is shown in Equation 8.1. The parameters for a fixed capacitor are constant.

$$C = \frac{\varepsilon A}{g} \tag{8.1}$$

where
ε = permittivity of material between the parallel plates (free space permittivity 8.85×10^{-12} F/M)
A = plate area
g = gap between the plates

273

TABLE 8.1
Transduction Methods

| | Output signals | |
Input signals	Mechanical	Electrical
Mechanical	Fluidics, acoustics	Piezoresistive
Electrical	Electrostatics, electromagnetic	Langmuir probe, transformer
Thermal	Thermal expansion	Pyroelectric
Magnetic	Magnetometer	Magnetoresistance
Chemical	ChemAbsorber	Ionization, ChemFET, ChemResistor

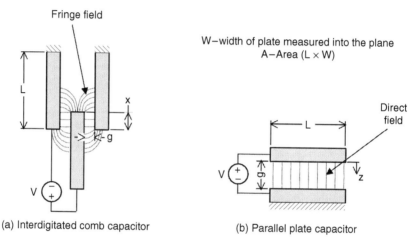

FIGURE 8.1 Parallel plate and interdigitated comb capacitor configurations.

For a variable parallel plate capacitor, the movable plate moves normally to the fixed plate as defined by the coordinate, z (Figure 8.1). The capacitance for the movable capacitor is

$$C = \frac{\varepsilon A}{g - z} \tag{8.2}$$

The energy, W, of a capacitor with a voltage, V, across the plates is given by Equation 6.80; this results in Equation 8.3 for this configuration:

$$W = \frac{1}{2} \frac{\varepsilon A}{g - z} V^2 \tag{8.3}$$

MEMS Sensors and Actuators

The electrostatic force between the plates can be determined by differentiating the energy function, W with respect to the coordinate in the direction of the force. In essence, this is the potential energy term of Lagrange's equation.

$$\frac{\partial W}{\partial z} = F = \frac{1}{2}\frac{\varepsilon A}{(g-z)^2}V^2 \tag{8.4}$$

For a parallel plate capacitor, the capacitance, Equation 8.2, is inversely proportional to the gap between the capacitors plates and the force, Equation 8.4, is inversely proportional to the gap between the capacitor plates squared. The capacitance and the force of the parallel plate capacitor are highly nonlinear. As z increases and the distance between the plates goes to zero, the electrostatic force becomes very large. Figure 8.2 shows the nonlinear parallel plate electrostatic force as the voltage is increased. If the movable parallel plate is attached to a spring, k, the equilibrium point for a capacitor-spring system for each voltage can be plotted at the intersection of the spring constant and force curve, which are marked with an "o." However, at the point at which the electrostatic force

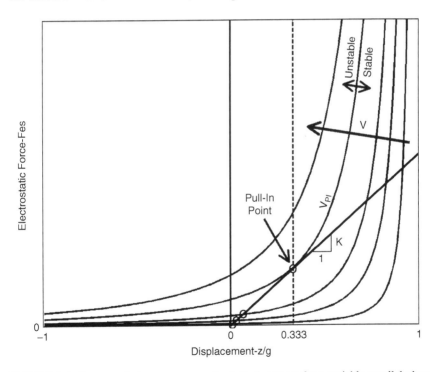

FIGURE 8.2 Electrostatic force vs. normalized displacement for a variable parallel plate capacitor.

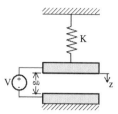

FIGURE 8.3 Spring and parallel plate capacitor.

curve and the spring constant curve are tangent, the electrostatic instability point, known as *pull-in*, occurs. At pull-in, the movable plate of the capacitor is unstable and moves to close the gap to zero. There is no stable intermediate point.

By analyzing the pull-in phenomena more closely, it can be seen that Equation 8.5 is the force balance between the spring and the electrostatic forces (Figure 8.3):

$$Kz = \frac{1}{2}\frac{\varepsilon A}{(g-z)^2}V^2 \tag{8.5}$$

Solve Equation 8.5 for the voltage squared, V^2:

$$V^2 = \frac{2Kz(g-z)^2}{\varepsilon A} \tag{8.6}$$

The pull-in phenomenon will occur when the derivative of voltage with respect to position is zero. The deflection, z, at which the derivative of voltage with respect to position is zero is the deflection at pull-in, Z_{PI}. This deflection at pull-in is shown in Equation 8.7 and illustrated in Figure 8.2. The corresponding voltage at which pull-in occurs, V_{PI}, is shown in Equation 8.8. Figure 8.4 shows the voltage vs. deflection curve for the parallel plate capacitor and the instability at pull-in. Note that the deflection at pull-in is a function of only geometric quantities (i.e., not a function of spring constant, area, etc.). The limited deflection before electrostatic instability (pull-in) occurs is a significant design constraint that must be considered in the design of sensors and actuators.

$$Z_{PI} = \frac{g}{3} \tag{8.7}$$

$$V_{PI} = \sqrt{\frac{8}{27}\frac{Kg^3}{\varepsilon A}} \tag{8.8}$$

MEMS Sensors and Actuators

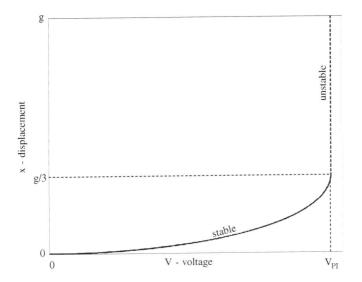

FIGURE 8.4 Voltage vs. deflection curve of a parallel plate capacitor.

Another significant effect in design involving a variable parallel plate capacitor is *electrostatic spring softening*. This effect can be explained by using a two-term Taylor series approximation for the electrostatic force term in Equation 8.9. The linear term of the Taylor series is the slope of the electrostatic force vs. deflection curve (Figure 8.2). The linear term, Equation 8.10, can also be viewed as a negative electrostatic stiffness, K_{es}, that will cancel the elastic stiffness at pull-in. The electrostatic spring softening can be used to soften or tune the spring stiffness electrostatically (Equation 8.11) or the resulting natural frequency of a design where this is critical.

$$Kz = \frac{1}{2} \frac{\varepsilon A}{(g-z)^2} V^2 \approx \frac{\varepsilon A V^2}{2g^2} + \frac{\varepsilon A V}{g^3} z \tag{8.9}$$

$$K_{es} = \frac{\varepsilon A V}{g^3} \tag{8.10}$$

$$(K - K_{es})z = \frac{\varepsilon A V^2}{2g^2} \tag{8.11}$$

Another frequently encountered configuration for a variable parallel plate capacitor is that in which the movable electrode has a fixed electrode on both

278   Micro Electro Mechanical System Design

sides with the same gap and voltage (Equation 8.12). The results for this configuration and others are summarized in Table 8.2.

$$Kz = \frac{1}{2}\frac{\varepsilon A}{(g-z)^2}V^2 + \frac{1}{2}\frac{\varepsilon A}{(g+z)^2}V^2 \qquad (8.12)$$

8.1.1.2 Interdigitated Comb Capacitor

The interdigitated comb variable capacitor configuration is shown in Figure 8.1. This configuration consists of a movable plate inserted between two fixed plates. The direct parallel plate electrostatic force and the fringe field force are the two types of electrostatic forces acting upon the movable plate. The motion of the movable plate is constrained so that motion can only be in the direction that inserts the movable plate between the two fixed plates and maintains the gaps between them. The width of the combs is w. The capacitance of this set of interdigitated combs is given in Equation 8.13. The "2" in Equation 8.13 is due to the capacitance of the fixed plates on either side of the movable plate.

The electrostatic force, Equation 8.14, is obtained as before by differentiating the electrostatic energy with respect to the displacement, x. It should be noted that because the force is not a function of displacement, this is a more controllable force. Figure 8.5 shows a plot of the electrostatic force vs. a normalized displacement for the interdigitated comb variable capacitor. The constant force contour lines are horizontal, indicating that they are not a function of displacement. If this variable capacitor is connected to an elastic spring, the spring and electrostatic force lines uniquely intersect for each voltage applied. There are no electrostatic instability phenomena for this configuration.

$$C = \frac{2\varepsilon w(L-x)}{g} \qquad (8.13)$$

$$F = \frac{\varepsilon w}{g}V^2 \qquad (8.14)$$

8.1.1.3 Electrostatic Actuators

The variable capacitors discussed earlier can be used for actuation or for sensing. Figure 8.6 shows an actuator implemented with interdigitated combs. Because electrostatics is only attractive, there are combs to actuate in both directions (i.e., positive and negative). The interdigitated comb actuator shown has a stroke of 17 µM in both directions with approximately a 25-µN force. The folded spring suspension allows motion in the comb meshing direction, but prevents the combs from moving laterally due to the parallel plate force of the comb fingers. The

MEMS Sensors and Actuators

TABLE 8.2
Variable Capacitor Configuration Capacitances, Force and Pull-In Displacement and Forces

	Parallel plate capacitor	Differential parallel plate capacitor	Interdigitated comb capacitor
Schematic			
Capacitance C	$\dfrac{\varepsilon A}{g-z}$	$\dfrac{\varepsilon A}{g-z} + \dfrac{\varepsilon A}{g+z}$	$\dfrac{2\varepsilon w(L-x)}{g}$
Electrostatic force F	$\dfrac{1}{2}\dfrac{\varepsilon A}{(g-z)^2}V^2$	0	$\dfrac{\varepsilon w}{g}V^2$
X_{PI}	$\dfrac{g}{3}$	0	Not applicable
V_{PI}	$\sqrt{\dfrac{8}{27}\dfrac{Kg^3}{\varepsilon A}}$	$\sqrt{\dfrac{1}{2}\dfrac{Kg^3}{\varepsilon A}}$	Not applicable

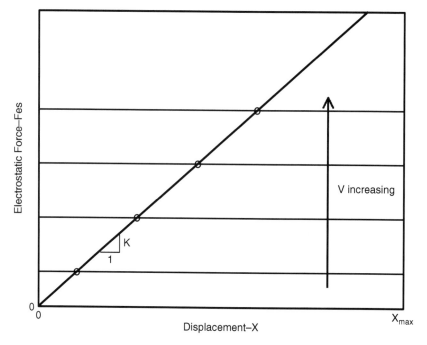

FIGURE 8.5 Electrostatic force vs. normalized displacement for an interdigitated comb variable capacitor.

parallel plate electrostatic forces are more powerful than the fringe field electrostatic forces of the combs. The actuation speed capabilities of electrostatic actuators are very high and limited only by the natural frequency of the suspension and the electrical time constant of the system.

In addition, the electrostatic force that will cause the combs to mesh and unmesh — the electrostatic field in the normal to the substrate direction — should be considered (Figure 8.7). The surface underneath the fixed and suspended combs acts as a ground plane that produces an asymmetric electrostatic field. The asymmetric field has more lines terminating on the upper surface of the suspended comb than the bottom surface, thus resulting in a levitation force pulling the suspended combs in the vertical direction [1,2]. For actuation that needs to be very precise, such as a vibratory gyroscope, this results in a wobble that must be minimized. For sensing, this will result in an offset requiring calibration.

The fundamental design trade-off for electrostatic actuation is the choice between parallel plate vs. interdigitated comb approaches. Parallel plate actuation can provide very high forces (~100 μN) with small stroke (~3 μm), but the force is highly nonlinear with instability within the displacement range. Interdigitated comb actuation provides a moderate level of force (~10 μN) with large strokes (~20 μm) with a very controllable force (i.e., no instabilities). In reality, both of these two types of electrostatic actuation are present in some degree in all realizable actuators.

MEMS Sensors and Actuators

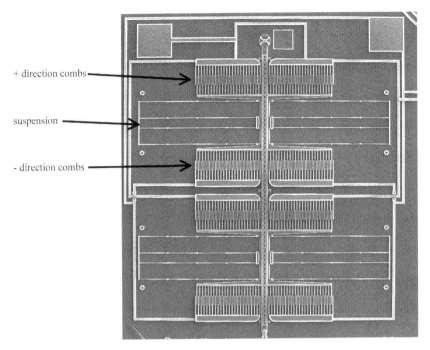

FIGURE 8.6 Electrostatic interdigitated comb actuators. (Courtesy of Sandia National Laboratories.)

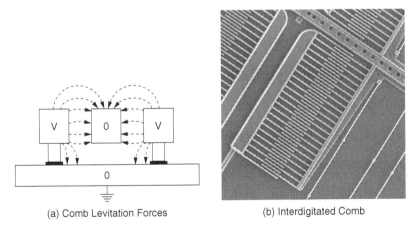

FIGURE 8.7 Interdigitated combs and levitation forces.

A number of very innovative approaches to electrostatic actuator design have been and will continue to be developed in order to attain specific performance metrics for various applications. For example, Figure 8.8 [3] shows an array of parallel plate electrodes utilized to attain high force, coupled with a compliant mechanism consisting of flexures to implement a leverage system to attain the

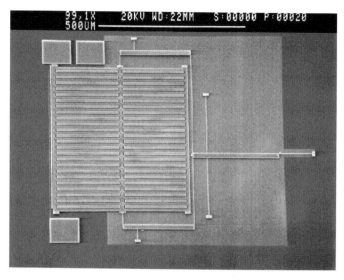

FIGURE 8.8 Parallel plate actuation using a leveraged displacement multiplier. (Courtesy of Sandia National Laboratories.)

desired actuator stroke. Figure 8.9 shows another actuator utilizing the same concept of using an electrostatic actuator with limited stroke coupled with a displacement multiplying mechanism [4]. The displacement multiplying mechanisms illustrated by these two examples are *compliant mechanisms*, which avoid the use of discrete joints with rubbing surface through the use of flexures. These

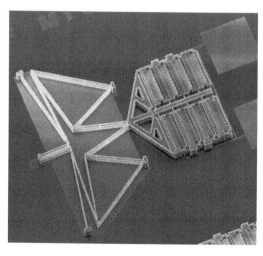

FIGURE 8.9 A compact electrostatic actuator with short stroke, high force, coupled with a displacement amplification system. (Courtesy of Sandia National Laboratory; S. Kota et al. *IEEE Int. Conf. Micro Electro Mechanical Syst.*, *MEMS 2000*, 164–169, January 2000, Miyazaki, Japan.)

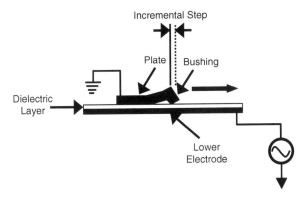

FIGURE 8.10 Scratch drive actuator schematic. (T. Akiyama and K. Shono, *JMEMS*, 2(3), 106–110, 1993.)

mechanisms can be designed via the use of structural and topology optimization methods [4] or via approximation methods that enable the use of traditional mechanism design techniques [5].

Another approach to actuator development is through the use of an *inchworm* type of electrostatically actuated mechanism [6,7] that couples the electrostatic forces and incremental motion via elastic flexing of a large electrostatic area plate. Different embodiments of this class of actuator are known as the *shuffle motor* [6] and the *scratch drive actuator* [7]. They are capable of high-force and large-scale motion with small step sizes within a very compact form factor. Figure 8.10 schematically shows the operation of a scratch drive actuator capable of small step sizes (~30 nm). This actuator operates by electrostatically flexing a plate that rotates a bushing, enabling the plate to move forward when the electrostatic force is released. Figure 8.11 shows several scratch drive actuators fabricated in a surface micromachine process.

The pull-in instability of parallel plate electrostatics is a significant constraint limiting the stable actuation range that electrostatic actuator designers have attempted to overcome. Stable repeatable actuation (positioning) throughout the operational range is required for devices such as analog optical mirrors. Several approaches have been explored to alleviate this situation, including:

- *Passive circuit components*. The inclusion of a fixed capacitor in series with the variable capacitor has been proposed [8]. The additional capacitor has the effect of extending the effective gap of the actuator and its stable range of operation. Parasitic capacitance is an issue, but the stable range of actuation can be extended [9]. The size of the additional capacitor is a complicating factor in actuator design. The use of an inductor to create a tuned LC circuit has also been proposed [10]; however, the practical use of an inductor severely limits this approach.
- *Closed loop control*. The use of closed loop control has been proposed and a switched capacitor implementation has been simulated [11] but

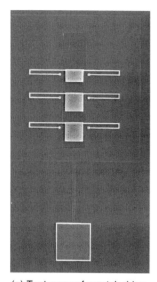
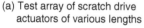
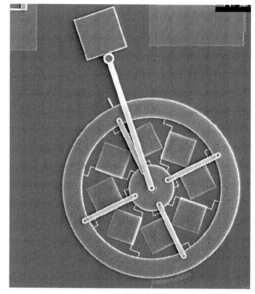

(a) Test array of scratch drive actuators of various lengths

(b) A rotary table driven by scratch drive actuators

FIGURE 8.11 Implementations of a scratch drive actuator. (Courtesy of Sandia National Laboratories.)

not demonstrated. This approach required an integrated electronics and MEMS fabrication process to implement. The utilization of this approach is very complicated and difficult.

- *Leveraged bending.* Leveraged bending [12] (Figure 8.12) is a simple technique to implement and has been demonstrated. This technique electrostatically actuates a beam close to its anchor and utilizes the leverage of the beam length to obtain the actuation deflection at another position on the beam. The stable deflection is obtained at the expense of greater beam length and actuation voltage.

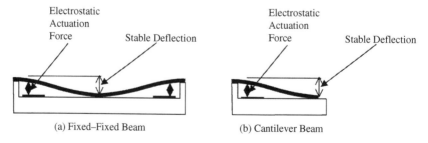

(a) Fixed–Fixed Beam   (b) Cantilever Beam

FIGURE 8.12 Leveraged bending schematic.

MEMS Sensors and Actuators

8.1.2 THERMAL ACTUATION

Thermal actuation uses the thermal expansion of materials (solid, liquid, gas) to achieve mechanical actuation. The thermal expansion of a solid material is characterized by coefficient of thermal expansion (CTE), α_T. Equation 8.15 states that the mechanical strain, ε, of a material is directly proportional to the temperature change, ΔT, where the proportionality constant α_T is the CTE. The CTE for a material has units of strain per change in temperature (1/°C). The CTE of a material is frequently a function of temperature, generally increasing with required temperature.

$$\varepsilon = \alpha_T \Delta T \tag{8.15}$$

Example 8.1

Problem: A 100-μm long beam is anchored to the substrate at both ends. The temperature of the beam is raised by 300°C. The beam is 2.5 μm thick by 4 μm wide and is made of polysilicon (assume $E = 160$ GPa and $\alpha_T = 2.5$ microstrain/°C). (a) Calculate the stress and strain in the beam. (b) Calculate the force and displacement produced in the beam. (c) Is buckling a concern for this beam?
  Solution: (a) $\varepsilon = \alpha_T \Delta T = (2.5 \times 10^{-6} \, °C^{-1}) \, (300°C) = 750 \, \mu\text{-strain}$

$$\sigma = E \, \alpha_T \, \Delta T = (1.6 \times 10^{11} \, \text{Pa}) \, (2.5 \times 10^{-6} \, °C^{-1}) \, (300°C) = 120 \times 10^6 \, \text{Pa}$$

(b) $F = A \, E \, \alpha_T \, \Delta T$
  $= (2.5 \times 10^{-6} \, \mu\text{m})(4 \times 10^{-6} \, \mu\text{m})(1.6 \times 10^{11} \, \text{Pa}) \, (2.5 \times 10^{-6} \, °C^{-1}) \, (300°C)$
  $= 1200 \, \mu\text{N}$
  $\Delta L = L \, \varepsilon = (100 \, \mu\text{m}) \, (0.000750) = 0.075 \, \mu\text{m}$

The thermal expansion of a solid produced high forces if constrained but a low amount of stroke. Thermal actuator design based on thermal expansion must be careful of "parasitic strains" in such things as anchors, which may easily negate the thermal strain. To achieve more change in length (stroke), a longer length beam is necessary, but as length is increased, buckling may occur.
  (c) The Euler column formula for the buckling of a beam is given by Equation 6.38. The beam can deflect laterally in two directions, but will fail due to buckling in the direction with the minimum area moment of inertia, I.

$$I = w \, t^3/12 = (4 \times 10^{-6} \, \mu\text{m}) \, (2.5 \times 10^{-6} \, \mu\text{m})^3/12 = 5.21 \times 10^{-24} \, \mu\text{m}^4$$

The force at which buckling can occur is calculated by the following Euler column formula. Buckling is very sensitive to the end conditions of the column. Table 6.2 lists the end condition constants for various column end conditions. The end condition constant, C, for a clamped–clamped beam is chosen to be 1.2. $C = 4$ would generally be too high because a clamped–clamped boundary con-

286   Micro Electro Mechanical System Design

dition is very difficult to achieve in practice. The low end condition constant in effect simulates the fixed–fixed boundary condition as slightly stiffer than a pinned–pinned boundary condition. Because the force produced by thermal expansion is greater than the critical buckling force, buckling will occur.

$$F_{cr} = C * \pi^2 * E * I / L^2 = 1.2 * \pi^2 * (1.6 \times 10^{11} \text{ Pa}) * (5.21 \times 10^{-24} \text{ } \mu m^4) /$$
$$(100 \times 10^6 \mu m)^2 = 987 \text{ } \mu N$$

Mechanical actuation due to the thermal expansion of a solid material can be accomplished by a device with two or more materials with different CTE, or by the thermal gradients within a device made of the one material.

A bimorph utilizes the CTE mismatch of materials to achieve mechanical deflection. Utilization of multiple layers of materials has also been used. The literature contains many examples of bimorph and multilayer cantilever beam thermal actuators [13–15]. Bimorph actuators are generally long thin beams with layers of materials with different CTE as illustrated in Figure 6.8. The CTE mismatch of the materials will produce a moment, M, in the beam when the temperature of the beam is changed, ΔT. The moment in the beam is calculated by integrating the thermal expansion forces through the thickness of the beam. Although this type of thermal actuator has limited stroke, it has been used in fluidic control valve applications.

Figure 8.13 shows two widely used thermal actuator approaches that utilize thermal gradients within a material to achieve mechanical actuation. The *lateral thermal actuator* [16] consists of a thin hot leg and a wide cold leg; these will create a difference in temperature due the differing current densities and ability to transfer heat. The actuator is heated by Joule heating due to current flow. The hot leg is heated to 400 to 600°C, and the actuator will deflect laterally. Figure 8.14 shows an SEM image of several of the lateral thermal actuators in parallel being used to close a latch. The hot leg–cold leg concept can be extended to vertical actuation by stacking the hot and cold legs atop each other. The actuator shown in Figure 8.14 is actuated by 7 to 10 V and achieves 8 μm of horizontal motion. These actuators can only move in one direction and are not reversible (i.e., positive vs. negative direction).

The *bent beam thermal actuator* [17] utilizes a different concept to achieve thermal actuation that is a reapplication of a structure for residual stress measurement [18,19] (discussed in Section 3.5.1). The bent beam thermal actuator is a symmetric structure consisting of a long thin beam canted at a small angle. When Joule heating is applied to the device, the bent beam expands and produces lateral motion. This beam is a mechanical amplifier of the small deflection produced by thermal expansion. The bent beam devices may be cascaded by running in parallel as well as by using additional bent beam structures for mechanical amplification (Figure 8.15).

A thermal actuator can produce high force (>100 μN) and a displacement of 10 μm or greater utilizing only low voltage ~ 10 V. However, these actuators can require significant power (~100 mW). Techniques have developed to minimize

MEMS Sensors and Actuators

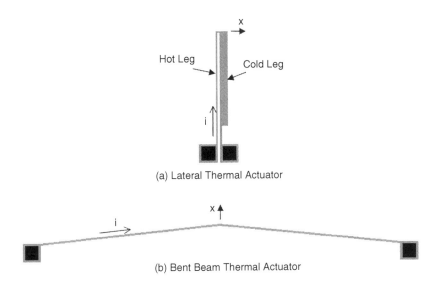

FIGURE 8.13 Electrothermal actuators.

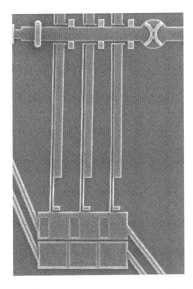

FIGURE 8.14 SEM images of electrothermal actuators. (Courtesy of Sandia National Laboratories.)

the impact of these power requirements [20]. The design of thermal actuators is a multiphysics problem [21] requiring thermal, structural, and electrical analysis. The heat transfer involved in a thermal analysis includes conduction and convection. Radiation heat transfer is generally not significant. The upper practical limit for temperature in the thermal actuator is approximately 600°C, above which material property changes such as localized plastic yielding and material grain

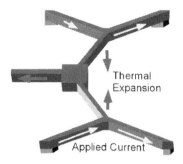
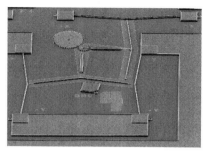

FIGURE 8.15 Cascaded bent-beam thermal actuators. (Courtesy of A. Oliver, Sandia National Laboratories.)

growth become an issue. Thermal actuators are also generally limited to a frequency response less than 1000 Hz because of the time constants associated with heat transfer.

Thermal actuation may also be used in a fluid medium. Figure 8.16 shows a schematic of thermal actuation used to vaporize ink that will eject an ink drop. Also, thermal actuation can be used for direct mechanical motion by thermal expansion of a fluid or change of phase of the fluid.

8.1.3 Lorentz Force Actuation

The Lorentz force is generated by a magnetic field, B, and current, i, in a conductor of length, L (see Figure 6.32 and Section 6.3.1). The Lorentz force, F_L, is given by Equation 8.16, where the F_L, i, and B are at right angles according to the right-hand rule implied by the cross-product. Lorentz force actuation may be applied to MEMS devices in a number of ways, as shown in Figure 8.17.

$$F_L = Li \times B \tag{8.16}$$

Figure 8.17a shows a microvalve application [22] that is actuated by the Lorentz force produced by the interaction of a DC current, i, and a magnetic field, B, created by a permanent magnet mounted adjacent to the valve. Alternatively, Figure 8.17b shows the generation of a cyclic Lorentz force that utilizes an AC current, i, interaction with a permanent magnet to produce the necessary AC actuation for a MEMS magnetometer [23]. This AC Lorentz force actuation scheme may also be used for a flexural plate wave device [24]. The Lorentz force in this case is used to produce a cyclic excitation force at the natural frequency of a bar packaged in a vacuum suspended at the nodes of the first mode of vibration. An AC current, i, is passed through the beam and will interact with the ambient magnetic field to produce a cyclic Lorentz force that stimulates the beam into resonance, which is detected, and a measure of the magnetic field.

MEMS Sensors and Actuators

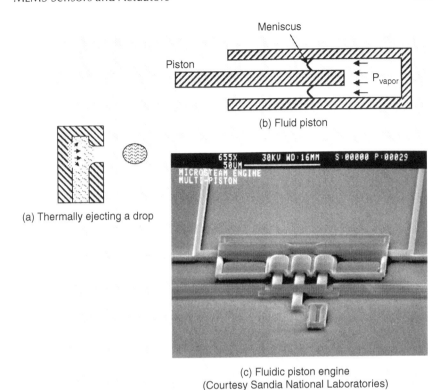

FIGURE 8.16 Thermal actuation in a fluid medium.

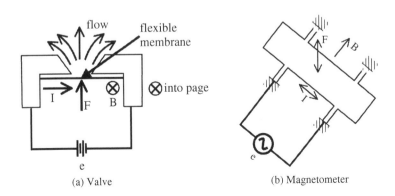

FIGURE 8.17 Examples of Lorentz force actuation for MEMS.

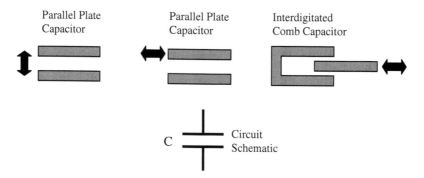

FIGURE 8.18 Examples of parallel plate and interdigitated capacitance structures.

8.2 MEMS SENSING

8.2.1 Capacitative Sensing

Electrostatic capacitance sensing is a frequently used transduction method for MEMS devices. MEMS fabrication techniques can readily produce parallel plate or interdigitated comb finger capacitors in a MEMS device that can move as a result of a physical variable (acceleration, pressure) excitation. Figure 8.18 shows schematically some variations of parallel plate and interdigitated capacitance structures. The relative motion of the plates can be in any direction (vertical or horizontal).

Differential capacitors are two capacitors that share a common electrode. Figure 8.19 shows a schematic of some differential capacitor structures, and Figure 8.20 shows some MEMS devices that employ a differential capacitor. Differential capacitors can be fabricated in such a manner that the two capacitors have nearly equal capacitance in the unperturbed position and change by equal and opposite amounts as the structure is deflected. The differential capacitor structure enables differential sensing, which can cancel many adverse or common mode effects to first order. Changing two capacitors by the same amount is a common mode effect to which a differential capacitor structure does not respond.

The size of the capacitors utilized in MEMS devices is small — generally a fraction of a picofarad. The variation of the nominal capacitance to be sensed to provide the dynamic signal of interest is in the femtofarad range or less. Also, undesirable stray capacitances called *parasitic capacitances* can interfere with capacitance sensing. Examples of parasitic capacitances may include the capacitance between a sense line and the substrate or the capacitance between two sense lines. Figure 8.21 illustrates the capacitances in a MEMS accelerometer, which include the direct capacitance, C_{pp}, as well as the fringe field, C_{f1}, C_{f2}, and parasitic capacitances, C_{pl}, C_{pu}, that exist between the upper plate, lower plate, and the substrate.

Frequently, the magnitudes of the parasitic capacitances are greater than the nominal and dynamic capacitance of the device, and the parasitic capacitances

MEMS Sensors and Actuators

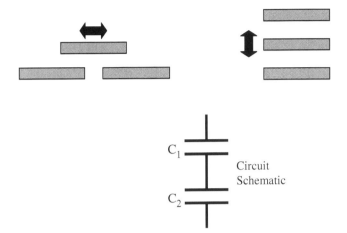

FIGURE 8.19 Differential capacitor schematic.

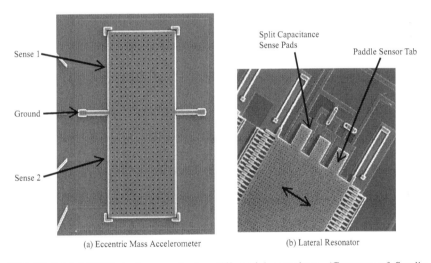

FIGURE 8.20 MEMS devices employing differential capacitors. (Courtesy of Sandia National Laboratories.)

may cause errors in sensing. The parasitic capacitances can be minimized by good layout practices, shielding, and direct on-chip integration of electronics. The parasitic capacitances will increase the overall electrical time constant of the device and limit the frequency at which it can be electrically charged and discharged; this will have an adverse effect on the sensor interface.

All of the various capacitance detection methods involve AC electrical excitation of a capacitance network, which will produce a voltage or frequency shift in the circuit that will be detected as a measure of capacitance change. In the case of a sensor, this is a measure of a physical variable such as acceleration or pressure. For example, a bridge circuit (Figure 8.22) can be excited by an AC voltage at a

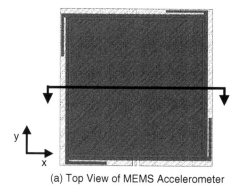

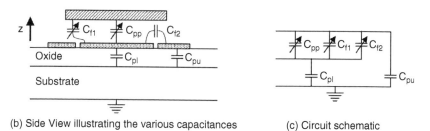

FIGURE 8.21 Schematic of capacitance in a MEMS device.

frequency, ω_c, and the output voltage of the bridge measured. The excitation frequency or carrier frequency, ω_c, is much higher than the frequency ω of the change in capacitance caused by physical excitation. The AC excitation, Equation 8.17, combines with the capacitance change, Equation 8.18, to produce an AM modulated signal at the output containing terms of the form of Equation 8.19.

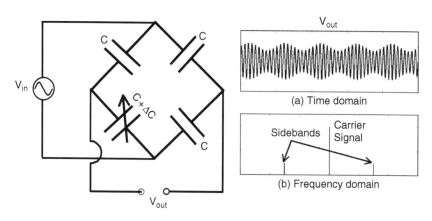

FIGURE 8.22 AC bridge circuit and AM modulation.

MEMS Sensors and Actuators

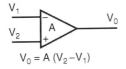

FIGURE 8.23 Operational amplifier.

Amplitude modulation (AM) is a multiplication of the carrier signal and the sensor signal in which the amplitude of the AC excitation (carrier frequency) is modified by the amplitude of a modulating signal (sensor signal). The sensor signal can be seen as the low-frequency envelope of the high-frequency excitation or carrier signal. A sensor that would use this type of excitation technique would also need to demodulate the signal to obtain the sensor signal. Demodulation methods will be discussed shortly.

$$V_{in} = A\sin(\omega_c t) \tag{8.17}$$

$$C + \Delta C = C(1+m)\sin(\omega_m t) \tag{8.18}$$

$$(1+m)\sin(\omega_m t)\sin(\omega_c t) = \\ \sin(\omega_c t) + \frac{m}{2}\cos(\omega_c - \omega_m)t - \frac{m}{2}\cos(\omega_c + \omega_m)t \tag{8.19}$$

The AC bridge shown in Figure 8.22 is not generally used by itself in MEMS applications due to the small gain and the size of the components required. However, two sense amplifier configurations are useful for capacitance measurements: a *voltage buffer* and the *integrator*. Both of these circuits use operational amplifiers as part of their circuitry.

An operational amplifier (Figure 8.23) is a circuit component that amplifies the difference between two input voltages to produce a single output, Equation 8.20. A few characteristics of an ideal operational amplifier should be known to facilitate circuit analysis:

- The differential gain is very high ($>10^5$).
- Input impedance is infinite (i.e., input current is zero).
- Output impedance is zero (i.e., acts like the output of an ideal voltage source).
- An operational amplifier has *common-mode rejection* (i.e., only the difference of $V1$ and $V2$ is amplified).

$$V_0 = A(V_2 - V_1) \tag{8.20}$$

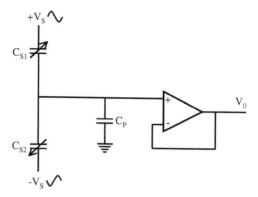

FIGURE 8.24 Voltage buffer schematic.

Figure 8.24 is a schematic of a voltage buffer that is sensing a differential capacitor, C_{S1} and C_{S2}. C_P is a parasitic capacitance, which models capacitance terms other than the sense capacitances. The sense capacitors have a high-frequency excitation signal (carrier signal) applied to them. The frequency of the excitation signal is one to two orders of magnitude higher than the expected frequency content of the variable capacitances induced by the physical variable being sensed. Because V_0 is feedback to the negative or inverting input of the operational amplifier, both amplifier inputs are at V_0.

Equation 8.21 is an expression of the conservation of charge at the negative operational amplifier input. Rearranging terms, the voltage transfer function is given by Equation 8.22. As can be seen from the transfer function, the gain is limited and cannot exceed unity. Because the excitation signal, V_S is multiplied by changing capacitances, C_{S1}, C_{S2}, in the transfer function, the output voltage, V_0 will be amplitude modulated. The parasitic capacitance also has a direct impact on the gain of the system. The sense capacitances are a nonlinear function of the displacement and the parasitic capacitance is a nonlinear function of voltage, which contributes to the distortion of the output voltage signal.

$$(V_S - V_0)C_{S1} - (V_0 + V_S)C_{S2} - V_0 C_P = 0 \tag{8.21}$$

$$\frac{V_0}{V_S} = \frac{C_{S1} - C_{S2}}{C_{S1} + C_{S2} + C_P} \tag{8.22}$$

The charge integrator is shown in Figure 8.25 and the transfer function in Equation 8.23. The charge integrator has better linearity and distortion performance than the voltage buffer. The parasitic capacitance does not directly appear in the transfer function of the charge integrator or affect its gain. V_0 is once again an amplitude-modulated signal. In practical implementations of the charge inte-

MEMS Sensors and Actuators

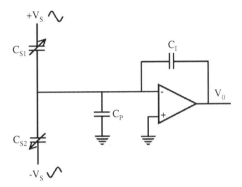

FIGURE 8.25 Charge integrator sense amplifier.

grator, a resistor in parallel with C_I is frequently used to provide a path to ground to prevent leakage current buildup on C_I.

$$\frac{V_0}{V_S} = \frac{C_{S1} - C_{S2}}{C_I} \tag{8.23}$$

The output voltage, V_0, of the voltage buffer or the charge integrator is an amplitude-modulated signal that needs to be *demodulated* to obtain the sensor signal of interest (the capacitance change due to physical variable). A conceptually simple method of demodulation is an *envelope detector*, which consists of two components: a rectifier and a low-pass filter (Figure 8.26). The rectifier, which may be implemented with a diode, will remove the positive or the negative portion of the low-frequency envelope of the carrier signal. The low-pass filter, whose frequency cutoff is set by the $R_1 C_1$ time constant, will remove the carrier signal and retain the analog signal of interest (Figure 8.27). The envelope detector is

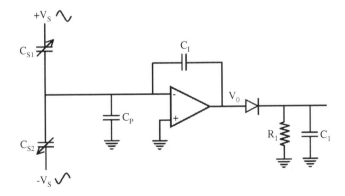

FIGURE 8.26 A charge integrator with an envelope detector to demodulate the signal.

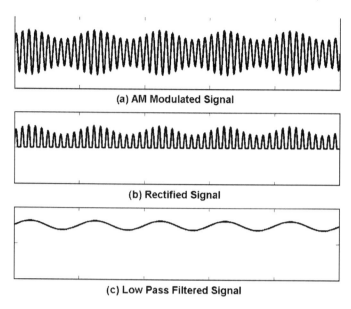

FIGURE 8.27 The signal processing stages involved in the envelope detector.

simply implemented and is limited by the nonlinear effects of diode rectification, which may lead to distortion. Other methods of demodulation such as *synchronous demodulation* are available, but they require more complex circuitry. Synchronous demodulation involves the multiplication of the AM signal with a phase-shifted signal at the same frequency as the carrier. The additional circuitry is involved with obtaining the carrier signal and mixing it with the AM signal.

With the advent of *switched capacitor* techniques, a new method of capacitance detection has arisen. Switched capacitor methods have two important advantages over the conventional voltage buffer or integrator described previously. A switched capacitor implementation depends upon the ratio of two capacitors, not their individual values. In microelectronic fabrication methods, it is possible to make a matched pair of anything on silicon, but it is very hard to make a component with a precise value. Also, the frequency of a filter implemented with the switched capacitor method can be tuned by merely changing the "clock" frequency of the switched capacitor circuit.

A classic example of the switched capacitor technique is the implementation of a resistor shown in Figure 8.28. A high-value resistor is expensive to implement in microelectronics because it is large. A resistor can be implemented very simply with two MOS switches and a capacitor using switched capacitor methods. The switches, S_1 and S_2, are controlled by nonoverlapping clocks so that when S_1 closes, S_2 is open. With S_1 closed and S_2 open, the capacitor is charged. When the switches reverse (i.e., S_1 open and S_2 closed), a charge, Δq, is transferred from V_1 to V_2. As this process is repeated by the cycling of the switches N times a current flow, i is established between the voltages $V1$ and $V2$ (Equation 8.24).

MEMS Sensors and Actuators

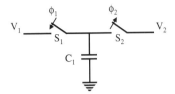

FIGURE 8.28 A switched capacitor resistor.

Recognizing that the left side of the equation is current, i, and the number of cycles, N, per time is the clock frequency, f_{clk}, the equation can be written as shown in Equation 8.25. The resistance, R, simulated by the switched capacitor technique is inversely proportional to the capacitor value, C_1, and the clock frequency, f_{clk}. The value of the resistance can be controlled externally just by adjusting the clock frequency. In essence, the switched capacitor resistor is a "bucket brigade" transferring charge from V_1 to V_2 as dictated by the clock.

$$\frac{\Delta q}{\Delta t} = C_1 (V_1 - V_2) \frac{N}{\Delta t} \qquad (8.24)$$

$$R = \frac{V_1 - V_2}{i} = \frac{1}{C_1 f_{clk}} \qquad (8.25)$$

Transferring and quantifying charge is necessary for capacitative sensing. Figure 8.29 is a schematic of an implementation of a switched capacitor integrator. The basic topology of the circuit is the same as for the integrator discussed previously; however, the voltage, V_S, is DC and four switches operate synchronously with the two nonoverlapping clocks, ϕ_1 and ϕ_2. The switches are implemented by CMOS transistors that are turned on or off by the clock, ϕ_1 and ϕ_2. The clock signals typically are externally supplied and are approximately 100 times faster than the analog signals from the sense capacitors, C_{S1} and C_{S2}.

The sequence of operation of the circuit is described as follows:

- ϕ_1 is high and ϕ_2 is low: S_1, S_2, and S_4 are closed and S_3 is open. V_0 is at 0 V and capacitor C_I is discharged. Capacitors C_{S1} and C_{S2} are charged by the voltage sources $+V_S$ and $-V_S$.
- ϕ_1 is low and ϕ_2 is high: S_1, S_2, and S_4 are open and S_3 is closed. The capacitors C_{S1} and C_{S2} are disconnected from the voltage sources $+V_S$ and $-V_S$ and connected to the operational amplifier. Because the operational amplifier cannot input current, the charge on C_{S1} and C_{S2} are transferred to capacitor C_I

The transfer function for the operation of the switched capacitor integrator is shown in Equation 8.26. The switched capacitor charge integrator operates by

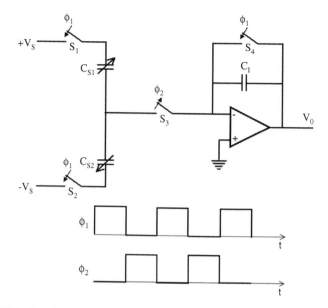

FIGURE 8.29 Switched capacitor implementation of a charge integrator.

a sequence of charging the sense capacitors and transferring the charge to the operational amplifier integration capacitor, C_I.

$$V_0 = \left(\frac{C_{S1} - C_{S2}}{C_I}\right) V_S \qquad (8.26)$$

8.2.2 Piezoresistive Sensing

8.2.2.1 Piezoresistivity

Piezoresistivity is a widely utilized phenomenon for MEMS sensors. The piezoresistive effect was first discovered by Lord Kelvin [27] in 1856 when he reported that certain metallic (iron, copper) conductors under mechanical strain exhibited a corresponding change in electrical resistance. This is the basic operating principle for the metal and foil strain gauges used for engineering measurements for many years. The piezoresistive effect in single-crystal silicon and germanium was first reported in 1954 [28]. The discovery of the piezoresistive effect in silicon had significant impact in the development of MEMS for the following reasons:

- Integration with MEMS devices and microelectronics is possible due to material compatibility.
- Integration of the piezoresistive material and the MEMS device allows good transmission of strain without hysteresis or creep.

MEMS Sensors and Actuators   **299**

- The piezoresistive effect in silicon is over an order of magnitude greater than in metals.
- MEMS fabrication processes allow good matching of resistors utilized in the Wheatstone bridge-sensing circuits.

The resistivity, ρ, of a semiconductor expressed by the Equation 8.27 is a function of the number of charge carriers, N, and their mobility, μ. The resistivity of a semiconductor such as silicon can be controlled by the concentration of an impurity that directly controls the number of charge carriers N. The addition of boron to silicon produces a p-type semiconductor material (positive charge holes are the majority carrier). An n-type semiconductor material (electrons are the majority carriers) can be created by the addition of arsenic or phosphorus as the impurities. For doping concentrations (charge carrier) varying from 10^{13} cm^{-3} to 10^{19} cm^{-3}, the resistivity of silicon varies over a wide range of approximately 500 Ω-cm to 5 mΩ-cm. The carrier mobility, μ, is a function of temperature and doping concentration. Mobility decreases with increasing temperature because lattice vibrations caused by increased temperature scatter the electrons. In silicon, hole mobilities are less than electron mobilities. For example, with a dopant concentration of 2.5×10^{-16} cm^{-3}, the electron mobility is $\mu_e \approx 1000$ cm^2/Vs compared to the hole mobility of $\mu_p \approx 500$ cm^2/Vs.

$$\rho = \frac{1}{qN\mu} \tag{8.27}$$

where

ρ = resistivity (Ω-cm)
q = electron charge (1.6×10^{-19} C)
N = number of charge carriers (cm^{-3})
μ = carrier (electron or hole) mobility (cm^2/Vs)

The piezoresistive effect is present in single-crystal and polycrystalline semiconductor materials. The analysis of piezoresistive single-crystal material will be discussed first, followed by a discussion of piezoresistivity in polycrystalline materials.

8.2.2.2 Piezoresistance in Single-Crystal Silicon

For single crystal silicon that has an anisotropic crystal structure of the cubic family, the electric field vector, $\{\mathbf{E}\}$, is related to the current density vector, $\{\mathbf{J}\}$, by a 3×3 resistivity matrix, $[\boldsymbol{\rho}]$ (Equation 8.28). (Note: boldface indicate matrix variables; $\{\}$ column or row matrices; and $[]$ square or rectangular matrices.) The $\{\mathbf{E}\}$, $\{\mathbf{J}\}$, and $[\boldsymbol{\rho}]$ matrix variables are direction dependent due to the anisotropic crystal structure.

$$\{E\} = [\rho]\{J\}$$

$$\begin{Bmatrix} E_x \\ E_y \\ E_z \end{Bmatrix} = \begin{bmatrix} \rho_{xx} & \rho_{xy} & \rho_{zx} \\ \rho_{xy} & \rho_{yy} & \rho_{yz} \\ \rho_{zx} & \rho_{yz} & \rho_{zz} \end{bmatrix} \begin{Bmatrix} J_x \\ J_y \\ J_z \end{Bmatrix} \quad (8.28)$$

For an unstressed crystal in the cubic family such as silicon, the resistivities along the <100> axis are identical and the off-diagonal terms of the resistivity matrix are zero:

$$\begin{Bmatrix} E_x \\ E_y \\ E_z \end{Bmatrix} = \begin{bmatrix} \rho & 0 & 0 \\ 0 & \rho & 0 \\ 0 & 0 & \rho \end{bmatrix} \begin{Bmatrix} J_x \\ J_y \\ J_z \end{Bmatrix} \quad (8.29)$$

The resistivity of a piezoresistive material is a function of stress that is also direction dependent due to the anisotropic crystal structure. Figure 8.30 shows a unit cell of material with the definition of the direction-dependent stress, electric field, and current density variables. The resistivity components of a stressed crystal can be written as the sum of the unstressed resistivity and the change in resistivity due to stress (Equation 8.30). The stresses applied to a unit cube of the crystal are the three normal stresses σ_{xx}, σ_{yy}, and σ_{zz}, and three shear stresses τ_{xy}, τ_{yz}, and τ_{xz}, as illustrated in Figure 8.30. The changes in resistivity, $\Delta\rho_{ij}$, due to stress in Equation 8.30 can be quantified by associating piezoelectric coefficients, π_{ij}, with every stress component (Equation 8.31). This approach produces a large number of coefficients; however, due to the symmetry of a cubic lattice, the number of piezoelectric coefficients reduces to three: π_{11}, π_{12}, and π_{44}. Equation 8.32 is the relationship between the change in resistivity, $\Delta\rho$, and stress, σ, modeled with the coefficients π_{ij}.

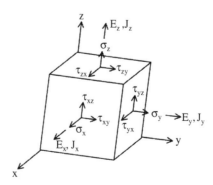

FIGURE 8.30 Unit cube illustrating the definition of the directional stresses (*s,t*), electric field (*E*), and current density (*J*) variables.

MEMS Sensors and Actuators

TABLE 8.3
Piezoresistive Coefficients for Single-Crystal Silicon at Room Temperature in <100> Orientation

Material	Dopant concentration (cm^{-3})	Resistivity (Ω-cm)	Piezoresistive coefficient (×10^{-11} Pa^{-1})		
			π_{11}	π_{12}	π_{44}
p-Type silicon	1.5×10^{15}	7.8	6.6	−1.1	138.1
n-Type silicon	4.0×10^{14}	11.7	−102.2	53.4	−13.6

Source: C.S. Smith, *Phys. Rev.*, 4, April, 1954.

The piezoresistive coefficients for <100> oriented silicon are shown in Table 8.3. The largest piezoresistance coefficients are $\pi_{11} = -102.2 \times 10^{-11}$ Pa^{-1} for n-type silicon and $\pi_{44} = +138.1 \times 10^{-11}$ Pa^{-1} for p-type silicon. The coefficients indicate the magnitude of the piezoresistive effect for a material. These coefficients are properties of a material and are affected by the temperature and doping of the material. The coefficients decrease with increasing temperature. The temperature sensitivity of the piezoelectric coefficients is a major concern for piezoelectric sensors. The piezoelectric coefficients will also decrease with increasing impurity (dopant) concentrations.

$$\begin{Bmatrix} \rho_{xx} \\ \rho_{yy} \\ \rho_{zz} \\ \rho_{xy} \\ \rho_{yz} \\ \rho_{zx} \end{Bmatrix} = \begin{Bmatrix} \rho \\ \rho \\ \rho \\ 0 \\ 0 \\ 0 \end{Bmatrix} + \begin{Bmatrix} \Delta\rho_{xx} \\ \Delta\rho_{yy} \\ \Delta\rho_{zz} \\ \Delta\rho_{xy} \\ \Delta\rho_{yz} \\ \Delta\rho_{zx} \end{Bmatrix} \tag{8.30}$$

$$\underbrace{}_{\substack{\text{resistivity} \\ \text{of the} \\ \text{stressed} \\ \text{material}}} \quad \underbrace{}_{\substack{\text{nominal} \\ \text{resistivity}}} \quad \underbrace{}_{\substack{\text{resistivity} \\ \text{change} \\ \text{due to stress}}}$$

$$\frac{1}{\rho}\{\Delta\rho\} = [\Pi]\{\sigma\} \tag{8.31}$$

$$\frac{1}{\rho}\begin{Bmatrix} \Delta\rho_{xx} \\ \Delta\rho_{yy} \\ \Delta\rho_{zz} \\ \Delta\rho_{xy} \\ \Delta\rho_{yz} \\ \Delta\rho_{zx} \end{Bmatrix} = \begin{bmatrix} \pi_{11} & \pi_{12} & \pi_{12} & 0 & 0 & 0 \\ \pi_{12} & \pi_{11} & \pi_{12} & 0 & 0 & 0 \\ \pi_{12} & \pi_{12} & \pi_{11} & 0 & 0 & 0 \\ 0 & 0 & 0 & \pi_{44} & 0 & 0 \\ 0 & 0 & 0 & 0 & \pi_{44} & 0 \\ 0 & 0 & 0 & 0 & 0 & \pi_{44} \end{bmatrix} \begin{Bmatrix} \sigma_{xx} \\ \sigma_{yy} \\ \sigma_{zz} \\ \tau_{xy} \\ \tau_{yz} \\ \tau_{xz} \end{Bmatrix} \tag{8.32}$$

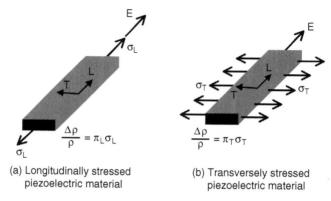

FIGURE 8.31 Longitudinally and transversely stressed piezoelectric material.

Equation 8.28, Equation 8.30, and Equation 8.32 can be combined to form Equation 8.33, which is an expression for the electric field in a stressed cubic crystal lattice such as silicon or germanium in the <100> orientation. The first term in this equation is the unstressed conduction term through the lattice; the second is the stressed conduction (piezoelectric effect) through the lattice. The second term is the same as the piezoelectric effect in metals such as wire or foil strain gauges. The last two terms are the piezoelectric effect associated with a stressed semiconductor lattice.

$$E_x = \underbrace{\rho J_x}_{\substack{\text{unstressed}\\\text{conduction}}} + \underbrace{\rho \pi_{11} \sigma_x J_x}_{\substack{\text{1st order}\\\text{piezoelectric}\\\text{effect}}} + \underbrace{\rho \pi_{12}\left(\sigma_y + \sigma_z\right) J_x + \rho \pi_{44}\left(J_y \tau_{xy} + J_z \tau_{xz}\right)}_{\substack{\text{stressed semiconductor lattice}\\\text{piezoelectric effect}}}$$
$$E_y = \rho J_y + \rho \pi_{11} \sigma_y J_y + \rho \pi_{12}\left(\sigma_x + \sigma_z\right) J_y + \rho \pi_{44}\left(J_x \tau_{xy} + J_z \tau_{yz}\right)$$
$$E_z = \rho J_z + \rho \pi_{11} \sigma_z J_z + \rho \pi_{12}\left(\sigma_x + \sigma_y\right) J_z + \rho \pi_{44}\left(J_x \tau_{xz} + J_y \tau_{yz}\right)$$

(8.33)

For generality, it is useful to express the piezoresistivity coefficients for arbitrary orientations. This is accomplished by defining longitudinal and transverse piezoresistivity coefficients, π_L and π_T, respectively. The longitudinal coefficient, π_L, is the case in which the stress and electric field are applied in the same direction. The stress and electric field are perpendicular for the transverse piezoresistivity coefficient, π_T. Figure 8.31 illustrates the concept of longitudinal and transverse directions for a bar of piezoelectric material that may possibly be used as a sensor.

If a transformation can be found to rotate the axes from the <100> orientation where the piezoelectric coefficients are known (Table 8.3) to the longitudinal–transverse axes, then the change in resistivity relationships, Equation 8.32, may be written as shown in Equation 8.34. It is also noted in Equation 8.34 that

MEMS Sensors and Actuators

TABLE 8.4
Longitudinal and Transverse Coefficients for Common Cubic Crystal Directions

Longitudinal direction	$l_1\ m_1\ n_1$	π_l	Transverse direction	$l_2\ m_2\ n_2$	π_t
(1 0 0)	1 0 0	π_{11}	(0 1 0)	0 1 0	π_{12}
(0 0 1)	0 0 1	π_{11}	(1 1 0)	$\dfrac{1}{\sqrt{2}}\ \dfrac{1}{\sqrt{2}}\ 0$	π_{12}
(1 1 1)	$\dfrac{1}{\sqrt{3}}\ \dfrac{1}{\sqrt{3}}\ \dfrac{1}{\sqrt{3}}$	$\dfrac{1}{3}(\pi_{11}+2\pi_{12}+2\pi_{44})$	(1 $\bar{1}$ 0)	$\dfrac{1}{\sqrt{2}}\ \dfrac{-1}{\sqrt{2}}\ 0$	$\dfrac{1}{3}(\pi_{11}+2\pi_{12}-\pi_{44})$
(1 $\bar{1}$ 0)	$\dfrac{1}{\sqrt{2}}\ \dfrac{-1}{\sqrt{2}}\ 0$	$\dfrac{1}{2}(\pi_{11}+\pi_{12}+\pi_{44})$	(1 1 1)	$\dfrac{1}{\sqrt{3}}\ \dfrac{1}{\sqrt{3}}\ \dfrac{1}{\sqrt{3}}$	$\dfrac{1}{3}(\pi_{11}+2\pi_{12}-\pi_{44})$
(1 1 0)	$\dfrac{1}{\sqrt{2}}\ \dfrac{1}{\sqrt{2}}\ 0$	$\dfrac{1}{2}(\pi_{11}+\pi_{12}+\pi_{44})$	(0 0 1)	0 0 1	π_{12}
(1 1 0)	$\dfrac{1}{\sqrt{2}}\ \dfrac{1}{\sqrt{2}}\ 0$	$\dfrac{1}{2}(\pi_{11}+\pi_{12}+\pi_{44})$	(1 $\bar{1}$ 0)	$\dfrac{1}{\sqrt{2}}\ \dfrac{-1}{\sqrt{2}}\ 0$	$\dfrac{1}{2}(\pi_{11}+\pi_{12}-\pi_{44})$

Source: B. Kloeck and N.F. De Rooij, in *Semiconductor Sensors*, S.M. Sze, Ed., John Wiley & Sons, New York, 1994.

the factional changes in resistivity and resistance are the same. Equation 8.34 is useful in calculations for piezoelectric sensor resistance changes.

$$\frac{\Delta\rho}{\rho} = \frac{\Delta R}{R} = \pi_L \sigma_L + \pi_T \sigma_T \tag{8.34}$$

The geometrical transformation necessary to rotate the piezoelectric coefficients from any orientation to the longitudinal–transverse directions is shown in Equation 8.35. The vector ($\mathbf{x}$, $\mathbf{y}$, $\mathbf{z}$) is the initial orientation and (L, T, z) are the longitudinal–transverse axes. The first two rows of the transformation matrix are the direction cosines of the longitudinal (l_1,m_1,n_1) and transverse (l_2,m_2,n_2) axes with respect to the original axes. This transformation information can be used in Equation 8.36 and Equation 8.37 to calculate the longitudinal and transverse piezoelectric coefficients. Table 8.4 contains the longitudinal and transverse coefficients for several frequently used orientations of the longitudinal and transverse axes.

$$
\left\{ \begin{array}{c} L \\ T \\ z' \end{array} \right\} = \left[\begin{array}{ccc} l_1 & m_1 & n_1 \\ l_2 & m_2 & n_2 \\ l_3 & m_3 & n_3 \end{array} \right] \left\{ \begin{array}{c} x \\ y \\ z \end{array} \right\} \tag{8.35}
$$

$$
\pi_L = \pi_{11} + 2\left(\pi_{44} + \pi_{12} - \pi_{11}\right)\left(l_1^2 m_1^2 + l_1^2 n_1^2 + m_1^2 n_1^2\right) \tag{8.36}
$$

$$
\pi_T = \pi_{12} - \left(\pi_{44} + \pi_{12} - \pi_{11}\right)\left(l_1^2 l_2^2 + m_1^2 m_2^2 + n_1^2 n_2^2\right) \tag{8.37}
$$

Gauge factor, G, is a term frequently used to describe the sensitivity of a piezoresistive sensor. The gauge factor is the ratio of the fractional change in resistance, $\Delta R/R$ and the strain, ε. The gauge factor for single-crystal silicon is approximately 100 and the gauge factor for a metal strain gauge is around 2.

$$
G = \frac{\Delta R}{R} \frac{1}{\varepsilon} \tag{8.38}
$$

8.2.2.3 Piezoresistivity of Polycrystalline and Amorphous Silicon

Polycrystalline silicon is composed of small silicon crystals separated by grain boundaries. Polycrystalline materials may show texture, which is a statistical measure of the crystal orientations within the polycrystalline material. Amorphous silicon has no crystalline structure at all. Deposition and patterning methods for polycrystalline and amorphous silicon are well developed, and their resistivity can be controlled by ion implantation with boron or phosphorus. The piezoelectric properties of polysilicon have been studied and found to be a promising piezoresistive material for sensors [32–34]. The total resistance in a polycrystalline material is a combination of the resistance of the grains and the resistance of the grain boundaries. This combination of effects can have significant impact on the resistivity and temperature-sensitivity properties of the material.

Polysilicon and amorphous silicon have an advantage of a high gauge factor compared to metal foil sensors; however, their gauge factors are significantly lower than that of single-crystal silicon and they strongly depend upon processing parameters. Table 8.5 is a comparison of the gauge factors for various piezoelectric materials. Piezoresistance coefficients for large-grained polysilicon can approach 60 to 70% of single-crystal silicon; however, for fine-grained (micromechanical) polysilicon, π_L is about seven times less than that of single-crystal silicon.

8.2.2.4 Signal Detection

A significant advantage of piezoresistive devices is that they generally do not require on-chip detection circuitry. Half- or full-bridge resistor bridges can pro-

MEMS Sensors and Actuators

TABLE 8.5
Approximate Gauge Factors for Several Piezoelectric Materials

Material	Gauge factor (G)
Single crystal silicon	100
Polysilicon	50
Amorphous silicon	30
Metal wire/foils	2

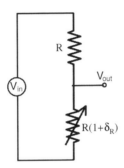

FIGURE 8.32 Voltage divider (half-bridge) circuit.

vide high sensitivity and first-order temperature compensation. Figure 8.32 shows a schematic of a half-bridge (i.e., voltage divider) circuit that could be used for piezoresistive signal detection. The variable resistance piezoresistor is denoted by the resistance $R(1 + \delta_R)$ indicated in the figure, where δ_R is the fractional change in the resistance due to strain (Equation 8.39). The output voltage for this circuit is given in Table 8.6 as well as a two-term Taylor series expansion around the small resistance change, δ_R. This equation indicates that the output voltage will have a DC offset of $V_{in}/2$ and a small nonlinearity due to the deleted terms from the Taylor series expansion. Neither of these attributes is desirable, but the half-bridge circuit is discussed for comparison to the Wheatstone bridge circuits, which can be visualized as a combination of two half-bridges.

$$\delta_R = \frac{\Delta R}{R} \qquad (8.39)$$

Three configurations of Wheatstone bridge circuits are presented in Figure 8.33 and the output voltages of these circuits are listed in Table 8.6. The three configurations, known as a single-active bridge, half-active bridge, and fully active bridge, utilize one, two, and four variable resistances, respectively.

A linearized output voltage relationship for the single- and half-active bridge circuits was obtained via a Taylor series expansion; therefore, both of these

TABLE 8.6
Resistance Signal Detection Circuits

Circuit	Output voltage	Comments
Half bridge (voltage divider)	$$V_{out} = \frac{R(1+\delta_R)}{R+R(1+\delta_R)}V_{in}$$ $$V_{out} \approx \left(\frac{1}{2}+\frac{\delta_R}{4}\right)V_{in}$$	Voltage offset; small nonlinearity
Single active Wheatstone bridge	$$V_{out} = \left(\frac{1}{2}-\frac{R(1+\delta_R)}{R+R(1+\delta_R)}\right)V_{in}$$ $$V_{out} \approx \frac{\delta_R}{4}V_{in}$$	Small nonlinearity
Half-active Wheatstone bridge	$$V_{out} = \left(\frac{\delta_R R}{R+R(1+\delta_R)}\right)V_{in}$$ $$V_{out} \approx \frac{\delta_R}{2}V_{in}$$	Small nonlinearity; better sensitivity; resistors must be well matched
Fully active Wheatstone bridge	$$V_{out} = \left(\frac{R(1+\delta_R)-R(1-\delta_R)}{2R}\right)V_{in}$$ $$V_{out} = \delta_R V_{in}$$	No offset; linear output; highest sensitivity; positive and negative resistance changes must be well matched

configurations have a small nonlinearity in their response. The output voltage gain for the single active-bridge is the same as the voltage divider, but the single active-bridge does not have an offset voltage. The half-active bridge circuit has an improved output voltage gain, but the two variable resistors must be matched. MEMS fabrication processes will tend to minimize the variations in the resistors.

The fully active bridge utilizes four variable resistors. Two resistors increase resistance with increasing strain, and the other two decrease resistance with increasing strain. By careful design in placement and orientation of the resistors, these criteria can be met with a single-crystal silicon piezoelectric approach. The fully active bridge has the largest output voltage gain and has a linear response.

8.2.3 ELECTRON TUNNELING

An extensive literature base exists on *tunneling tip* methods of transduction. The method was initially used in the scanning tunneling microscope [38] (STM), which has been used in material science research such as the study of atomic

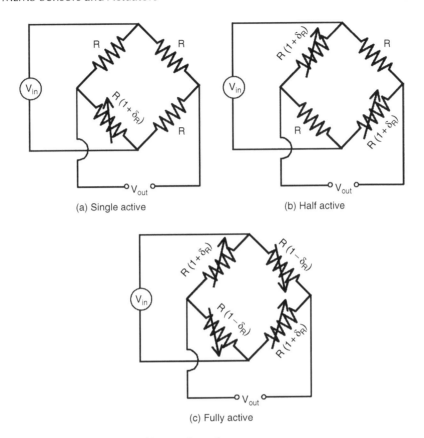

FIGURE 8.33 Wheatstone bridge configurations.

scale surface structure. This method of transduction has also been used for infrared (IR) [39–41], magnetometer [42], and accelerometer [39,43] sensors. Electron tunneling can provide an extremely sensitive method of position transduction.

Electron tunneling is a phenomenon in which a current is passed across a narrow gap (Figure 8.34). Classically, a gap of finite size would pose a barrier to current flow. However, for sufficiently small gaps (~10Å), the probabilistic nature of quantum mechanics becomes apparent. In quantum mechanics, when a particle comes to a barrier that it does not have enough energy to penetrate, the wave function dies off exponentially. However, if the gap is small enough, the wave function will predict a significant probability of finding the particle on the other side of the gap. Therefore, if the gap is small enough, a tunneling current will exist even though a break has occurred in the circuit. The electron tunneling current is described by:

$$I \propto V \exp\left(-\alpha\sqrt{\Phi s}\right) \quad (8.40)$$

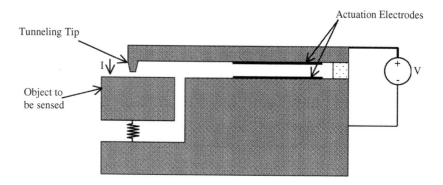

FIGURE 8.34 Electron tunneling transduction schematic.

where

I = tunneling current
s = gap
Φ = height of the tunneling barrier
V = bias voltage ($V \ll \Phi$)
α = conversion factor (1.025 Å^{-1} $ev^{-1/2}$)

Typical values of Φ and s are 1 eV and 10 Å. A typical tip bias voltage, V, is only a few 100 mV and a typical tunneling current, I, is 1 nA. Electron tunneling is an extremely sensitive method of position transduction in which the current can vary by a factor of three for each Å change in gap separation, s. Tass et al. [6] reported that a change of electrode displacement of 0.003 nm resulted in a 1% change in tunneling current for a tunneling gap of 1 nm with gold electrodes in air.

Figure 8.34 is a schematic of an electron tunneling tip transducer. The tunneling tip and opposing surface must be metalized with a thin layer of metal such as 100 Å of gold, which is adequate for this purpose. The tunnel effect is not extremely sensitive to tip geometry. One reported tunneling tip was a 50-μm pyramid with a 1 to 5-μm radius of curvature; even a 5-μm mesa will suffice. This makes the fabrication of the tunneling tip more tractable.

Because the tunneling tip is so close to the surface of a moving mass or membrane to be measured, the gap must be controlled by feedback during operation. This can be accomplished by measuring the tunneling current and applying correction signals an actuator to control the position of the tunneling tip or the moving mass. Because the tunneling tip is small, it can be controlled with minimal effort.

8.2.4 SENSOR NOISE

Noise represents a fundamental limit to the performance of a sensor or a control system. Therefore, noise is a significant aspect of the design of MEMS sensors.

MEMS Sensors and Actuators

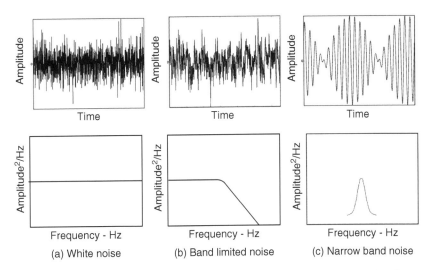

FIGURE 8.35 Time history and PSD of white, band-limited, and narrow-band noise.

Noise in a sensor can arise from many sources, ranging from very small temperature-induced vibrations of the sensor atoms to electronic noise produced by the discrete nature of electrons.

Noise is a random signal that requires some preliminary definitions and metrics to facilitate further discussion. A noise signal can be described in the time domain via plots of amplitude vs. time or in the frequency domain via plots of amplitude or phase vs. frequency (Figure 8.35). For this discussion of noise, frequency will be expressed in units of hertz (Hz). An important function for noise analysis is the *power spectral density function (PSD)*, $S(f)$, which is a function of frequency and has units of amplitude squared per hertz. For example, the PSD function can be describing a voltage signal that would be a plot of V^2/Hz vs. hertz.

Alternatively, for an acceleration signal, the plot would be acceleration squared per hertz (i.e., g^2/Hz). *The PSD provides an indication of how the "power" of the signal is distributed over frequency.* The term "power" is used because the function can have units of voltage squared or displacement squared (indicative of power) per hertz. Figure 8.35 shows the relationship between a signal represented in time domain vs. the frequency domain via a PSD. The detailed mathematical signal-processing techniques [44] required to make this transition between the time domain and frequency domain are beyond the scope of this book.

The difference in the frequency content in the time domain signals of Figure 8.35 is apparent. A noise signal that has equal frequency content over all frequencies is known as *white noise*; otherwise, a signal that has limited frequency content is called *band limited*. As signals pass through different instruments such as amplifiers, filters, actuators, or sensors, the frequency content of the signal is altered.

The *mean square* of the signal, $\bar{v}^2$, can be obtained from the PSD by integrating over frequency (Equation 8.41). The mean square of a signal is a scalar, which is a useful metric indicative of the magnitude of the signal.

$$\bar{v}^2 = \int_0^\infty S(f) df \qquad (8.41)$$

In noise analysis, it is often necessary to evaluate the magnitude of a noise signal, which is a combination of several other signals. Noise signals are frequently uncorrelated; this allows the sum of the mean square amplitude of the uncorrelated signals to be added to obtain the mean square amplitude of the total:

$$\overline{v_{total}^2} = \sum_{i=1}^n \overline{v_i^2} \qquad (8.42)$$

In sensors, a desired sensor signal, v_s, is proportional to a quantity to be measured, and sensor noise, v_n, is an unwanted signal produced by the noise sources of the system. The *signal to noise ratio* (SNR) is a metric of the relative amount of signal and noise present in a sensor output. SNR is defined as the ratio of the mean square signal, v_s^2, to the mean square noise, v_n^2 (Equation 8.43). Due to the typical magnitude of SNR, the magnitude is often expressed in decibels (dB) as:

$$SNR = \frac{\overline{v_s^2}}{\overline{v_n^2}}$$

$$SNR = 10 \log \left(\frac{\overline{v_s^2}}{\overline{v_n^2}} \right) dB \qquad (8.43)$$

When signals and noises are applied to systems (Figure 8.36), it is frequently desired to calculate the response of the system. For linear systems, a frequency response function, FRF, which can be calculated from the basic principles or measured, defines the response of the system vs. frequency due to inputs.

The PSD of the system output can be calculated by the product of the magnitude of the FRF squared and the PSD of the input (Equation 8.44). Because these are functions of frequency, the product is performed at each frequency. The mean square system response is calculated by integrating Equation 8.44 over frequency similar to Equation 8.41.

$$S_o(f) = |H(f)|^2 S_i(f) \qquad (8.44)$$

MEMS Sensors and Actuators

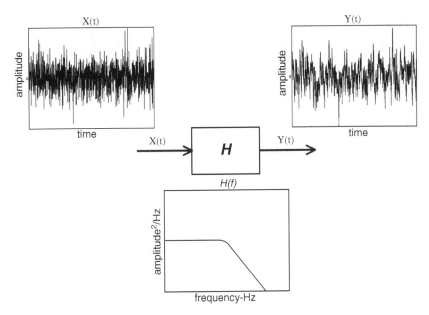

FIGURE 8.36 A signal, X, connected to a linear system, H, producing an output signal, Y.

$$\overline{v_o^2} = \int_0^\infty |H(f)|^2 S_i(f) df \qquad (8.45)$$

8.2.4.1 Noise Sources

Noise in a sensor can arise from a number of mechanical and electrical effects. Several of the most frequently encountered types of sensor noise are discussed next. Table 8.7 summarizes the types of noise, noise models, and mechanisms.

Shot noise is associated with direct current flow across potential barriers present in devices such as p–n diodes and bipolar transistors. The passage of each current carrier (electrons or holes) across the p–n junction in diodes and transistors is a random event dependent upon the carrier having sufficient energy. What appears as a continuous external current is actually a large number of discrete pulses. The time constant associated with the passage of carriers across the potential barrier is extremely small; therefore, the PSD of the shot noise current can be modeled as *white*, which is valid well beyond the range of practical electronic circuits. Equation 8.46 is the amplitude of the shot noise current PSD, where q is an electron charge and I_{DC} is the direct current flow.

$$S_i^{shot}(f) = 2qI_{DC} \qquad (8.46)$$

Flicker noise is a low-frequency noise component arising from the capture and release of charge carriers by *trap* sites produced by crystal defects or con-

TABLE 8.7
Summary of Noise Models and Mechanisms

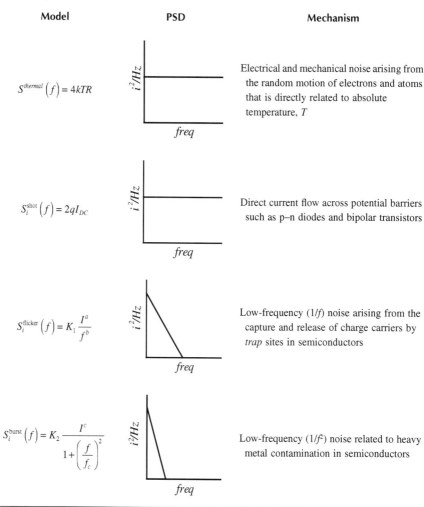

Model	PSD	Mechanism
$S^{thermal}(f) = 4kTR$	(flat)	Electrical and mechanical noise arising from the random motion of electrons and atoms that is directly related to absolute temperature, T
$S_i^{shot}(f) = 2qI_{DC}$	(flat)	Direct current flow across potential barriers such as p–n diodes and bipolar transistors
$S_i^{flicker}(f) = K_1 \dfrac{I^a}{f^b}$	(1/f)	Low-frequency (1/f) noise arising from the capture and release of charge carriers by *trap* sites in semiconductors
$S_i^{burst}(f) = K_2 \dfrac{I^c}{1+\left(\dfrac{f}{f_c}\right)^2}$	(1/f²)	Low-frequency (1/f^2) noise related to heavy metal contamination in semiconductors

tamination in semiconductors. The trap sites capture and release carriers randomly with time constants that are primarily low frequency. The flicker noise current PSD is modeled by Equation 8.47, which shows a 1/f dependency in amplitude vs. frequency; this is why this type of noise is frequently called *1/f noise*. The constants a, b, and K_1 in Equation 8.47 are primarily fabrication process and device dependent. Once these parameters are determined for a particular process and type device, they may be used for prediction of noise in other similar devices made in that process.

MEMS Sensors and Actuators

313

$$S_i^{flicker}(f) = K_1 \frac{I^a}{f^b} \tag{8.47}$$

where

I = direct current
K_1 = constant for a particular device and process
a = constant in the range 0.5 to 2
b = constant typically about unity

Another type of low-frequency noise, termed *burst or popcorn noise*, is sometimes found in integrated circuits and discrete transistors. The mechanism is not fully understood, but is believed to be related to heavy metal contamination. Equation 8.48 is a model of burst noise that contains constants that need to be determined for the specific application. Burst noise can also occur with multiple frequency components (f_c), which can be seen in an experimentally obtained PSD of the noise.

$$S_i^{burst}(f) = K_2 \frac{I^c}{1 + \left(\dfrac{f}{f_c}\right)^2} \tag{8.48}$$

where

I = direct current
K_2 = constant for a particular device and process
c = constant in the range 0.5 to 2
f_c = a particular frequency for the noise process

Thermal noise is noise that can be generated in electrical and mechanical components; it arises from a completely unique mechanism — the random motion of electrons and atoms, which is directly proportional to the absolute temperature, T. Thermal noise is generated in these systems through their energy dissipation mechanisms (i.e., resistance for electrical or damping for mechanical). Because dissipation mechanisms provide a way for energy to leave an electrical or mechanical system, they also provide a way for energy to enter. The energy entering the system is the thermal vibration of electrons and molecules. The association linking the paths for energy dissipation and entrance is expressed by the fluctuation–dissipation theorem [47,48]. Thermal noise is also known as *Johnson noise* in electrical systems and *Brownian noise* in mechanical systems. The PSD for thermal noise is white. This is the most fundamental noise limitation for a sensor.

$$S^{thermal}\left(f\right) = 4kTR \tag{8.49}$$

where

I = direct current
k = Boltzmann's constant (1.38×10^{-23} J/K)
T = absolute temperature (K)
R = dissipation constant (electrical resistance or mechanical damping coefficient)

8.2.5 MEMS Physical Sensors

Physical sensors are one of the biggest application areas for MEMS technology. In this section, an overview of three of the most widely used types of physical sensors and their implementation in MEMS technology will be presented.

Accelerometers and gyroscopes are inertial sensors that measure acceleration and rotation rate, respectively. The performance grades of these types of sensors are found in Table 8.8 [50], which presents the bias stability and cost range for the four performance grades. The performance ranges from strategic grade inertial instruments used on strategic missiles and submarines to instrument grade instruments used in automotive and commercial applications. Currently, MEMS inertial sensors are available in the tactical and instrument grade inertial sensors.

One of the first commercial application areas of MEMS technology was pressure sensing. Today, pressure sensing is one the largest commercial applications of MEMS technology. The implementation of pressure sensors with MEMS technology for different pressure/vacuum regimes will be presented.

8.2.5.1 Accelerometer

Accelerometers are one of the most frequently utilized physical sensors for detecting and measuring motion. Accelerometers have found applications ranging from measurement and control to inertial navigation. MEMS implementations of accelerometers have found a large commercial market in automotive airbag deployment

TABLE 8.8
Inertial Instrument Performance Grades

Performance grade	Accelerometer bias stability	Gyroscope bias stability	Cost
Strategic	<1 µg	<0.0001°/h	<$10,000,000/unit
Navigation	10–50 µg	0.001–0.01°/h	<$100,000/unit
Tactical	0.1–1 mg	1–10°/h	<$30,000/unit
Instrumentation	10–100 mg	30–100°/h	$250–$2000/axis

Source: M.R. Daily, Defense Manufacturing Conference, *DMC '99*, Nov. 1999.

MEMS Sensors and Actuators

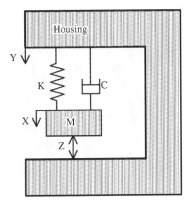

FIGURE 8.37 Accelerometer components.

systems [52]. The basic configuration of an accelerometer is the same for all of these applications. Three items are the basic components of an accelerometer:

- Inertial mass
- Suspension
- Sensing element

The suspension supporting the inertial mass will deflect under acceleration due to D'Alembert's principle [53]. A sensing element will transduce the deflection of the suspension to an electrical signal. The transduction can be accomplished by a number of means, but the most common utilize piezoresistive, piezoelectric, or capacitance means. Piezoresistive and capacitance sensing are discussed in this chapter.

Figure 8.37 schematically depicts an accelerometer. The motions of the housing and inertial mass are denoted by the coordinates Y and X, respectively, which are absolute displacements. The coordinate Z is the relative displacement of the inertial mass relative to the housing, which is related by Equation 8.50. Z is the variable that could be transduced to an electrical signal. Equation 8.51 is a force balance on the inertial mass including spring, K, and damper, C, forces.

Using Equation 8.50 and rearranging Equation 8.51 results in Equation 8.52. This shows that the acceleration input to the housing, $\ddot{Y}$, is related to the dynamics of the relative displacement of the housing and inertial mass, Z. The relative displacement, Z, is a quantity that can be measured within the case through means such as the displacement of a beam, deflection of a piezoelectric crystal, or capacitance change of the mass relative to the housing.

$$Z = X - Y \tag{8.50}$$

$$M\ddot{X} = -C\left(\dot{X} - \dot{Y}\right) - K\left(X - Y\right) \tag{8.51}$$

$$M\ddot{Z} + C\dot{Z} + KZ = M\ddot{Y} \tag{8.52}$$

Equation 8.52 is a second-order differential equation, which was discussed in Section 6.2.1. This type of equation will have a resonance at the system's natural frequency, ω_n. Resonance is a condition in which the spring forces balance the inertia forces and the damping force controls the amplitude. With no damping, the amplitude of the system would theoretically become infinite. Figure 6.24 shows the frequency response function of a second-order system with the maximum response occurring at resonance. Figure 6.23 shows the time response of the second-order system for various amounts of damping. The response varies from oscillatory for $\zeta < 1$ to nonoscillatory for $\zeta \geq 1$.

The design of an accelerometer depends upon the system damping. The amount of damping will determine the dynamic response and the Brownian noise and thus the noise floor of the sensor. Macroscale accelerometers are designed to have a damping ratio of $\zeta \cong 0.7$.

This type of system will have a fast response with very small overshoot. The accelerometer needs to sense signals that contain a combination of many frequencies. Two important metrics for accelerometer design are *amplitude distortion* and *phase distortion*. To prevent amplitude distortion, the accelerometer transfer function must amplify the signals of different frequencies equally. This means that the magnitude of the frequency response function must be flat in the operating range, which occurs in the low-frequency range of the frequency response function (Figure 8.38). For no phase distortion, the phase of the harmonic components of the signal must increase linearly with frequency (Equation 8.53). This will shift the harmonic components in time equally [54]. A damping ratio of $\zeta \cong 0.7$ almost perfectly eliminates phase distortion, and restricting the operating range to approximately $0 < \omega/\omega n \leq 0.2$ also minimizes amplitude distortion. For most MEMS accelerometer designs, the natural frequency of the inertial mass-suspension system is at least an order of magnitude higher than the highest frequency signal to be sensed.

$$\varphi = \frac{\pi}{2} \frac{\omega}{\omega_n} \tag{8.53}$$

The mechanical sensitivity of an accelerometer, S_M, is the relationship between the relative deflection of the inertial mass and case, Z, and the input acceleration, $\ddot{Y}$. Because the operating range of the accelerometer is at low frequency, where $\omega \approx 0$, an acceleration input to the system may be approximated by a constant acceleration balanced by the suspension (Equation 8.54). Thus, the mechanical sensitivity, S_M, of an accelerometer is shown in Equation 8.55.

$$KZ = M\ddot{Y} \tag{8.54}$$

MEMS Sensors and Actuators

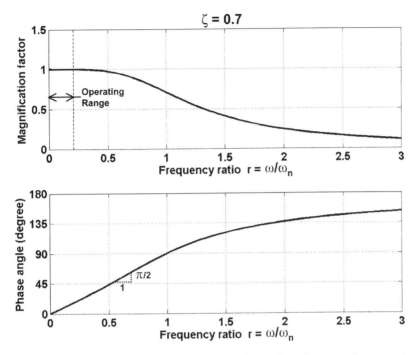

FIGURE 8.38 Frequency response magnitude and phase of a spring-mass-damper system with $\zeta = 0.7$.

$$S_M = \frac{M}{K} = \frac{1}{\omega_n^2} \tag{8.55}$$

The accelerometer described thus far is an *open-loop sensor* consisting of the inertial mass-suspension system, position sensing, and amplification signal-conditioning elements. Figure 8.39a is a block diagram of an open-loop accelerometer for which, the greater the acceleration $\ddot{Y}$ input is, the greater the relative displacement, Z, which will be transduced into an electrical signal. Linearity is an important sensor quality because of calibration and signal-conditioning issues. However, some of the transduction means such as capacitance sensing are non-linear with increasing displacement. This can be mitigated by limiting the acceleration input range of the sensor, but also has the adverse effect of limiting the sensor dynamic range. Generally, an open-loop accelerometer is satisfactory for applications in which the dynamic range is less than 5000:1 and the scale factor error can be 0.1% or greater. This is generally the case for instrument-grade accelerometers; however, for accelerometers used in inertial navigation, this is not sufficient.

An alternative approach is to maintain the inertial mass in the undeflected or zero position during acceleration. This will require a control loop with a force

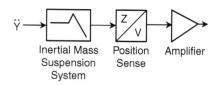

(a) Open-Loop Accelerometer Schematic

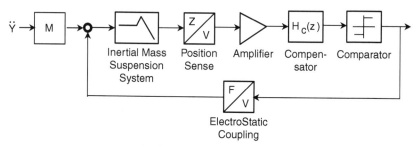

(b) Closed-Loop Accelerometer Schematic

FIGURE 8.39 Open-loop and closed-loop accelerometer system block diagram.

actuator to maintain the inertial mass position. Figure 8.39b shows a block diagram for a *closed-loop* accelerometer. A closed-loop sensor involves some additional items compared to an open-loop sensor. The former will need an actuator to apply force to the inertial mass to maintain its position, as well as a control compensator to maintain the closed-loop stability of the system. The additional electronics involved in a closed-loop sensor may be implemented digitally or by analog. Analog control electronics will continuously vary the feedback signal to maintain the inertial mass position; however, this will require the scale factors of the position sensing, amplifiers, and force feedback system to be known precisely over the entire operating range, which is challenging.

A digital implementation of a closed-loop accelerometer utilizing a method such as a $\Sigma\Delta$ modulator [55] can alleviate many of the concerns of a purely analog implementation. For example, the nonlinear relationship between voltage and force in electrostatic actuators is eliminated by quantizing the feedback signal to 1 bit and encoding only the sign of the proof mass displacement from the undeflected (nominal) position. This is the function of the comparator in Figure 8.39b. Linearity is assured because only two displacement/force levels are sensed or generated. Due to the time delay involved in position sensing, comparator, and force actuation, the closed-loop system must contain a compensator, H_c, to maintain stability [56]. This digital control loop will operate at a frequency much higher (~ two to three orders of magnitude) than the inertial mass-suspension natural frequency or the frequency content of the signal to be measured. An example of an implementation of this approach is presented in Lemkin and Boser [57].

A number of trade-offs are associated with open-loop vs. closed-loop or analog vs. digital implementation of an accelerometer. The sensitivity and bandwidth of the open-loop accelerometer are related to the natural frequency,

MEMS Sensors and Actuators 319

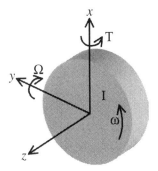

FIGURE 8.40 Precession of a rotating body.

ω_n, of the suspension. The closed-loop accelerometer implementation can make the sensitivity–bandwidth trade-off somewhat independently. Digital implementations can also reduce the impact of nonlinearities (e.g., electrostatics) on the sensor. However, a closed-loop accelerometer implementation will be limited by the capacity of the force actuation to balance the inertial mass for high-input accelerations.

8.2.5.2 Gyroscope

A gyroscope is an inertial instrument capable of sensing rotation that can be implemented in a number of ways. Invented by Leon Foucault in 1852, the first gyroscope was based on the angular momentum of a spinning wheel. The angular momentum, H, is the product of the mass moment of inertia, I, and the angular velocity, ω, of the wheel (Equation 8.56). Due to Newton's laws of motion, the angular momentum of a body will remain unchanged unless acted on by a torque, T (Equation 8.57). If a torque is applied in the same axis as the angular velocity, the effect is to accelerate or decelerate the rotating body, which is denoted by the first term of Equation 8.57.

However, if the torque is applied orthogonal to the spin axis, the rotating body will *precess*, Ω, denoted by the second term of Equation 8.57. These effects are illustrated in Figure 8.40. The cross-product in the second term generates the interesting gyroscopic effects (i.e., Ω, H, and T are related by the right-hand rule). Precession or the moments generated by precession are the effect utilized by this form of gyroscope as a measure of angular rate.

The spinning wheel gyroscope is used to implement a class of high-performance gyroscopes for inertial navigation as well as other lower performance applications. Because the fabrication of this type of gyroscope requires precision bearings, machining, drive motors, and electronics, it is very costly. However, in the 1950s inertial navigation for missiles, aircraft, and submarines came to rely on this type of gyroscope.

$$H = I\omega \qquad (8.56)$$

$$T = \frac{dH}{dt} + \Omega \times H \qquad (8.57)$$

In the 1980s and 1990s when MEMS technology was reaching the stage of maturity sufficient for application to gyroscopic sensing, several avenues were pursued. The development of a MEMS spinning mass gyroscope was initially inhibited due to the lack of low-friction bearings and the significant stiction and adhesion forces at the microscale. However, promising research on the development of an electrostatic levitated spinning mass MEMS gyroscope is proceeding [58,59]. This is an ambitious approach because of the necessity of closed-loop control to stabilize the levitation, in addition to driving the spinning mass and sensing its deflections due to precession. This approach is currently in the research stages and no MEMS spinning mass gyroscope commercial products are available.

Optical rotation rate sensors based upon the *Sagnac effect* have also been developed [60]. Georges Sagnac discovered this effect in 1913 while performing a modification of the Michelson–Morley experiment [61]. An optical gyroscope utilizing the Sagnac effect can be implemented with two counter-rotating light beams circulating around an optical path of radius, R, where the optical path is rotating with angular velocity, Ω (Figure 8.41). The Sagnac effect can be observed by the time difference, Δt, between the clockwise and counterclockwise beams striking a detector that is in and rotating with the optical path. If the optical path is not rotating, the optical signal traveling in either direction will complete the path at the same time. However, if the optical path is rotating clockwise, as shown in Figure 8.41b, the optical signal traveling in the same direction as the rotation will have a slightly longer distance to travel than the optical signal traveling in the opposite direction.

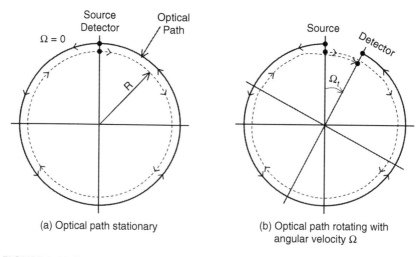

FIGURE 8.41 Sagnac effect on two counter-rotating beams of light.

MEMS Sensors and Actuators

The tangential speed of the rotating optical path is $v = \Omega R$. The initial separation of the start and end points of the optical signals is $2\pi R$ or, if one allows the signal to circulate N times around the path, $N2\pi R$. The time difference in the arrival of the signals due to the Sagnac effect can be calculated as shown in Equation 8.58, where c is the speed of light in the optical path medium. The Sagnac effect time interval is very small. For example, the measurement the Earth's rotation rate (i.e., 15°/h) with a 1-km long optical path will produce a Sagnac effect of only $\Delta t = 3.3 \times 10^{-9}$ sec. A short time interval such as this can be resolved by phase shift effects of the optical signals.

$$\Delta t = N2\pi R \left(\frac{1}{c-v} - \frac{1}{c+v} \right) \approx \frac{N4\pi R^2 \Omega}{c^2} \qquad (8.58)$$

The basic configuration schematically described in Figure 8.41 can be implemented with fiber optics with multiple turns (N) to increase path length and the Δt or phase shift measured as an indication of rotation rate, Ω. The Sagnac effect is the basis of a number of optical rotation rate sensors such as the interferometric fiber-optic gyro (IFOG).

In 1982, a micro-optical-gyro (MOG) concept utilizing MEMS and microelectronic fabrication techniques was patented [60] and initial development pursued by Northrup [60]. MOGs utilize wave guides etched into the substrate by MEMS etching techniques. This initial effort by Northrup to produce an MOG was discontinued. However, other organizations [63] are still pursuing this concept, but this approach does not currently have an MOG commercial product available.

Another approach for rotation rate sensing lies in the dynamics of vibrating mechanical systems. The fact that vibrating objects are sensitive to rotation has been known since 1890. The initial concept for an implementable *vibratory gyroscope* was based on the vibration of a metal tuning fork [64,65]. By the 1960s, engineers were seeking alternatives to the spinning mass gyroscope due to its size, fragility, and expense. Subsequent technology developments enabled the realization of a functioning vibratory gyroscope [66–68]. The vibratory gyroscope was also later discovered to be the mechanism utilized by biological systems such as a fly's ability to sense angular rotation [69].

Vibratory gyroscopes are based on *Coriolis acceleration*, which is an acceleration produced due to the changing direction in space of the velocity of the body relative to the moving system. For example, Figure 8.42 shows the Coriolis acceleration, $A_{coriolis}$, produced on a body moving around an axis with a fixed angular velocity, Ω, and moving radially with a velocity V as well. The Coriolis acceleration is defined by Equation 8.59. The detection of the deflection of an object due to Coriolis acceleration is the basis for a vibratory gyroscope.

$$A_{coriolis} = 2\Omega \times V \qquad (8.59)$$

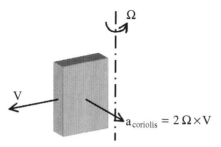

FIGURE 8.42 Coriolis acceleration on a moving body in a rotating system.

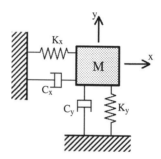

FIGURE 8.43 Single mass gyroscope schematic.

A vibratory gyroscope is composed of a resonator that will oscillate a body along one axis and measure the orthogonal movement or force on the body due to Coriolis acceleration. Figure 8.43 is a schematic of a plate being driven along the x axis, the rotation rate to be measured; Ω is along the z axis and the Coriolis acceleration response is sensed along the y axis. Equation 8.60 and Equation 8.61 are the equations of motion (force balance) for the body in the drive (x) and sense (y) axes, respectively. These are a system of coupled second-order equations coupled via Coriolis acceleration terms. The physical mechanism for a vibratory gyroscope is the transfer of energy from one resonator axis to another via the Coriolis acceleration coupling. The suspension for this device can have a unique natural frequency, ω_x, ω_y and a unique damping ratio for ζ_x, ζ_y for each axis (Equation 8.62).

$$\ddot{x} + 2\zeta_x \omega_x \dot{x} + \omega_x^2 x = \frac{1}{M} F_x - 2\Omega \dot{y} \tag{8.60}$$

$$\ddot{y} + 2\zeta_y \omega_y \dot{y} + \omega_y^2 = 2\Omega \dot{x} \tag{8.61}$$

$$\omega_y = \sqrt{\frac{K_y}{M}} \qquad \omega_x = \sqrt{\frac{K_x}{M}} \tag{8.62}$$

MEMS Sensors and Actuators

The relative positioning of the suspension natural frequencies is a gyroscope design decision. Frequently, the sense direction natural frequency, ω_y, is approximately 10% less than the drive direction natural frequency, ω_x. This will provide a modest mechanical gain without significant bandwidth or phase shift reductions. The damping ratio of the mass in the x and y axes depends on the orientation of the mass relative to the substrate, which will determine the damping mechanism involved (e.g., squeeze film vs. lateral shear damping).

The implementation of the gyroscope will require the mass to be driven in the x axis by the force F_x. For many MEMS designs, F_x is electrostatic, such as an interdigitated electrostatic comb drive. The drive amplitude, x, must be maintained very accurately because any variation will directly contribute an error into the sense direction amplitude (Equation 8.60) and the gyroscope output. For this reason, the drive axis amplitude is controlled by an automatic gain control feedback loop.

Because the oscillatory drive portion of the gyroscope (Equation 8.60) is fixed to a high degree of accuracy by the gain control loop, Equation 8.61 governs the dynamics of the gyroscope response. Because the x axis (drive axis) is an oscillator, the response of the y axis (sense axis) will also be oscillatory (Equation 8.63). The Coriolis term that is the input to Equation 8.61 is twice the product of the angular rate and the velocity of the x axis oscillator, which produces a *modulated* signal. Therefore, the gyroscope output will need to be *demodulated* to extract the rotation rate signal. Section 8.2.1 contains a discussion of modulation and demodulation of signals.

The velocity, $\dot{x}$, of the drive signal that is the input to the Coriolis term of Equation 8.61 is simple harmonic motion, which will be zero at the extremes of motion of the driven mass and a maximum as the mass passes through the undeflected position. The mass x displacement and the Coriolis force, which contains an x velocity term, have a 90° phase difference; therefore, the y displacement due to the Coriolis force will also have a 90° phase difference. These signals are said to be in *quadrature*. This will lead to an oval deflection path (symmetric about the x axis) of the mass shown in Figure 8.44a when the gyroscope is subject to a constant rotation rate. With a zero rotation rate, the mass deflection pattern will not deflect in the y direction and oscillate entirely along the x axis as shown in Figure 8.44b.

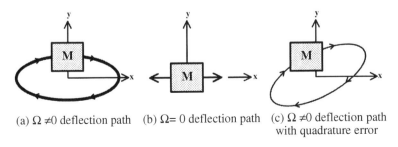

(a) $\Omega \neq 0$ deflection path   (b) $\Omega = 0$ deflection path   (c) $\Omega \neq 0$ deflection path with quadrature error

FIGURE 8.44 Gyroscope mass deflection response due to Coriolis acceleration.

324   Micro Electro Mechanical System Design

However, if mass or stiffness imbalances exist in the system dynamics as indicated in Equation 8.64, the mass deflection pattern will be as shown in Figure 8.44c. These subtle imbalances in the vibration of the sense mass produce a deflection in the y direction known as *quadrature error*, which contaminates the Coriolis signal, which is the measure of rotation rate. The effects of quadrature error can be negated by a quadrature error cancellation [70,71] scheme involving the use of electrostatic actuators with properly phased signals to cancel the imbalance, or by synchronous detection methods, which take advantage of the quadrature relationship to extract the Coriolis signal.

$$x = Xe^{j\omega t} \qquad y = Ye^{j(\omega t + \varphi)} \tag{8.63}$$

$$\ddot{y} + \delta_M \ddot{x} + 2\zeta_y \omega_y \dot{y} + \omega_y^2 + \delta_K x = 2\Omega \dot{x} \tag{8.64}$$

The first silicon integrated micromachined vibratory gyroscope was described by O'Connor and Shupe in 1981 [72]. In the ensuing years, development of a MEMS gyroscope was spurred by the lure of a low-cost instrument that could be mass produced. Efforts to produce single resonator gyroscopes schematically are shown in Figure 8.43. However, the one configuration that has been employed for macroscale and MEMS vibratory gyroscopes is the *tuning fork gyro* (TFG) (Figure 8.45). The TFG consists of two plates that are driven in an antiphase manner (i.e., both plates move outward and inward relative to the center axis). The rotational field will cause the plates to move perpendicular to the substrate in opposite directions. This configuration enables differential sensing, which will allow common mode signals such as external accelerations to be rejected. The use of two masses vibrating in antiphase also causes momenta to cancel locally and make the gyroscope less sensitive to mounting. The two masses may have coupled or separate suspensions.

Two MEMS TFGs have been successfully developed for commercial applications by Draper Laboratories [73,74] and Analog Device [75]. These are examples of tactical and instrument grade *MEMS* gyroscopes. The applications for a gyroscope such as these include tactical grade navigation, platform stabilization, automobile skid control, and stabilization.

8.2.5.3 Pressure Sensors

The method of pressure sensor implementation differs depending on whether the pressure to be sensed is greater than atmospheric or less than atmospheric (vacuum). Currently, the greatest commercial application for a MEMS pressure sensor is for the greater than atmospheric regime, which is the focus here. However, methods of implementing a vacuum sensor will first be briefly discussed.

The method of implementation of a vacuum sensor depends on the operating range. In the 1 to 2000 mtorr range, vacuum may be sensed by measuring the

MEMS Sensors and Actuators

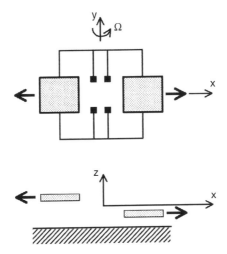

FIGURE 8.45 Tuning fork gyro (TFG) schematic.

thermal conductivity of the ambient gas. However, at vacuums of 1 mtorr to 10^{-8} torr, ionization of the gas may be used as a measure of vacuum.

A *Pirani gauge*, which measures vacuum, can be implemented by measuring the thermal conductivity of the ambient gas. Thermal conductivity absolute pressure gauges are widely used in vacuum systems [77]. Thermal conductivity proportional to pressure can be implemented by a heated filament suspended in a gas (Figure 8.46). The resistance of the suspended filament is a function of temperature, which is related to the surrounding pressure and thermal conductivity. The mechanism for this relationship has been studied in detail [78]. The MEMS implementation of a Pirani gauge is presented in Mastrangelo and Muller [79].

At very low pressures (1 mtorr to 10^{-8} torr), gas ionization may be used as a measure of vacuum. An *Ionization gauge* emits electrons from a cathode; these are accelerated toward an anode plate. Positive ions are created by the elec-

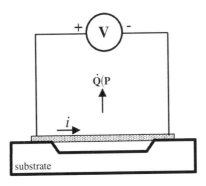

FIGURE 8.46 Pirani gauge microbridge implementation schematic.

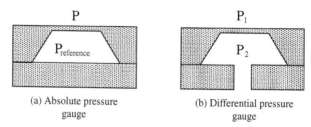

(a) Absolute pressure gauge   (b) Differential pressure gauge

FIGURE 8.47 Pressure sensor schematic for greater than atmospheric pressures.

tron–gas collisions. The current on the anode is proportional to the absolute pressure (vacuum) of the gas.

Most pressure sensors for greater than atmospheric pressure utilize a deformable diaphragm. The deflection of the diaphragm is the measure of pressure, which can be sensed by capacitative, piezoresistive, or optical means. Figure 8.47 is a schematic of typical pressure sensors. The pressure sensor may measure absolute pressure that has a vacuum or a reference pressure on one side of the diaphragm. Alternatively, a gauge pressure or differential pressure sensor would have one side of the diaphragm vented to atmosphere or to another pressure that would be a reference for the measurement.

The pressure diaphragm is generally rectangular or circular in shape. Pressure sensors fabricated with bulk micromachining methods are generally rectangular, due to the anisotropic etching techniques utilized. The fabrication sequence of the bulk micromachined Motorola MPX200 pressure sensor is shown in Figure 8.48. This sensor has a square, single-crystal diaphragm that is 1448 μm in length and 26.5 μm thick. A four-terminal X-ducer™ [81] shear strain gauge technology is used to read pressure.

The shape of surface micromachined pressure sensors is not restricted because they are photolithograhically defined. Figure 8.49 shows a surface micromachined pressure sensor that has a 2-μm thick diaphragm and is 200 μm in diameter. Circular diaphragms may have an advantage over square or rectangular ones due to the absence of stress concentrations at the corners.

The applied pressure for the diaphragm-based pressure sensors is determined by the deflection of the diaphragm. Section 6.1.6 discusses the relationships for small deformations of a flat plate in bending with no initial built-in stresses due to residual stress. Figure 6.15 and Figure 6.1.6 provide the maximum deflection and stress for a rectangular and circular diaphragm with fixed boundary conditions (i.e., no deflection or rotation) at the boundary. The pressure on the diaphragm is directly proportional to the applied pressure for rectangular and circular diaphragms. For the case of a diaphragm with large built-in stress or large deflections, the direct proportionality is no longer true. In general, it is desirable to use a diaphragm with a linear relationship with pressure because calibration and measurement are simpler. Deflection of the diaphragm may be sensed via capacitance or piezoresistive sensing (discussed in this chapter).

MEMS Sensors and Actuators

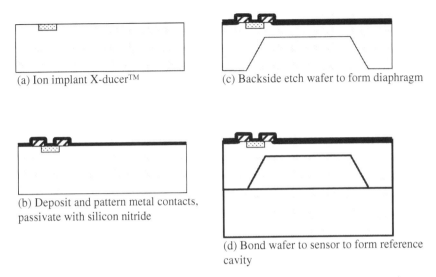

(a) Ion implant X-ducer™

(b) Deposit and pattern metal contacts, passivate with silicon nitride

(c) Backside etch wafer to form diaphragm

(d) Bond wafer to sensor to form reference cavity

FIGURE 8.48 A schematic of the fabrication sequence for the bulk micromachined Motorola MPX200 pressure sensor. (After W.P. Eaton et al., *Proc. SPIE*, 3514, 431–438, Sept. 1998.)

FIGURE 8.49 Surface micromachined pressure sensor. (Courtesy of Sandia National Laboratories; W.P. Eaton et al., *Proc. SPIE*, 3514, 431–438, Sept. 1998.)

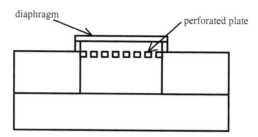

FIGURE 8.50 Microphone with a perforated backplate. (After W.P. Eaton and J.H. Smith, Micromachined pressure sensors: review and recent developments, *Smart Mater. Struct.*, 6, 530–539, 1997.)

Pressure sensors are generally used for low-frequency pressure measurement. Microphones are similar to pressure sensors, but microphones sense a dynamic pressure signal. The frequency response and mechanical sensitivity of ordinary pressure sensors are typically inadequate for use as a microphone due to the acoustic resistance and squeeze film damping between the diaphragm and stationary cavity of the sensor. The damping effects can be mitigated by perforating a stationary plate or venting so that the air can escape to a larger chamber [82] (Figure 8.50). These modifications will allow a flat frequency response over a broader range. Piezoresistive [84] and piezoelectric [85] microphones have been reported. A micromachined hydrophone with a frequency response of 2 kHz has also been reported [86,87]; it is based on a capacitance-sensed microphone filled with a compressible fluid.

8.2.6 CHEMICAL SENSORS

Types of chemical sensors vary widely. Figure 8.51 is a schematic of the basic elements of a chemical sensor. The chemical to be sensed, analyte, interacts with a chemically sensitive layer, which will produce an effect that can be transduced to an electrical signal. The chemically sensitive layer and the method of transduction to an electrical signal are application specific and must be tailored to the analyte of interest. The chemical reaction that occurs between the analyte and the chemically sensitive layer may be *reversible* or *irreversible*. Some reversible chemical reactions may be reversed simply by removing the chemical to be sensed, thus causing it to dissociate from the sensitive layer. Others may be reversed through the addition of heat, which will cause the analyte to detach with no net change in the chemically sensitive layer. An irreversible reaction will cause the sensitive layer to be consumed, which limits the sensor lifetime. Biological sensors are similar in approach to chemical sensors except that the sensitive coating may include biological materials such as an antigen. A detailed review of chemical and biological sensors may be obtained in Taylor et al. [88], Cass [89], and Ko [90].

The effect produced by the chemical reaction between the analyte and the chemically sensitive layer is varied and application specific; these include:

MEMS Sensors and Actuators

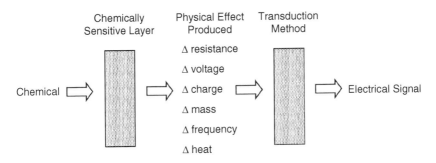

FIGURE 8.51 Elements of a chemical sensor.

- *Electrochemical.* These effects include changes in charge, electric potential, or current; they can be directly transduced to an electrical output. Examples of chemical sensors that measure the electrochemical effect are chemically sensitive resistors, FETs, and capacitors.
- *Heat.* Calorimetric chemical sensors are based upon the measurement of the heat produced by the analyte reacting with the chemically sensitive layer. The heat produced by the reaction is directly related to the analyte concentration. Calorimetric sensors for the detection of glucose, gases, etc. are documented in the literature.
- *Optical.* Optical chemical sensors utilize optical absorption, transmission, or luminescence resulting from the analyte interacting with the chemically sensitive layer. These chemical sensors can be highly sensitive but can be limited by the optical properties at the wavelength of interest. Optical sensors can be used to sense pH, oxygen, glucose, etc., depending on an appropriate choice of the chemically sensitive material.
- *Mass change.* The mass change resulting from the chemical reaction of the analyte and the chemically sensitive layer may be detected by highly sensitive acoustic wave devices [91]. These devices generate a high-frequency wave on the surface of a piezoelectric crystal, which is coated with the chemically sensitive material. As the wave passes through the material with adsorbed analyte molecules, their effect on the velocity of the acoustic wave is detected. The acoustic wave devices are split into two categories: bulk acoustic wave (BAW) and surface acoustic wave (SAW). SAW devices generally operate at frequencies above 50 MHz. A well-known member of the BAW device category is the quartz microbalance (QMB), which is generally 10 to 15 mm in size, operates between 10 and 30 MHz, and can detect mass changes as low as 10^{-9} to 10^{-10} g/cm². Chemical sensing can also be implemented with resonating beams [92,93].

A unique application of chemical sensing is the *electronic nose* [94]. This is a combination of multiple chemical sensor detection and pattern recognition to

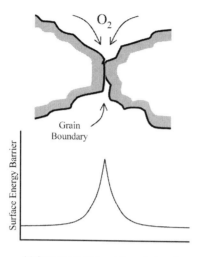 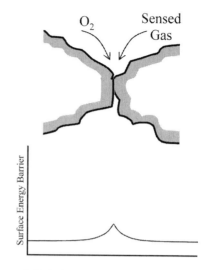

(a) Oxygen adsorption at the grain boundary produces a surface energy barrier

(b) Reduction reaction at the grain boundary with the gas to be sensed reduces the surface energy barrier

FIGURE 8.52 Operation of the Taguchi gas sensor.

achieve the recognition of complex odors. The electronic nose has found application in a spectrum of markets such as the food processing industry for the recognition of food freshness, beverages, and perfumes.

8.2.6.1 Taguchi Gas Sensor

The Taguchi gas sensor (TGS), a very successful commercially available [95] sensor for a wide variety of gases, was first developed by N. Taguchi in 1971. Applications include alcohol breath analyzers; automatic cooking controls; combustible gases (methane, propane, CO, hydrogen, etc.); volatile organic vapors (alcohol, ketone, esters, benzols, etc.); and others. This sensor is a solid-state sensor composed of a sintered metal oxide (SnO_2), which detects gases through an increase in electrical conductivity when reducing gases are adsorbed on the sensor's surface. The sensing material, tin dioxide (SnO_2), is a polycrystalline material consisting of crystals (grains) embedded in an amorphous matrix of the material. The sensors operate as follows (illustrated in Figure 8.52):

- The metal oxide is heated to 300 to 400°C in air.
- Oxygen is adsorbed on the crystal surface with a negative charge. Donor electrons in the crystal are transferred to the adsorbed oxygen, which results in positive charges in a space charge layer in the crystal. A surface potential that is a potential barrier to electron flow (i.e., resistance increases) is produced.

MEMS Sensors and Actuators

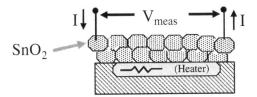

FIGURE 8.53 Schematic of a Taguchi gas sensor.

$$\frac{1}{2}O_2 + SnO_2 \rightarrow O^-ad(SnO_2) \qquad (8.65)$$

- An analyte molecule reacts with the adsorbed oxygen to release electrons; this decreases the resistance. Equation 8.66 shows this reaction for CO. This is a reversible reaction that is subsequently repeated.

$$CO + O^-ad(SnO_2) \rightarrow CO_2 + SnO_2 \qquad (8.66)$$

A schematic of the TGS is shown in Figure 8.53. The sensor incorporates a heater to heat the metal oxide to 300 to 400°C, which consumes a few hundred milliwatts. The temperature increases the chemical reaction rate and speeds the sensor response time. The metal oxide can be doped to improve the sensitivity and selectivity of the sensor to specific gases [97]. In addition to tin dioxide (SnO2), zinc oxide (ZnO) has been used for chemical sensing [98].

8.2.6.2 Combustible Gas Sensor

A number of combustible gas-sensing devices have been implemented with field effect transistors, capacitors, resistors, and diodes. These devices are based on the adsorption of hydrogen in catalytic metal films [99–102]. For example, a Pd–Ni alloy will adsorb hydrogen molecules on the surface, where the atoms are dissociated and free to diffuse into the bulk. The dissolved hydrogen atoms, which are typically located at octahedral interstices in the face-centered cubic crystal, act as additional impurity scattering sites. This results in an increase in the resistance [103,104] or a change in threshold voltage of a transistor whose gate is coated with the catalytic material.

Figure 8.54 shows an example of a hydrogen combustible gas detector that is implemented with this type of technology and developed at Sandia National Laboratories. This sensor used a differential measurement of resistance, which required a passivated filament and a filament coated with a catalytic material (Pt).

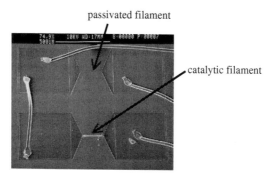

FIGURE 8.54 Hydrogen combustible gas detector. (Courtesy of Sandia National Laboratories.)

QUESTIONS

1. An actuator consisting of a 1-N/m spring suspension supports an electrostatic array that will have a stroke of 3 μm to perform an actuation task (e.g., close contacts). The actuation needs to occur at 30 V or less. This is a digital actuator with only two positions: open and closed. Design an actuator to achieve these specifications. What type of electrostatic actuator (parallel plate, interdigitated comb) was chosen and why? What are the details of your design (i.e., number of combs, electrode area. etc.)?

2. An actuator consisting of a 1-N/m spring suspension supports an electrostatic array that will have a maximum stroke of 10 μm to perform an actuation task. This is an analog actuator that needs to be controlled at a number of intermediate position between 0 and 10 μm. Design an actuator to achieve these specifications. What type of electrostatic actuator (parallel plate, interdigitated comb) was chosen and why? What are the details of your design (i.e., number of combs, electrode area, etc.)?

3. An actuation method for a resonator that operates at 10 kHz with a 2-μm peak-to-peak stroke needs to be designed. Assume the total suspension spring constant is 1 N/m. What type of actuation would you choose and why?

4. Design a thermal actuator to achieve 7 μm of total travel. Assume that you are using a surface micromachining process with a 2-μm layer thickness. Assume that you will heat your material to 400°C for actuation.
   a. Explain the advantages and disadvantages of your design.
   b. How much force is produced at 0-μm stroke?
   c. How much force is produced at 7-μm stroke?

5. A plate 500 × 500 × 2 μm will move ±1 μm due to acceleration input. The natural frequency of the plate and suspension is 8 kHz. If 10 nm

MEMS Sensors and Actuators

of plate motion can be sensed, what is the corresponding acceleration input? What is the nominal capacitance of the suspended plate? What is the change in capacitance corresponding to 10-nm motion? Choose a capacitance sense amplifier configuration and explain your reasons.

6. How many piezoresistive coefficients describe a piezoresistor in single-crystal silicon? How many piezoresistive coefficients describe a piezoresistor in polycrystalline silicon? If the number of coefficients differs, explain.

REFERENCES

1. W.C. Tang, Electrostatic comb drive for resonant sensor and actuator applications, Ph.D. dissertation, University of California, Berkeley, November 1990.
2. W.C. Tang, M.G. Lim, R.T. Howe, Electrostatic comb drive levitation and control method, *JMEMS*, 1(4), 170–178, December 1992.
3. J.J. Sniegowski, J.H. Smith, An application of mechanical leverage to microactuation, *Int. Conf. Solid-State Sensors Actuators, TRANSDUCERS'95*, 2, 364–367, 1995.
4. S. Kota, J. Hetrick, L. Zhe, S. Rodgers, T. Krybowski, Synthesizing high-performance compliant stroke amplification systems for MEMS, *IEEE Int. Conf. Micro Electro Mechanical Syst., MEMS 2000*, 164–169, January 2000, Miyazaki, Japan.
5. L.L. Howell, *Compliant Mechanisms*, John Wiley & Sons, New York, 2001.
6. N. Tas, I. Wissink, I. Sander, T. Larnmerink, M. Elwenspoek, Modeling, design and testing of the electrostatic shuffle motor, *Sensors Actuators*, 70, 171, 1998.
7. T. Akiyama, K. Shono, Controlled stepwise motion in polysilicon microstructures, *JMEMS*, 2(3), 106–110, 1993.
8. J.I. Seeger, S.B. Crary, Stabilization of electrostatically actuated mechanical devices, *Transducers '97*, 1133–1136, 1997.
9. E.K. Chan, R.W. Dutton, Electrostatic micromechanical actuator with extended range of travel, *JMEMS*, 9(3), 321–328, September 2000.
10. J.M. Kyynarainen, A.S. Oja, H. Seppa, Increasing the dynamic range of a micromechanical moving-plate capacitor, *Analog IC Signal Proc.*, 29, 61–70, 2001.
11. J.I. Seeger, B.E. Boser, Dynamics and control of parallel-plate actuators beyond the electrostatic instability, *Transducers '99*, 474–477, 1999.
12. E.S. Hung, S.D. Senturia, Extending the travel range of analog-tuned electrostatic actuators, *JMEMS*, Vol. 8, No. 4, 497–505, December 1999
13. G. Greitmann, R. A. Buser, Tactile Microgripper for Automated Handling of Microparts, *Sensors and Actuators A*, 53, 410–415, 1996.
14. J. Buhler, J. Funk, O. Paul, F.P. Steiner, H. Baltes, Thermally actuated CMOS micromirrors, *Sensors Actuators A*, 46–47, 572–575, 1995.
15. H.P. Trah, H. Baumann, C. Doring, H. Goebel, T. Grauer, M. Mettner, Micromachined valve with hydraulically actuated membrane subsequent to a thermoelectrically controlled bimorph cantilever, *Sensors Actuators A*, 39, 169–176, 1993.
16. J.H. Comtois, V.M. Bright, Applications for surface-micromachined polysilicon thermal actuators, *Sensors Actuators A*, 58, 19–25, 1997.
17. L. Que, J. Park, Y. Gianchandani, Bent-beam electro-thermal actuators for high force applications, *Proc. IEEE MEMS '99*, 31–36, 1999.

18. Y.G. Gianchandani, K. Najafi, Bent-beam strain sensors, *JMEMS*, 5(1), 52–58, March 1996.
19. P.M. Zavracky, G.G. Adams, P.D. Aquilino, Strain analysis of silicon-on-insulator films produced by zone melting recrystallization, *JMEMS*, 4(1), 42–48, March 1995.
20. J.T. Butler, V.M. Bright, W.D. Cowan, Average power control and positioning of polysilicon thermal actuators, *Sensor Actuators*, 72, 88–97, 1999.
21. C.D. Lott, T.W. McLain, J.N. Harb, L.L. Howell, Modeling the thermal behavior of a surface-micromachned linear-displacement thermomechanical microactuator, *Sensor Actuators*, 101, 239–250, 2002.
22. D. Bosch, B. Heimhofer, F. Muck, H. Seidel, U. Thumser, W. Welser, A silicon microvalve with combined electromagnetic/electrostatic actuation, *Sensors Actuators A*, 37–38, 684–692, 1993.
23. D.K. Wickenden, R.B. Givens, R. Osiander, J.L. Champion, D.A. Oursler, T.J. Kistenmacher, MEMS-based resonating xylophone bar magnetometers, *Proc. SPIE*, 3514, 350–358, 1998.
24. M.A. Butler, S.J. Martin, J.J. Spates, and M.A. Mitchell, Magnetically excited flexural plate wave devices, *Transducers '97*, 1031–1034, 1997.
25. J.T. Kung, H-S Lee, R.T. Howe, A digital readout technique for capacitive sensor applications, *IEEE J. Solid-State Circuits*, 23(4), 972–977, August 1988.
26. M.A. Lemkin, Micro accelerometer design with digital feedback control, dissertation, University of California, Berkeley, 1997.
27. W. Thomson (Lord Kelvin), The electro-dynamic qualities of metals, *Phil. Trans. Royal Soc.*, 146, 733, London, 1856.
28. C.S. Smith, Piezoresistive effect in germanium and silicon, *Phys. Rev.*, 94, 42–49, April, 1954.
29. B. Kloeck, N.F. De Rooij, Mechanical sensors, in *Semiconductor Sensors*, S.M. Sze, Ed., chapter 4, John Wiley & Sons, New York, 1994.
30. S. Middelhoek, S.A. Audet, *Silicon Sensors*, Academic Press, New York, 1989.
31. Y. Kanda, A graphical representation of the piezoresistance coefficients in silicon, *IEEE Trans. Electron Device*, ED-29, 64–70, 1982.
32. P.J. French, A.G.R. Evans, Polycrystalline silicon strain sensors, *Sensors Actuator*, 8, 219–225, 1985.
33. P.J. French, A.G.R. Evans, Piezoresistance in polysilicon and its applications to strain gauges, *Solid-State Electron.*, 32(1), 1–10, 1989.
34. V.A. Gridchin, V.M. Lubimsky, M.P. Sarina, Piezoresistive properties of polysilicon films, *Sensors Actuators A*, 49, 67–72, 1995.
35. W.P. Eaton, Surface micromachined pressure sensors, Ph.D. dissertation, University of New Mexico, May, 1997.
36. W.P. Eaton, J.H. Smith, Planar surface-micromachined pressure sensor with a subsurface, embedded reference pressure cavity, *Proc. SPIE*, 2882, 259–265, 1996.
37. H.L. Stalford, C. Apblett, S.S. Mani, W.K. Schubert, M. Jenkins, Sensitivity of piezoresistive readout device for microfabricated acoustic spectrum analyzer, *Proc. SPIE*, 5344, 36–43, 2004.
38. G. Binnig, H. Rohrer, Scanning tunneling microscopy, *IBM J. Res. Develop.*, 30, 355, 1986
39. T.W. Kenny, W.J. Kaiser, J.K. Reynolds, J.A. Podosek, H.K. Rockstand, E.C. Vote, S.B. Waltman. Electron tunnel sensors, *J. Vac. Sci. Technol.* A 10(4), Jul/Aug, 1992.

MEMS Sensors and Actuators

40. T.W. Kenny, W.J. Kaiser, J.A. Podosek, H.K. Rockstand, J.K. Reynolds, E.C. Vote, Micromachined tunneling displacement transducers for physical sensors, *J. Vac. Sci. Technol.* A, 11(4), July/August, 1993.

41. T.W. Kenny, W.J. Kaiser, S.B. Waltman, J.K. Reynolds, Novel infrared detector based on a tunneling displacement transducer, *Appl. Phys. Lett.* 59(15), 7 October 1991.

42. D. DiLella, L.J. Whitman, J.J. Colton, T.W. Kenny, W.J. Kaiser, E.C. Vote, J.A. Podosek, L.M. Miller, A micromachined magnetic-field sensor based on an electron tunneling displacement transducer, *Sensors Actuators* 86, 8–20, 2000.

43. P.M. Zavracky, B. McClelland, K. Warner, J. Wang, F. Harley, B. Dolgin, Design and process considerations for a tunneling tip accelerometer, *J. Micromech. Microengineering*, 6(3), 352–358, September 1996.

44. J.S. Bendat, *Principles and Applications of Random Noise Theory*, Krieger Publishing Company, Malabar, FL, 1977.

45. A. van der Ziel, *Noise: Sources, Characterization, Measurement*, Prentice Hall, Englewood Cliffs, NJ, 1970.

46. P.R. Gray, R.G. Meyer, *Analysis and Design of Analog Integrated Circuits*, 3rd ed., New York, Wiley, 1984.

47. T.B. Gabrielson, Fundamental noise limits in miniature acoustic and vibration sensors, *NASC-91113–50*, 31 December 1991.

48. T.B. Gabrielson, Mechanical-thermal noise in micromachined acoustic and vibration sensors, *IEEE Trans. Elec. Dev.*, 40(5), 903–909, May 1993.

49. H.B. Callen, T.A. Walton, Irreversibility and generalized noise, *Phys. Rev.* 83(1), 34–40, 1951.

50. M.R. Daily, MEMS manufacturing challenges for inertial measurement unit applications, Defense Manufacturing Conference, DMC '99, Miami, FL, November 1999.

51. N. Yazdi, F. Ayazi, K. Najafi, Micromachined inertial sensors, *Proc. IEEE*, 86(8), 1640–1659, August 1998.

52. Analog Devices Inc., http://www.analog.com/.

53. D.T. Greenwood, *Principles of Dynamics*, Prentice Hall, Englewood Cliffs, NJ, 1965.

54. W.T. Thomson, *Theory of Vibration with Applications*, Prentice Hall, Inc., Englewood Cliffs, NJ, 1981.

55. R. van de Plassche, *Integrated Analog to Digital and Digital to Analog Converters*, Kluwer Academic Publishers, Dordrecht, 1994.

56. C. Lu, M. Lemkin, B.E. Boser, A monolithic surface micromachined accelerometer with digital output, *IEEE J. Solid State Circuits*, 30(12), Dec. 1995.

57. M. Lemkin, B.E. Boser, A micromachined fully differential lateral accelerometer, *IEEE Custom I.C. Conf.*, 315–318, 1996.

58. C. Shearwood, C.B. Williams, P.H. Mellor, R.B. Yates, M.R.J. Gibbs, A.D. Mattingley, Levitation of a micromachined rotor for application in a rotating gyroscope, *Electron. Lett.*, 31(21), 1845–1846, 1995.

59. C. Shearwood, K.Y. Ho, C.B. Williams, H. Gong, Development of a levitated micromotor for application as a gyroscope, *Sensors Actuators*, 83, 85–92, 2000.

60. A. Lawrence, *Modern Inertial Technology*, Springer-Verlag, Heidelberg, 1998.

61. G. Sagnac, Sur la preuve de la réalité de l'éther lumineux par l'expérience de l'interférographe tournant, *C.R. Acad. Sci.*, 95, 1410–1413, 1913.

62. A.W. Lawrence, Thin film laser gyro, U.S. Patent 4,326,803, issued 27 April 1982.

336   Micro Electro Mechanical System Design

63. C. Monovoukas, A.K. Swiecki, F. Maseeh, Integrated optical gyroscopes offering low cost, small size and vibration immunity, *Proc. SPIE*, 3936, 293–300, 2000.
64. J. Lyman, E. Norden, Rate and attitude indicating instrument, U.S. Patent 2,309,853, issued April 10, 1941.
65. J. Lyman, Angular velocity responsive apparatus, U.S. Patent, 2,513, 340, issued October 17, 1945.
66. W.H. Quick, Theory of the vibrating string as an angular motion sensor, *Trans. ASME, J. Appl. Mech.*, 31, 523–534, September 1964.
67. D. Boocock, L. Maunder, Vibration of a symmetric tuning fork, *J. Mech. Eng. Sci.*, 11(4), 1969.
68. G.W. Hunt, A.E.W. Hobbs, Development of an accurate tuning-fork gyroscope, symposium on Gryos, *Proc. Inst. Mech. Eng.*, 179(3E), 1964–1965.
69. W.P. Chan, F. Prete, M.H. Dickinson, Visual input to the efferent control of a fly's gyroscope, *Science*, Vol. 280, No. 5361, 289–292, April 10, 1998.
70. W.A. Clark, Micromachined vibratory rate gyroscopes, Ph.D. dissertation, University of California, Berkeley, Fall 1997.
71. W.A. Clark, R.T. Howe, R. Horowitz, Surface micromachined Z-axis vibratory rate gyroscope, *Tech. Dig. Solid-State Sensor Actuator Workshop*, 83–287, June 1996.
72. J.M. O'Connor, D.M. Shupe, Vibrating beam rotation sensor, U.S. Patent 4,381,672, issued May 3, 1983.
73. B. Greiff, T. Boxenhorn, T. King, L. Niles, Silicon monolithic gyroscope, *Transducers '91, Dig. Tech. Papers*, 966–969, 1991.
74. A. Kourepenis, J. Borenstein, J. Connelly, R. Elliott, P. Ward, M. Weinberg, IEEE 1998 Position Location and Navigation Symposium, *PLANS 98*, 1–8, 1998.
75. J.A. Green, S.J. Silverman, J.F. Chang, S.R. Lewis, Single-chip surface micromachined integrated gyroscope with 50°/h Allan deviation, *IEEE J. Solid-State Circuits*, 37(12), December 2002.
76. T. Juneau, A.P. Pisano, Micromachined dual input axis angular rate sensor, *Tech. Dig. Solid-State Sensor Actuator Workshop*, 299–302, June, 1996.
77. J.H. Leck, *Pressure Measurement in Vacuum Systems*, Chapman & Hall, London, 1964.
78. H.V. Ubisch, On the conduction of heat in rarefied gases and its manometric application, *Appl. Sci. Res.*, A2, 364–430, 1948,
79. C.H. Mastrangelo, R.S. Muller, Microfabricated thermal absolute-pressure sensor with on-chip digital front-end processor, *IEEE J. Solid State Circuits*, 26(12), 1998–2007, December 1991.
80. W.P. Eaton, J.H. Smith, D.J. Monk, G. O'Brian, T.F. Miller, Comparison of bulk and surface micromachined pressure sensors, *Proc. SPIE*, 3514, 431–438, September 1998.
81. Motorola, Inc., *Motorola Pressure Sensors, Sensor Device Data/Handbook*, 4.5, 1997.
82. J. Bergqvist, F. Rudolf, J. Maisano, F. Parodi, M. Rossi, A silicon condenser microphone with a highly perforated backplate, *Transducers '91*, 266–269, 1991.
83. W.P. Eaton, J.H. Smith, Micromachined pressure sensors: review and recent developments, *Smart Mater. Struct.*, 6, 530–539, 1997.
84. R. Schellin, M. Strecker, U. Nothelfer, G. Schuster, Low pressure acoustic sensors for airborne sound with piezoresistive monocrystalline silicon and electrochemically etched diaphragms, *Sensors Actuators A*, 46–47, 156–160, 1995.

MEMS Sensors and Actuators

85. R.P. Pied, E.S. Kim, D.H. Hong, R.S. Muller, Piezoelectric microphone with on-chip CMOS circuits, *JMEMS*, 2(3), 111–120, September 1993.

86. J. Bernstein, A micromachined condenser hydrophone, *IEEE Solid State Sensor Actuator Workshop, Hilton Head '92*, 161–165, 1992.

87. J. Bernstein, M. Weinberg, E. McLaughlin, J. Powers, F. Tito, Advanced micromachined condenser hydrophone, *IEEE Solid State Sensor Actuator Workshop, Hilton Head '94*, 73–77, 1994.

88. R. Taylor, A.D. Little, J.S. Schultz, *Handbook of Chemical and Biological Sensors*, IOP Publishing, Bristol, 1996.

89. A.E.G. Cass, *Biosensors: A Practical Approach*, Oxford University Press, Oxford, 1990.

90. W.H. Ko, Frontiers in solid state biomedical transducers, *Proc. IEEE Int. Electron Devices Meeting*, 112–115, 1985.

91. D.S. Ballantine, R.M. White, S.J. Martin, *Acoustic Wave Sensors*, Academic Press, New York, 1997.

92. T. Ono, X. Li, H. Miyashita, M. Esashi, Mass sensing of adsorbed molecules in sub-picogram sample with ultrathin silicon resonator, *Rev. Sci. Instrum.*, 74(3), 1240–1243, March 2003.

93. D. Lange, A. Koll, O. Brand, H. Baltes, CMOS chemical microsensors based on resonant cantilever beams, *Proc. SPIE*, 3328, 233–243, March 1998.

94. J.W. Garder, P.N. Barlett, *Electronic Noses: Principles and Applications*, Oxford University Press, Oxford, 1999.

95. Figaro Engineering Inc.: http://www.figarosensor.com/.

96. K.D. Schierbaum, U. Weimar, W. Gopel, U. Komalowski, Conductance, work function and catalytic activity of SnO2 based gas sensors, *Sensor Actuators*, 3, 205–214, 1991.

97. N. Yamazoe, Y. Kurokawa, T. Seiyama, Effects of additives on semiconductor gas sensors, *Sensors Actuators*, 4, 283–289, 1983.

98. H. Nanto, T. Kobayaski, N. Dougami, M. Habara, H. Yamamoto, E. Kusano, A. Kinbara, Y. Douguchi, Smart chemical sensors using ZnO semiconducting thin films for freshness detection of foods and beverages, *Proc. SPIE*, 3328, 418–427, March 1998.

99. I. Lundstrom, S. Shivaraman, C. S. Svensson, L. Lundkvist, A hydrogen-sensitive MOS field effect transistor, *Appl. Phys. Lett.*, 26, 55, 1975.

100. M. Armgarth, C. Nylander, A stable hydrogen sensitive Pd gate metal oxide semiconductor capacitor, *Appl. Phys. Lett.*, 39, 91, 1981.

101. R.C. Hughes, W.K. Schubert, Thin films of Pd/Ni alloys for detection of high hycrogen concentrations, *J. Appl. Phys.*, 71, 542, 1992.

102. P.F. Ruths, S. Askok, S.J. Fonash, J.M. Ruths, A study of Pd/Si MIS Schottky barrier diode hydrogen detector, *IEEE Trans. Electron Devices*, ED-28, 1003, 1981.

103. H. Conrad, G. Ertl, E. E. Latta, Adsorption of hydrogen on palladium single crystal surfaces, *Surface Sci.*, 41, 435, 1974.

104. P. Wright, The effect of occluded hydrogen on the electrical resistance of palladium, *Proc. Phys. Soc.*, 63A, 727, 1950.

9 Packaging

Once the fabrication of the MEMS device has been completed, parts are obtained and assembled, protected, and tested. These parts can consist of one or more die resulting from a surface micromachined or bulk micromachined process or, alternatively, an object such as a gear, rotor, or seismic mass resulting from a bulk micromachined or LIGA process. The parts obtained from a MEMS fabrication process are rarely the end *product*. A number of steps need to be accomplished once the MEMS fabrication is done in order to produce a MEMS product such as a sensing device (e.g., accelerometer, pressure sensor); an optical device (e.g., the TI DLP™); or a switch. The major issues for packaging of MEMS devices include handling, thermal budget, mechanical stress, encapsulation, and device motion requirements.

9.1 PACKAGING PROCESS STEPS

The objective of packaging is multifaceted. Packaging operations may consist of some postfabrication processes required for a device to function properly as well as assembly of the MEMS device into a next level assembly or final product that provides a function for the end user. The package containing the MEMS device needs to provide the following functions:

- *Mechanical support*. The package must mechanically support or contain the MEMS device so that it can function alone or within another system. The package will physically protect the MEMS device.
- *Interconnection*. The package must provide for communication between the microscale connection of a MEMS device and the macroscale connection that will be used to function or interface with the device. The connections may encompass a variety of physical phenomena such as electrical, optical, fluidic, biologic, etc.
- *Environment control*. The package must control the environment necessary for the MEMS device to function properly throughout its lifetime. The necessary environmental controls may include thermal management, particulate contamination, or ambient atmosphere control (i.e., humidity, atmosphere, and atmospheric pressure).

Similar to MEMS fabrication processing, packaging has its roots in and leverages the microelectronics infrastructure; however, MEMS packaging fre-

340　　　　　　　　　　　　　　　　　Micro Electro Mechanical System Design

quently has different requirements from those of microelectronics. The packaging processing sequence will be broken down into the following steps, which will be discussed in detail in this section:

- *Postfabrication processing.* Additional processing must occur to enable the functionality of the basic MEMS device or part. These processes may include dicing, releasing, and coating the device. Exactly which of these postfabrication processes is performed and its sequence is very application specific. This process sequence is frequently called the *back end of the line* (BEOL).
- *Package selection and design.* The package provides the functions of mechanical support, interconnection, and environmental control for the MEMS device that it contains. Depending upon the application or the point in the MEMS device development cycle (i.e., prototype, product development, or product), the package that may be utilized could be a simple IC package or a specifically designed package for the intended product. Basic options and considerations will be discussed.
- *Die attach.* The MEMS die is fixed in the package to provide support and interconnection.
- *Wire bond and sealing.* This section will discuss the connection from the MEMS die to the package to provide interconnect between the micro- and macrosize regimes. The package will be sealed to protect the MEMS die from handling or contamination. The sealed package can provide an inert environment for the MEMS die to prevent corrosion or humidity, which can affect device performance. Alternatively, the package may also provide a vacuum that may be required for some MEMS devices.

9.1.1 POSTFABRICATION PROCESSING

Postfabrication or back end of the line (BEOL) processing consists of a variety of processes required to enable the MEMS device to be fully useful. The selection of the processes to be used is very application specific and influenced by the fabrication technology utilized (surface micromachining, bulk micromachining, LIGA). The following processes will be discussed in this section:

- *Release*: the removal of the sacrificial layers at the end of fabrication of a surface micromachined device.
- *Drying*: the removal of the etchant and rinsing solutions utilized in the release process to yield a dry MEMS device.
- *Coating*: a coating applied to the MEMS device to further enable or enhance its functionality — for example, a thin coating applied to lubricate or reduce friction in the MEMS device. Alternatively, a biological coating could be applied to enable a biological sensor, or a coating could be applied to enhance reflectivity of an optical device.

Packaging

- *Assembly*: assembly of MEMS die or parts to form the next level of assembly.
- *Encapsulation*: a MEMS die may be encapsulated to enhance its protection.

9.1.1.1 Release Process

The release process of surface micromachining is discussed in Chapter 3. The exact nature of the release process will depend upon the material system used in the surface micromachine process. It is generally desired to have a release process that will remove the sacrificial layer as quickly as possible with minimal impact on the structural layer material that remains. For example, the SUMMiT™ process utilizes polysilicon as the structural material and silicon dioxide as the sacrificial material with hydrofluoric acid release chemistry. Alternatively, the Texas Instruments DMD™ device utilizes aluminum as the structural material and photoresist as the sacrificial material released with oxygen plasma etch. The release etch would ideally produce no particles that would inhibit the MEMS device function and produce a free-standing structure.

9.1.1.2 Drying Process

For release processes that involve liquid etchants, drying is a critical process in the BEOL sequence. The drying of a MEMS device is crucial because the fabrication processes typically involve thin layers (e.g., ~2 to 4 µm) and small gaps, which give rise to a stiction phenomenon in which the layers can be adhered to each other or to the substrate. Dyck et al. [1] provide a review and discussion of the stiction literature. Stiction is chiefly due to the surface–volume ratio scaling of MEMS devices (Chapter 4), which magnify the effects of surface forces [2,3] such as surface tension in liquid–vapor interfaces. In the ordinary drying process after the release etch, liquid evaporates and capillary forces due to surface tension arise as a result of the liquid–vapor interface. The surface forces at this interface can be avoided by using processes that avoid the liquid–vapor transition, such as supercritical drying or freeze drying.

The *supercritical drying* process proposed by Mulhern [4] avoids the liquid–vapor interface by transferring the liquid via the supercritical phase. Carbon dioxide is used because of its low critical temperature and pressure (i.e., T_c = 31.1°C, p_c = 72.8 atm). During the drying process, no liquid–vapor interface exists, thus, no capillary forces are present.

Freeze drying removes the liquid by freezing it using evaporation cooling in a vacuum chamber. The solid (frozen) material is then removed by sublimation. Freeze drying was first applied by Guckel [5,6] in a surface micromachined device released with a hydrofluoric (HF) acid etch. The wafer is immersed in a water–methanol mixture after the liquid release etch. The liquid is then frozen by evaporative cooling in a vacuum chamber. The methanol is added to avoid supercooling the water and thus causing it to freeze too rapidly. Then the solid water–methanol is removed by sublimation at 0.15 mbar.

9.1.1.3 Coating Processes

Coating a MEMS device may be required to enable its functionality for a particular application or to enhance the yield of the device. The types of coatings that may be used are as wide ranging as applications for MEMS are. Examples of coatings include surface passivation; antistiction coatings [7,8]; optical coatings [9–12]; biocompatibility [13]; and biologically sensitive [14] or chemically sensitive [15–17] coatings. Selected examples are discussed next.

- *Antifriction/antistiction coatings.* Self assembled monolayer (SAM) coatings have been developed to aid in the reduction of stiction and friction in MEMS devices. Stiction is the adhesion of compliant MEMS surfaces when the restoring forces are unable to overcome the interface forces (e.g., van der Waals, capillary forces, electrostatic forces). The SAM coatings frequently consist of long chain molecules of various chemistries that can produce oriented hydrophobic monolayers on a MEMS surface; this aids stiction reduction.
- *Optical coatings.* MEMS optical devices frequently require a highly reflective surface for the optical wavelength of interest. Gold (Au) is a useful optical reflective coating in the infrared and visible spectrums. Depending upon the surface to be coated, an adhesion layer may be required to ensure good adherence to the surface. For example, Picard et al. [9] investigated the use of chromium (Cr) and titanium (Ti) as an adhesion layer for gold deposition. However, for other wavelengths, alternative coatings may be necessary. For example, a molybdenum–polysilicon (Mo/Si) multilayer coating [10–12] is necessary in the extreme ultraviolet (EUV) spectrum ($\lambda = 13.4$ nm). The Au/Ti, Au/Cr, and Mo/Si coatings are generally deposited by evaporation or sputtering.
- *Biologically and chemically compatible coatings.* MEMS-based fluid systems may handle biological and chemical materials. For the MEMS system to function properly, the surfaces of the MEMS devices must not chemically or biologically react with the sample materials because that would produce materials that may foul the system [13] with reacted chemicals or biological elements.
- *Biologically or chemically sensitive coatings.* Biological and chemical sensors require portions of the MEMS device to be coated with sensitive materials that are used to produce an effect that can be measured. For example, the effect to be measured may be a mass change on the treated portion of a vibrating beam. The biologically sensitive coating may be an antigen applied to a vibrating sensor [14]. A chemically sensitive coating [15–17] may be applied to a structure, which will adsorb a particular chemical that will change the resistance of the material.

Packaging

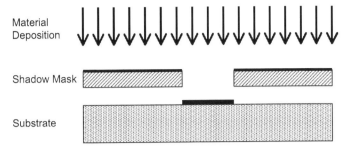

FIGURE 9.1 Shadow mask utilized in an evaporation or sputtering deposition process to control the deposition location of material.

Coatings are very application specific, but the MEMS designer must be aware of the possible impact these coatings may have on the device and its fabrication. The effects that a coating may have are quite varied; several of these effects are discussed next.

- *Electrical shorts.* Many optical coatings are electrically conductive and are deposited by evaporation or sputtering. It is desirable to apply the optical coating to the optical surface; it may be patterned via a lift-off process (see Section 2.6.2). Alternatively, the deposition of the optical coating may be controlled via a *shadow mask* utilized in the evaporation or sputtering deposition (Figure 9.1). The shadow mask forms a barrier that permits deposition only in selected areas. Also, the optical device design may be *self-shadowing*, as illustrated in Figure 9.2. A self-shadowing design uses an overhanging structure to cause breaks in the conductive coating so that the device is not electrically shorted. This device is difficult to design, but can prevent fine features from being electrically shorted. The shadow mask approach

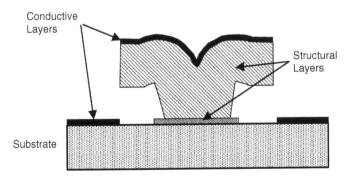

FIGURE 9.2 Self-shadowing MEMS design will prevent electrical shorting of conductive blanket deposited films.

is limited by the precision with which the mask can be aligned, whereas the self-shadowing approach is only limited by the precision of the lithography process utilized and the amount of underspray in the deposition process.

- *Residual stress.* The coating residual stress is a very important concern especially for optical coatings. Significant engineering effort is involved in the process development of a low-stress coating process [9]. A metric for the performance of optical devices is flatness of the coated optical surface. The process parameters of deposition rates and material alloys are significant in engineering the coating processes. The MEMS optical design and fabrication process is frequently influenced by the necessity to produce mirror surfaces that can withstand the residual stresses inherent in the MEMS fabrication process and the applied coatings. Frequently, this pushes the MEMS design to have thicker, reinforced, or smaller mirror surfaces to accommodate the residual stress effects.
- *Particulate.* Particles may be generated by a coating that may interfere with the device operation by shorting or jamming. If this is a possibility, the MEMS designer may be able to shield the critical portions of the design from these effects.
- *Coating out-gassing.* Coatings may produce gases when exposed to an elevated temperature. These gases may impede device operation by adverse chemical reactions or changing the pressure or vacuum in the package. These problems may be circumvented by a change of coating material or use of chemical *getters* [18] to remove the offending gases. The use of a chemical getter may be needed due to outgasing of the applied coatings; also, package leakage and outgasing of the die attach material may also be a significant source of undesirable internal package gases. Getters are selective chemical scavengers designed to capture undesirable substances. They are available for hydrogen, oxygen, moisture, and particulate materials [19]. An early use of chemical getters was the removal of oxygen from triode vacuum tubes, which caused the filaments to oxidize and fail. A solution to this problem was the use of active metals that reacted more quickly with the oxygen than the tungsten filament of the triode.
- *Coating thermal environment.* The thermal environment involved in the deposition of the coating may have an impact on the MEMS material properties used if the temperature is sufficiently high. If the coating is applied to a MEMS device that contains microelectronic properties, the thermal environment may affect the electronic properties or melt the interconnect layers.
- *Coating stability.* Some coatings may degrade at elevated temperature. This could be the chemistry of the coating changes that makes a SAMS coating for stiction ineffective or the corrosion of a mirror coating. Many times these effects will impose a limit on subsequent temperature processing that the MEMS device can withstand.

Packaging

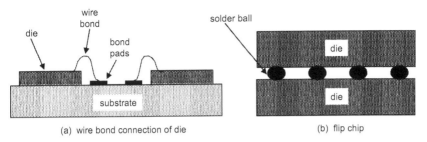

FIGURE 9.3 Illustration of direct wire bonding of die and flip-chip bonding to form a multichip module.

9.1.1.4 Assembly

A number of items fall under the umbrella of assembly. Assembly can include assembly of MEMS or microelectronic die for a next level of assembly; assembly of MEMS parts; or erection of MEMS structures. As with most things in the area of MEMS, the technologies that will be discussed under this topic are in various stages of development and maturity.

- *Assembly of die.* MEMS or microelectronic die can be assembled together to form a greater assembly [20]. Interconnection between die can be made via a number of methods, such as wire bonding between die mounted on a common substrate; however, *flip-chip* bonding has significant advantages (Figure 9.3). In flip-chip bonding, the die are bonded together via small solder balls (<200 μm) without the use of bond wires. This approach enables a small package with intimate electrical contact, which mitigates the electrical parasitic of bond wire resistance and capacitance. This method enables a multichip module (MCM) in which multiple chips are stacked using flip-chip bonding and connections through the wafer to enable the vertical connections [21]. This approach allows improved system performance and small geometric form factor.

  The range of capabilities enabled by MEMS devices has prompted research in techniques in which MEMS components fabricated in different technologies can be combined via a packaging method such as flip-chip bonding to enable a viable solution or higher performance. For example, Michalicek et. al. [22] demonstrated the transfer of a surface micromachined device assembly to another substrate (Figure 9.4) to increase functionality; this allowed electronic control of the MEMS devices. This approach could also be utilized to create RF devices (e.g., switches, filters, variable capacitors) utilizing silicon surface micromachined devices removed from the host silicon wafer; that is not suited to RF applications to a better suited substrate.

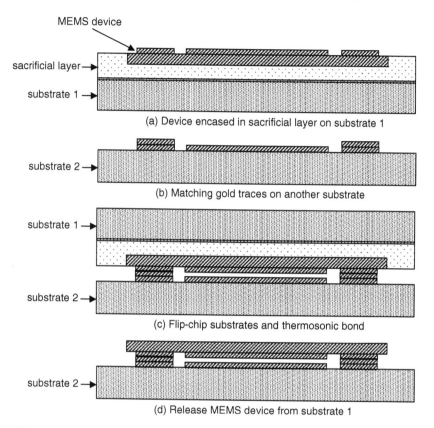

FIGURE 9.4 Method for transferring MEMS devices via flip-chip assembly. (Modified from M.A. Michalicek et. al., *Proc. SPIE*, 3878, 68–79.)

- *Erecting structures.* The erection of MEMS structures is still a research topic, but the development of technology and devices to produce truly three-dimensional structures may be enabling to systems in the future. The actual assembly may be performed by on-chip actuators or residual stress devices on the chip, or manually assembled via a probe. Figure 9.5 shows erected vertical mirrors, which have been manually erected via an assisting truss structure, and a mirror erected at an angle via residual stressed beams. The erection of MEMS structures may also be enabled by surface tension of solder [23] or fluids [24,25], which would eliminate the need for manual assembly, actuators, or aiding truss structures. The use of properly arranged and designed receiving sites with surface tension forces may enable automated assembly.
- *Assembly of MEMS parts.* Some MEMS fabrication technologies will produce parts (e.g., bulk micromachining, LIGA) that will require assembly to form a functioning MEMS device. Figure 9.6 shows a LIGA mechanism that required the manual assembly of the various

Packaging

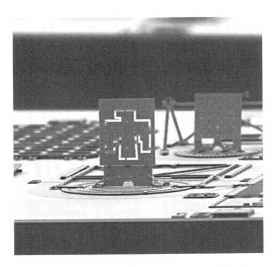

FIGURE 9.5 Erected MEMS mirrors utilizing manual assembly via a truss structure and residual stress. (Courtesy of Sandia National Laboratories.)

parts (e.g., gears, linkages). Manual assembly is usually not an acceptable method except for the very small lot sizes typically encountered for research and development prototypes. The area of microassembly is currently in the very early stages of development; thus, no standard procedures, standards, or tools are readily available. However, research has been conducted in parallel and serial methods of microassembly [26–31,33]. Serial microassembly utilizes *robotic visual servoing* (RVS) [30] to locate and manipulate parts. The RVS approach to serial microassembly is ideal for small lot sizes and diverse parts to assemble. As the lot sizes become larger, fixtures for the parts become a viable alternative. Also, parallel assembly of the parts is the path that would be needed for large lot sizes. Figure 9.7 shows the equipment utilized for parallel assembly of a shaft and gear at the wafer scale. Serial and parallel assembly utilized the available techniques of robotic assembly; however, much of the equipment is unique to this size scale.

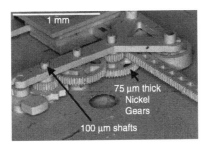

FIGURE 9.6 Assembled LIGA mechanism. (Courtesy of Sandia National Laboratories.)

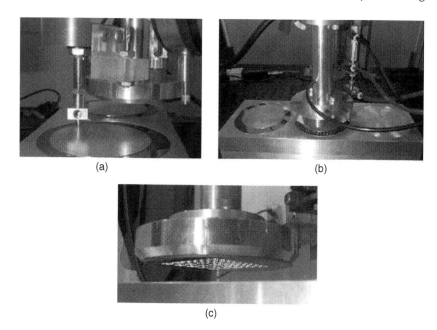

FIGURE 9.7 Parallel assembly of LIGA components. (a) The pin insertion tool is picking up a 386-μm diameter pin. The pin is placed in the wafer in the background. (b) The wafer of gears is being placed on the wafer of pins. (c) View of the wafer of gears before being placed on the pins. (Courtesy of Intelligent Systems and Robotics, Sandia National Laboratories.)

9.1.1.5 Encapsulation

Packaging can represent as much as 70% of the cost of a MEMS product. Manufacturability, which is the result of batch fabrication of hundreds to thousands of devices at a time on a single wafer, is one of the reasons that integrated circuits are so successful. One of the reasons that packaging can represent such a significant fraction of the total cost is the need to handle individual die. MEMS packaging can achieve greater manufacturability by avoiding the handling of individual die in the release, dry, and coat processes; this can be achieved by performing wafer level release, dry, and coat processes and encapsulation to protect the MEMS device through the rest of the packaging process.

The encapsulation will provide a protective shell over the sensitive MEMS parts while feed-throughs are passed through the package to connect to other components or to pass an electrical signal (Figure 9.8). The feed-throughs are needed to transfer signals to or from sensors, actuators, and circuitry. The feed-throughs must have low parasitics (e.g., resistance, capacitance) and they must be sealed to avoid any leakage. The encapsulation illustrated in Figure 9.8a is achieved by bonding an encapsulating substrate to the MEMS wafer. A reliable bonding process is essential to achieving a good seal. A variety of bonding

Packaging

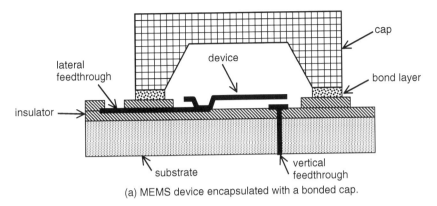

FIGURE 9.8 Encapsulation of a MEMS device with a wafer capsule or a thin-film shell. Vertical and horizontal feed-throughs for signals are as indicated.

processes (e.g., silicon-glass bonding, glass frit bonding, eutectic or solder bonding) has been utilized [34].

To achieve a wafer-scale encapsulation, the processes of release, dry, coat, and encapsulation must be compatible. A significant concern will be the impact of the temperature of the encapsulation process upon the MEMS coatings. Sandia National Laboratories [32] has developed a wafer-scale encapsulation that utilizes anodic bonding to join the encapsulating wafer to the MEMS wafer (Figure 9.9). The temperature of the bonding process is the most important parameter in determining if the SAM coatings survive. The SAM coatings utilized in this process require a temperature of less than 320°C to survive functionally. These glass-capped wafers can then be run through a standard wafer saw. The result of the Sandia research indicates that anodic bonding is sufficiently strong for microsystem packaging applications.

An alternative approach to encapsulation utilizing wafer bonding is a thin-film shell package (Figure 9.8b). Thin-film shell encapsulation is attractive due to the small area encapsulation that can be achieved. A variety of techniques is available for thin-film encapsulation, which is compatible with wafer-level processing. However, thin-film materials can be physically fragile. The thin-film encapsulation can be accomplished with organic materials (i.e., epoxies, silicones,

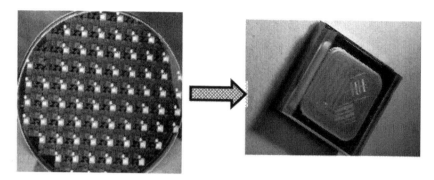

FIGURE 9.9 A machined pyrex anodically bonded over a released 150-mm SUMMiT™ wafer and diced device with cap. (Courtesy of Sandia National Laboratories.)

or polymers such as polyamides or Parylene) or inorganic materials (i.e., silicon nitride or silicon carbide). The organic materials are a challenge due to the possibility of moisture penetration and attack by harsh environments. The inorganic materials are a challenge for use in encapsulation because of the temperature required for deposition and the nonconformal nature of the film.

9.1.2 Package Selection/Design

Once the postfabrication processing is accomplished, a selection of a suitable package, based upon the intended use of the MEMS device, must be made. For a prototype research and development application requiring only electrical connectivity, a microelectronic package may be used. Figure 9.10 shows a schematic of the basic features of an electronic package. Figure 9.11 shows a few examples of frequently utilized IC packages. More detailed compilations of IC packages available may be obtained from Web sites listed in References 35 through 38.

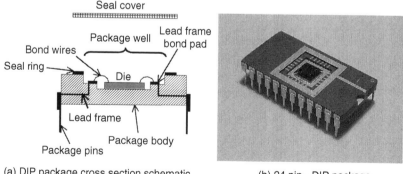

(a) DIP package cross section schematic (b) 24 pin - DIP package

FIGURE 9.10 Schematic of a microelectronic package denoting the lead frame, well, and seal cover.

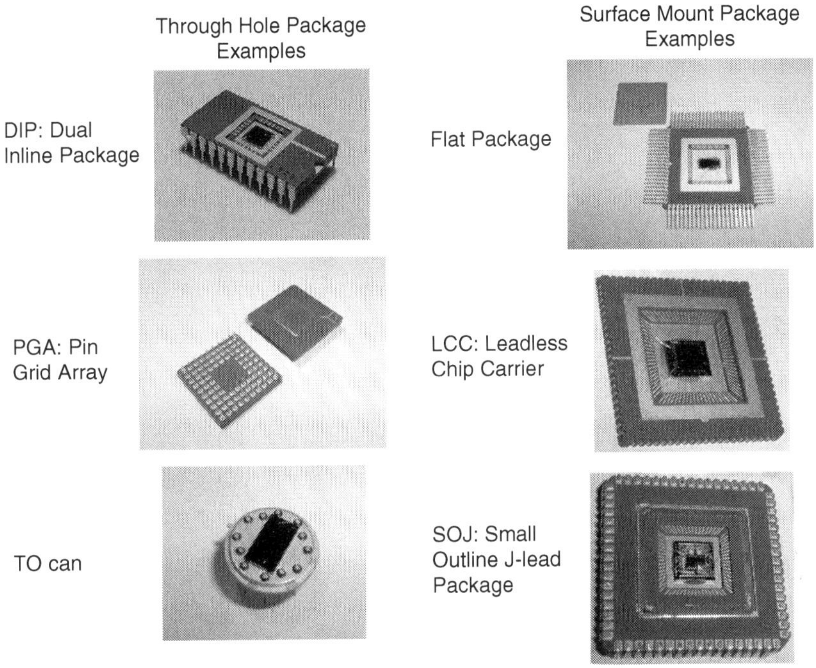

Packaging

351

Through Hole Package Examples

Surface Mount Package Examples

DIP: Dual Inline Package

Flat Package

PGA: Pin Grid Array

LCC: Leadless Chip Carrier

TO can

SOJ: Small Outline J-lead Package

FIGURE 9.11 Examples of THD and SMD packages.

The two basic types of connections are the *through-hole device* (THD) and *surface mount device* (SMD) packages. The THD package interfaces with a printed circuit board with pins inserted in the plated through-holes of the circuit board and soldered in place. The SMD package connections are more varied. Surface mount technology packages enable tight spacing of connections out of the package, which will allow more I/O (input–output) per unit area of the package. Figure 9.12 illustrates the gull-wing, J-lead, and solder-ball grid types of connection utilized in surface mount technology to facilitate a dense package connection interface. The ball grid array (BGA) package is a relatively new style of package that enables a smaller parasitic (i.e., capacitance, resistance) than the traditional lead frame and bond wire approach to packaging. The BGA style of package facilitates high connection density (e.g., 100s to 2000 connections in some cases) and high-frequency packaging interconnection.

The most advanced methods of packaging involve use of the microsolder ball for attaching chips to each other, a package, or a circuit board. The use of a solder ball eliminates the traditional wire bond that provides a low parasitic connection. The following will briefly touch on the packaging technologies that utilize solder ball attachment.

Flip-chip or *bump bonding* uses small solder balls (e.g., 0.1 to 0.2 mm) that are directly attached to metallized pads on the bottom of a chip or die. The solder

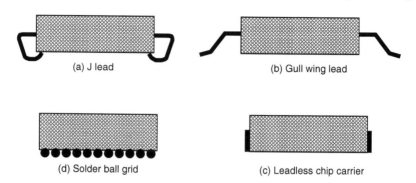

FIGURE 9.12 Example of connection lead types utilized in surface mount device packages.

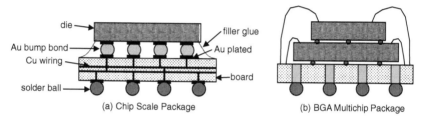

FIGURE 9.13 Chip scale package and multichip module package schematic.

used is typically a 97% Pb/3% Sn. The common flip-chip process is called C4 (controlled collapse chip connection) and was patented by IBM in 1969 [39].

Chip scale packaging (CSP) belongs to the BGA family of packages. The chip die bonding is no longer done by wire bonding, but is accomplished by using gold bump bonding between the chip die and the package base substrate. The package base consists of multilayers, which provide signal and ground planes of copper. A filler material or glue is between the chip die and the package base to provide additional attachment and stress relief. The solder ball count for this type of package is 40 to 200 balls. The CSP package area is 80% occupied by the die. The *multichip module* (MCM) is the ability to integrate several chip die into one package. This approach to packaging facilitates small size and short interconnection paths between die. The solder ball is utilized once again for chip–chip connection. Figure 9.13 illustrates the CSP and MCM package concepts.

9.1.3 Die Attach

Die attach is the traditional method for attaching a die to a package. The goal of a die attach process is to provide a 100% void-free attachment between the die and package, which would minimize heat resistance, and a die attach material that has no outgas, which may contaminate a MEMS device.

Two approaches for die attach are *epoxy adhesive* and *eutectic* die attach. The epoxy die attach utilizes epoxies with different fillers depending upon the desired conductivity (silica filler for insulating, silver for conducting) and cure

Packaging 353

temperature (typically, <150°C). An issue with epoxy die attach may be outgas during and after curing; this may contaminate a MEMS device. This contamination can affect MEMS devices by causing stiction and have an impact on device reliability. Research has developed a low outgas epoxy [40] that may meet the requirements for MEMS devices. The epoxy may be automatically dispensed in a pattern in the package; the die handled by a vacuum pick-up will insert the die onto the epoxy pattern in the package die attachment area in the package well. Appropriate space on the device layout will need to be provided to facilitate the automated vacuum pickup.

The eutectic die attach layer, such as gold-silicon (AuSi) or tin-lead (SnPb), is reflowed to provide a mechanically strong bond that conducts heat and electricity well. The consideration for use of a eutectic die attach is the high-temperature processing (183°C for SnPb or 379°C for AuSi). The eutectic die attach does not have the outgas concerns of the epoxy die attach.

9.1.4 Wire Bond and Sealing

Wire bonding is the oldest method for connecting a chip or die to the inner lead frame of the package. Gold wires with 25 to 50 µm diameter or aluminum wires with 25 µm diameter are commonly used. The bonding of the wires occurs by application of heat and ultrasonic energy to form a *thermosonic bond*; application of ultrasonic energy alone to form an *ultrasonic* bond (e.g., a type of cold weld); or application of heat alone and compression to form a *thermocompression* bond.

A ball bond tool implements a thermocompression bond by pressing heated gold balls onto metallized pads. A torch heats a gold wire tip above the melting point (~400°C), which allows surface tension to form a ball on the end of the wire. The tool with a mold on the end presses the ball onto the metallized bond pad. The bond is formed by mechanical pressure and diffusion of the bonding material (i.e., gold ball–metallized bond pad).

A wedge bond tool implements an ultrasonic bond by feeding a gold or aluminum wire, which is pressed onto the metallized bond pad and ultrasonic energy (~50 kHz) applied. Figure 9.14 is a schematic the operation of a ball and wedge bond tool. Figure 9.15 shows a wire bond performed by a ball and a wedge bond tool.

Sealing the package well is the final step in packaging a microsystem. Sealing the package cavity can be accomplished by use of a eutectic solder. The use of eutectic solder such as AuSn requires high temperature to braze the package lid. Alternatively, epoxy can be used to attach the package lid; this requires a much lower temperature, but it produces a nonhermetic seal and has the issue of outgassing discussed earlier. Glass frit seals are often used to attach a ceramic lid to a ceramic package, but this requires temperatures higher than brazing.

9.2 PACKAGING CASE STUDIES

This section will discuss three variations of packaging of interest to a MEMS designer or project:

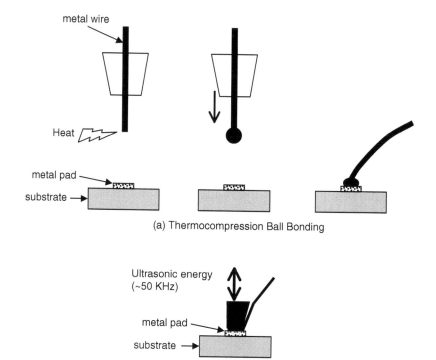

FIGURE 9.14 Schematic of a ball and wedge bond tool.

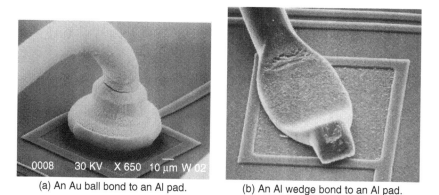

(a) An Au ball bond to an Al pad.   (b) An Al wedge bond to an Al pad.

FIGURE 9.15 Example ball and wedge bond.

Packaging   355

- *R&D prototype packaging.* R&D prototype packaging is a brief discussion of the packaging required for a preliminary prototype that will facilitate initial test and evaluation of a device. This example will utilize commercial microelectronic packaging for the test and evaluation of a MEMS device that require electrical signals only. This approach will generally not be the final product package required, but the use of commercial electrical packages facilitates research and development of a prototype. The initial considerations necessary for this type of packaging will be discussed.
- *DMD packaging.* This will review some of the packaging considerations for the Texas Instruments digital mirror device (DMD), which is one of the premier MEMS success stories. This is an example of a package for a commercial product that needs electrical and optical feed-throughs in the package.
- *Sandia electrical-fluidic packaging (EMDIP™).* The section will review the packaging considerations for MEMS fluidic research devices. These devices require electrical as well as fluidic package feed-throughs. This was a packaging development effort to achieve an in-house standard electrical-microfluidic package that would facilitate ongoing research efforts.

9.2.1 R&D Prototype Packaging

This section will discuss packaging issues for packaging a MEMS device in a standard electrical package for preliminary test and evaluation of a MEMS prototype device. A list of some simple common-sense rules that will greatly aid the ability to package a device is developed. These *MEMS packaging layout rules*, which are also applicable for more advanced packages and applications, follow:

- Adequately sized bond pads. Utilize adequately sized bond pads on the MEMS layout. A typical size is 120×120 micron bond pads. Figure 9.16 shows the size of a bond pad and a wedge bond tool that illustrates the need for adequately sized bond pads.
- Provide space for vacuum pickup. Manual manipulation of die with tweezers is not a manufacturable process due to the possibility of particle generation. Vacuum pickup or collets are used to pickup a die. The pick-up must be separated from MEMS structures to prevent the possibility of damage. A 300-μm space from any structure is a typical recommendation, but this depends upon the tools used.
- Adequate bond pad spacing. The bond pad should be spaced around the outsize periphery of the die. A standard layout, which may facilitate testing and multiple uses of probe cards, is recommended. Figure 9.17 shows a standard layout of bond pads for a 24-pin DIP for a SUMMiT™ module.

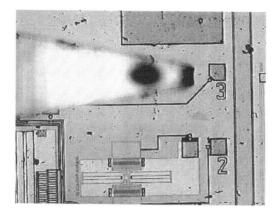

FIGURE 9.16 Wedge bond tool and a bond pad to illustrate the need for adequately sized bond pad. (Courtesy of A. Oliver, Sandia National Laboratories.)

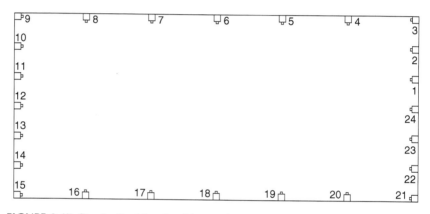

FIGURE 9.17 Standardized bond pad layout for a SUMMiT MEMS module. (Courtesy of Sandia National Laboratories.)

- No MEMS devices or conducting traces outside the bond pad perimeter. This will minimize electrical parasitics or damage during the wire-bonding process.
- Minimize length of bond wires. This will prevent sagging or sensitivity of the package to acceleration. Keep the length of the bond wires less than 0.1 in. (2.5 mm).
- No crossed bond wires. This will preclude shorting of the bond wires due to handling.
- Obtain specific packaging specification and rules from the packaging engineer. For exact specifications of the packaging issues that apply to a specific device, consult the packaging engineer, who is knowledgeable about the tools and processes available at the facility.

Packaging

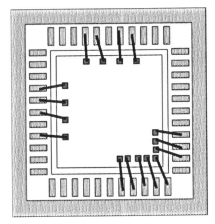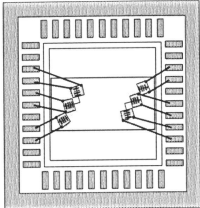

FIGURE 9.18 Example of a wire bond layout that violates the MEMS packaging layout rules. (Courtesy of Sandia National Laboratories.)

Figure 9.18 illustrates two types of layouts that may be problematic for packaging. In Figure 9.18a, the bond pads are bunched in a corner; this will produce difficulties for the wire bonding tool. The wire bond lengths shown in Figure 9.18b are long, closely spaced, and cross over the MEMS device.

9.2.2 DMD Packaging

The digital mirror device (DMD) developed by Texas Instruments, Inc. has been accepted by the data projector market and is making significant inroads into the high-definition home entertainment market as well. The DMD is a large arrayed MOEMS device (shown in Figure 9.19). This section will discuss the packaging requirements for the DMD [41,42] — an application of a MEMS device that requires electrical and optical feed-throughs in the package for functionality. The packaging design drivers are cost, image quality, and product life.

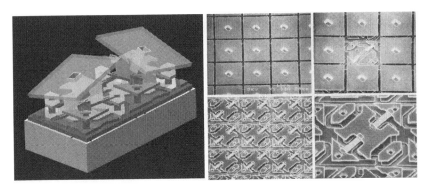

FIGURE 9.19 The Texas Instruments DMD. (Courtesy of Texas Instruments.)

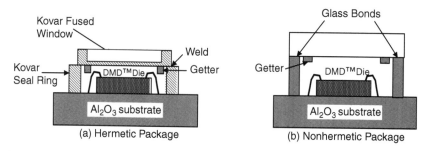

FIGURE 9.20 Schematic of hermetic and nonhermetic package designs. (After J. Faris and T. Kocian, *Tex. Instrum. Tech. J.*, 15(3), 87–94, 1998, and A. Kunzman et al., *Proc. SPIE*, 4207, 1–10, 2000.)

FIGURE 9.21 Texas Instruments HD1 package. (Courtesy of Texas Instruments.)

The first production versions employed the hermetic packaging approach [41] shown in Figure 9.20a. For the hermetic package, the device is packaged on a ceramic substrate with an optical quality window seam welded to the substrate; this requires a glass–metal fusing process. The window–substrate seam weld and the glass–metal fusing process are cost drivers.

To address the cost issues during the design for the product HD1 (Figure 9.21) for the high-definition home entertainment market, a nonhermetic package approach was used (Figure 9.20b). The device is still packaged on a ceramic substrate and the window lid is directly bonded to the substrate; however, this eliminated the window–substrate seam weld and the glass–metal fusing, which greatly reduced the cost. The seam weld process was replaced with an adhesive dispense and curing process.

The window qualities are very important for the image quality of the device. The qualities of number of window defects, DMD array placement accuracy, and parallelism with respect to the system optics are the leading factors impacting image quality, with number of window defects most significant. Because the array of mirrors, DMD, is the image plane, the gap height, h, between the mirror array and the window has a great impact on the effect of window defects on image quality (i.e., the defects are out of the focus). Studies performed by Texas Instru-

Packaging 359

ments have shown that "at a gap height of 1.0 mm, all window defects smaller than 24 µm are undetectable by the human eye" [42]. For the HD1 package, the gap height was set at 1.5 mm, which is greater than that in previous designs. The increased gap height does not have an impact on other parameters such as mirror lubrication. The nonhermetic package uses Borofloat glass vs. the more expensive Corning 7056 previously used.

The DMD array is sensitive to contamination and moisture, which are affected by a nonhermetic package permeation rate. The permeation rate into the package is controlled by the window adhesive layer, which depends upon adhesive type, temperature, bond line width, bond line thickness, and uniformity. The permeation rate is also affected by the internal package pressure as well as the external environment pressure, temperature, and humidity. The decision to use a nonhermetic package was based upon empirically based models and verified by accelerated package tests. A 20-year lifetime is achieved under nominal conditions.

The new package design further consists of a one-piece heatsink and stud with a compliant thermal pad and a spring clip, which increases the manufacturability of the package.

9.2.3 ELECTRICAL-FLUIDIC PACKAGING

A group of researchers at Sandia National Laboratories has developed a two-level packaging scheme for electromicrofluidic applications [44,45]. The motivation for this work was to develop a standardized electromicrofluidic packaging method that would interface to and enhance research in surface micromachined microfluidic devices. The researchers were striving to achieve the following attributes with their packaging technology:

- Adaptable to a variety of electromicrofluidic applications
- Producible and inexpensive in quantity
- Layered assembly
- Optimal fluid compatibility with materials used
- Protect delicate device and wire bonds from the environment
- Hermetic seal
- Compatible with optical devices
- Accommodate environmental sampling
- Accommodate cooling channels
- Modular and easy to handle, test, and ship
- Serviceable (modules can be detached and replaced)
- Accommodate multiple independent fluidic and electrical connections

The EMDIP™ (electromicrofluidic dual in-line package) was developed as a first-level package for electromicrofluidic devices fabricated in silicon (Figure 9.22). The EMDIP is constructed in layers. The base has the same electrical leads as a standard 24-pin DIP package and will plug into a standard socket. The base has eight 1-mm diameter holes that provide the fluidic interface to the second-

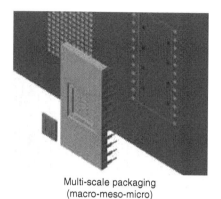

Multi-scale packaging
(macro-meso-micro)

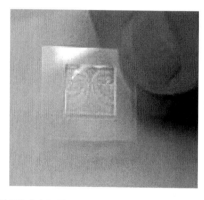

Polymer based microfluidic
interconnects

FIGURE 9.22 Exploded view of the 24-pin, eight-channel EMDIP™. (Courtesy of Sandia National Laboratories.)

level package (discussed later). The lead frame provides electrical connection to the DIP package pins and also provides metal pads along the periphery of the DIP package well that can be wire bonded to the top side of the electromicrofluidic die. The fluidic manifold is formed by a double-sided adhesive tape, which forms the flow channel that transitions from a 200-μm Bosch etched hole through the electromicrofluidic die to a 1-mm through-hole in the DIP package. The mesoscale manifold forms the top of the fluidic channel and provides holes that mate with the 200-μm Bosch holes in the die. Another piece of double-sided adhesive tape adheres the die to the mesoscale manifold.

The use of adhesive tape prevents adhesive flowing into the fluidic channels and forming a blockage. The curved transition from 200 μm to 1 mm eliminates sharp corners and prevents the formation of stagnant fluid pockets. The layered construction is amenable to methods other than adhesive tape, such as thermoplastics. The adhesive tape is the application-specific portion of the EMDIP that can be specialized to accommodate the number of fluidic channels up to a maximum of eight.

The fluidic printed wiring board (FPWD) is the second-level package designed to interface to multiple EMDIP modules as well as standard electronic

FIGURE 9.23 Exploded view of the fluidic printed wiring board with a standard 24-conductor electrical connector and an eight-channel fluidic connector. (Courtesy of Sandia National Laboratories.)

components. Figure 9.23 is an exploded view of the FPWB, electrical connector, and fluidic connector. The materials used to construct the FPWB are high-temperature thermoplastics or glass. The FPWB consists of two parts: the channel board and the cover board. The channel board contains the fluidic channel and the cover board forms the channel lid. The fluidic connector, which was developed by Peter Krulevitch and Willam Benett [43] and patented by Lawrence Livermore National Laboratories, connects to the channel board with a 3.4-mm channel. The 1-mm holes in the cover board are designed to mate with the fluidic connections of the EMDIP. The cover board also contains electrical traces, which route electrical connections from the EMDIP to the electrical connector. The interface between the EMDIP and FPWB make electrical as well as fluidic connections.

9.3 SUMMARY

Packaging is a significant fraction of the cost of any MEMS product and requires consideration concurrently with the device design and fabrication. Packaging provides the basic functions of mechanical support, interconnection, and environmental control for the MEMS device. Although the exact processes required for packaging a device are application specific, they may span a wide range, which may include any or all the following:

- Die separation
- Release
- Drying
- Coating
- Assembly
- Encapsulation
- Die attachment
- Wire bond and sealing

The requirement for the development of a packaging process is in many ways the same as fabrication process development. The interaction of the packaging processes with each other and with the fabricated MEMS device is a major issue for consideration. For example, the temperature profile required for a packaging process step, such as die attach or outgassing of the die attach material, can affect the integrity of a previously applied coating or the MEMS device.

QUESTIONS

1. What are the packaging functions?
2. What are the functions of the release and drying processes and what type of fabrication processes may utilize these?
3. What are the issues for dicing a surface micromachine wafer after release?
4. What are the concerns for MEMS devices regarding the method of die attach?
5. What three methods can be utilized to prevent electrical shorting due to the deposition of a conductive coating?
6. What is the purpose of encapsulation? Describe how encapsulation is utilized in the DMD.
7. Why would a MEMS device possibly be concerned about the temperature profile of the packaging process? Which packaging process steps may have a significant temperature profile?
8. Describe three MEMS applications that may require the device to have a coating. Describe the coating and issues that may have an impact on the MEMS device.
9. What are two types of wire bond that may be used? Describe them.
10. Of which design layout considerations for packaging should a MEMS designer be aware? Why is each important?
11. What are the problems with the wire bond layout shown in Figure 9.24?

FIGURE 9.24 View of a package well, die, and wirebonds.

Packaging **363**

12. Research the packaging of a commercial MEMS device and discuss the issues that drove the package design.

REFERENCES

1. C.W. Dyck, J.H. Smith, S.L. Miller, C.L. Russick, C.L.J. Adkins, Supercritical carbon dioxide solvent extraction from surface micromachined micromechanical structures, *Proc. SPIE Symp. Micromachining Microfabrication*, 2879, 225–235, Austin, TX, 1996.
2. J.N. Israelachvili, *Intermolecular and Surface Forces*, Academic Press, New York, 1992.
3. R. Legtenberg, A.C. Harrie, J.E. Tilmans, M. Elwenspoek, Stiction of surface micromachined structures after rinsing and drying: model and investigation of adhesion mechanisms, *Sensors Actuators*, 43, 230–238, 1994.
4. G.T. Mulhern, D.S. Soane, R.T. Howe, Supercritical carbon dioxide drying of microstructures, *Proc. 7th Int. Conf. Solid-State Sensors Actuators, Transducers '93*, Yokohama, 296–299, 1993.
5. H. Guckel, J.J. Sniegowski, T.R. Christenson, S. Mohney, S.F. Kelly, Fabrication of micromechanical devices from polysilicon films with smooth surfaces, *Sensors Actuators*, 20, 117–122, 1989.
6. H. Guckel, J.J. Sniegowski, T.R. Christenson, R. Raissi, The application of fine-grained, tensile polysilicon to mechanically resonant transducers, *Sensors Actuators*, 21, 346–351, 1990.
7. U. Srinivasan, M.R. Houston, R.T. Howe, Alkyltrichlorosilane-based self-assembled monolayer films for stiction reduction in silicon micromachines, *J. MEMS*, 7(2), 252–260, June 1998.
8. X.Y. Zhu, J.E. Houston, Molecular lubricants for silicon-based microelectromechanical systems (MEMS): a novel assembly strategy, *Triboloby Lett.*, 7, 87–90, 1999.
9. Y.N. Picard, D.P. Adams, O.B. Spahn, S.M. Yalisove, D.J. Dagel, J. Sobczak, Low-stress, high-reflectivity thin films for MEMS mirrors, *Proc. Mater. Res. Soc. Symp.*, San Francisco, CA, 729, 113–118, 2002.
10. S.S. Andreev, N.N. Salaschchenko, L.A. Suslov, A.N. Yablonsky, S.Y. Zuev, Stress reduction of Mo/Si multilayer structures, *Nuc. Inst. Methods Phys. Res.*, 470, 162–167, 2001.
11. P.B. Mirkarimi, E.A. Spiller, D.G. Stearns, V. Sperry, S.L. Baker, An ion-assisted Mo-Si deposition process for planarizing reticle substrates for extreme ultraviolet lithography, *IEEE J. Quantum Electron.*, 37(12), 1514–1516, December 2001.
12. T. Chasse, H. Neumann, B. Ocker, M. Scherer, W. Frank, F. Frost, D. Hirsch, A. Schindler, G. Wagner, M. Lorenz, G. Otto, Mo/Si multilayers for EUV lithography by ion beam sputter deposition, *Vacuum*, 71, 407–415, 2003.
13. T.W. Schneider, L.E. Aloi, R.C. White, Control of protein adsorption for MEMs and microfluidic applications on silicon surfaces using silane based self-assembled monolayers of an oligo(ethylene oxide) derivative, *Proc. SPIE*, 4205, 128–134, 2001.
14. W.H. Ko, Frontiers in solid state biomedical transducers, *Proc. IEEE Int. Electron Devices Meeting*, 112–115, 1985.
15. Figaro Engineering Inc., http://www.figarosensor.com/.

364   Micro Electro Mechanical System Design

16. R.C. Hughes, W.K. Schubert, Thin films of Pd/Ni alloys for detection of high hycrogen concentrations, *J. Appl. Phys.*, 71, 542, 1992.

17. H. Nanto, T. Kobayaski, N. Dougami, M. Habara, H. Yamamoto, E. Kusano, A. Kinbara, Y. Douguchi, Smart chemical sensors using ZnO semiconducting thin films for freshness detection of foods and beverages, *Proc. SPIE*, 3328, 418–427, March 1998.

18. K. Gilleo, M. Previti, Getters: molecular scavengers for packaging, International Symposium on Advanced Packaging Material Properties and Interfaces, Braselton, GA, March 2001; also www.hdi-online.com, January 2001.

19. STAYDRY™, Cookson Semiconductor Packaging Materials, Alpharetta, GA.

20. J.T. Butler, V.M. Bright, J.H. Comtois, Multichip module packaging of microelectromechanical systems, *Sensors Actuators A*, 70, 15–22, 1998.

21. S. Linder, H. Baltes, F. Gnaedinger, E. Doering, Fabrication technology for wafer through-hole interconnections and three-dimensional stacks of chips and wafers, *Proc. IEEE Micro Electro Mech. Syst. (MEMS '94)*, 349–354, January 1994, Japan.

22. M.A. Michalicek, W. Zhang, K.F. Harsh, V.M. Bright, Y.C. Lee, Micromirror arrays fabricated by flip-chip assembly, *Proc. SPIE*, 3878, 68–79, 1999.

23. K.F. Harsh, V.M. Bright, Y.C. Lee, Solder self-assembly for three-dimensional microelectromechanical systems, *Sensors Actuators*, 77, 237–244, 1999.

24. H.J. Yeh, J.S. Smith, Fluidic self assembly for the Integration of GaAs light-emitting diodes on Si substrates, *IEEE Photonics Technol. Lett.*, 6, 706–708, 1994.

25. U. Srinivasan, D. Liepmann, R.T. Howe, Microstructure to substrate self-assembly using capillary forces, *JMEMS*, 10(1), March 2001.

26. J.T. Feddema, T. Christenson, Parallel assembly of high aspect ratio microstructures, *Proc. SPIE*, 3834, 153–164, September 1999.

27. A. Singh, D.A. Horsley, M.B. Cohen, A.P. Pisano, R.T. Howe, Batch transfer of microstructures using flip chip bump bonding, *Transducers '97*, Chicago.

28. J.T. Feddema, C.G. Keller, R.T. Howe, Experiments in micromanipulation and CAD-driven microassembly, *Proc. SPIE*, 3202, 98–107, 1998.

29. J.T. Feddema, R.W. Simon, Visual servoing and CAD-driven microassembly, *IEEE Robotics Automation Mag.*, 5(4), 18–24, December 1998.

30. S. Hutchinson, G.D. Hagar, P.I. Corke, A tutorial on visual servo control, *IEEE Trans. Robotics and Automation*, 12(5), 651–670, October 1996.

31. M.B. Cohn, R.T. Howe, Wafer to wafer transfer of microstructures using break away tethers, U.S. Patent No. 6,142,358, issued November 7, 2000.

32. L.E.S. Rohwer, A.D. Oliver, M.V. Collins, Wafer level micropackaging of MEMS devices using thin film anodic bonding, *Mat. Res. Soc. Proc.*, 729, 229–234, 2002.

33. M.B. Cohn, K.F. Bohringer, J.M. Noworolski, A. Singh, C.G. Keller, K.Y. Goldberg, R.T. Howe, Microassembly technologies for MEMS, *Proc. SPIE*, 3514, 2–16, September, 1998.

34. K. Najafi, Micropackaging technologies for integrated microsystems: applications to MEMS and MOEMS, *Proc. SPIE*, 4979, 1–19, 2003.

35. http://www.amkor.com — wafer bumping, MEMS.

36. http://www.necel.com — general packing information.

37. http://www.oki.com/semi/english — IC packages, materials.

38. http://www.intel.com — IC packaging, environmental conditions.

39. L.S. Goodman, Geometric optimization of controlled collapse interconnections, *IBM J. Res. Dev.* 13(3), 251–265, 1969.

Packaging

40. M.N. Nguyen, M.B. Grosse, Low moisture polymer adhesive for hermetic packages, *IEEE Trans. Comp, Hybrids, Manf. Tech.*, 15, 964–971, 1992.

41. J. Faris, T. Kocian, DMD™ Packages — evolution and strategy, *Tex. Instrum. Tech. J.*, 15(3), 87–94, 1998.

42. A. Kunzman, J. O'Connor, D. Segler, T. Migl, K. Bell, Advancing the DMD™ device for high-definition home entertainment, *Proc. SPIE*, 4207, 1–10, 2000.

43. W. Benett, P. Krulevitch, A flexible packaging and interconnect scheme for microfluidic systems, *Proc. SPIE*, 3606, 111, 1999.

44. P. Galambos, G. Benavides, M. Okandan, M. Jenkins, D. Hetherington, Precision alignment packaging for microsystems with multiple fluid connections, *IMECE 2001 Proc.*, IMECE 2001/MEMS-23902, November 2001.

45. G.L. Benavides, P.C. Galambos, Electromicrofluidic packaging, Sandia National Laboratories Report, SAND2002-1941, June, 2002.

46. G.L. Benavides, P.C. Galambos, J.A. Emerson, K.A. Peterson, R.K. Giunta, R.D. Watson, Packaging of electromicrofluidic devices, U.S. Patent 6,443,179, issued September 3, 2002.

47. G.L. Benavides, P.C. Galambos, J.A. Emerson, K.A. Peterson, R.K. Giunta, R.D. Watson, Packaging of electromicrofluidic devices, U.S. Patent 6,548,895, issued April 15, 2003.

10 Reliability

10.1 RELIABILITY THEORY AND TERMINOLOGY

Reliability is the *ability of a device or system to perform a required function for a specified amount of time*. The function and level of reliability required of a device are obviously application specific. A sensor vital to an aircraft flight control system whose failure can mean loss of life is an example of a high-reliability application. Alternatively, a switch in a piece of commercial electronics may fail and be an annoyance, but it is hardly a tragedy. Since World War II, reliability techniques have been widely applied to military and commercial devices and systems; this has led to the development of the field of reliability. Examples of the reliability literature can be found in the journals listed in Table 10.1.

The reliability of a system may depend upon many things, such as the subsystems (e.g., individual electrical, mechanical, or MEMS devices); packaging; power supply; software; or cooling systems. The overall reliability of a system is complex because the interrelations between the different subsystems can combine in complex ways to cause a system failure. Alternatively, the system could be designed so that subsystems can compensate for failures in other aspects of the system. For example, software could be designed to monitor the system and avoid a system failure by utilizing other subsystems to perform a function.

However, this chapter focuses on reliability at the device level — specifically, the reliability of MEMS devices. Data to access the reliability of a device are obtained by operating a large number of devices under normal operating conditions and noting when failure occurs. Three items need to be defined unambiguously to formulate the reliability experiment properly:

- *Method of operation*: a detailed definition of the method of device operation — for example, a switch is the device to be tested for reliability. An application of voltage (10 V/0 V) to close or open a switch at 100-Hz frequency using a square wave signal.
- *Definition of failure*: definition of what constitutes failure of this device — for example, resistance of the switch contact > 1 kΩ when the 10-V operate signal is applied or the resistance < 1 MΩ when the 0-V signal to open the switch is applied. This statement of failure denotes that contacts are fouled so that adequate continuity when closed is not present or sufficient isolation is not present when the switch is open. This condition may be present when the switch mechanism has failed

367

368   Micro Electro Mechanical System Design

TABLE 10.1
List of Reliability Journals

Journal	Publisher
Microelectronic Reliability	Elsevier Publishers
IEEE Transactions on Reliability	IEEE
Reliability Engineering & System Safety	Elsevier Publishers
Risk Analysis	Blackwell Science

---

so that the contacts do not open or close when the appropriate signal is applied.

- *Sample size*: a large enough sample size utilized in the reliability test so that a statistically meaningful result can be obtained. Reliability data are by their nature nondeterministic because numerous factors can affect the outcome of the reliability experiment. The nondeterministic factors include manufacturing variations (e.g., deposition, etch, patterning); material variations; and environmental variations (e.g., humidity, particulate, material interactions, shock, vibration, thermal). These variations cannot be predicted; therefore, the reliability data must be considered *random* and analyzed with statistical methods.

These three items define how to operate the device, what constitutes failure, and the number of items to test. The number of cycles that have elapsed when the failure occurs defines when the failure occurs, which can have *units of time* or *cycles of operation*. In the example discussed earlier, the cycles of operation were open–close switch operation cycles. For other devices, this could possibly be cycles of force, voltage, or pressure application.

Figure 10.1 shows an example of how reliability data may be recorded and analyzed. Figure 10.1a graphically shows the record of the raw data (i.e., device designation number vs. the time of failure of that device). In this case, the time of failure is denoted in units of 10^6 h of operation. This plot by itself is not unduly meaningful. Distribution plots of when the failures occur can provide more insight.

Figure 10.1b is a *cumulative failure distribution* plot, which plots the total number of failures that occurred up to a specified operation time. The cumulative failure distribution plot can be derived from the raw data plot (Figure 10.1a). For example, if the first failure occurred at ~0.02×10^6 h of operation, the total number of devices that failed up to that time is one. Proceeding along the time-to-failure axis of Figure 10.1a and denoting the total number of devices that have failed up to that time is then plotted on the ordinate (i.e., total number of device failures) of Figure 10.1b. For example, a total of five devices have failed by 0.5×10^6 h of operation. The last device in the sample set failed at ~0.95×10^6 h of operation; therefore, the total number of device failures at that time is ten (the sample size).

Reliability

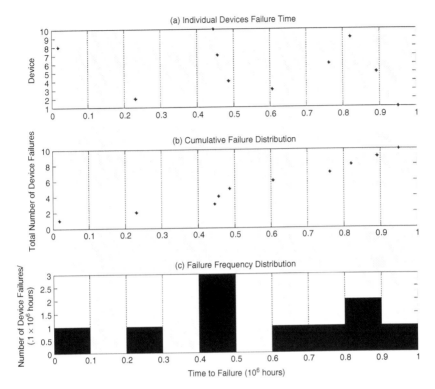

FIGURE 10.1 Device failure and failure distribution plots utilizing a ten-device sample set. (a) Individual device failure vs. time. (b) Cumulative failure distribution. (c) Failure frequency distribution.

The total number of device failures in a cumulative distribution is always 0 at $t = 0$, and the cumulative distribution is equal to the sample size as $t \Rightarrow \infty$. Instead of dealing with numbers of devices for the cumulative distribution, the plot can be put on a probability basis by normalizing by the total number of devices tested. In this case, the plots would be known as the cumulative probability distribution function (CPDF). It is assumed that there are no inoperative devices at $t = 0$ (i.e., no samples dead on arrival). For purposes of this discussion, the initially inoperative devices would factor into the fabrication yield instead of the reliability of the device. The cumulative failure distribution shows the minimum time to failure of a device, the maximum time of failure of all the devices, and how device failures were distributed between those limits. The slope of the cumulative failure distribution shows the device failure rate over the duration of the test data.

A frequently used unit in reliability test data is the FIT (failure in time). A FIT is defined as one failure per billion (10^9) device hours or operations. This definition of the FIT unit was chosen because failure rates are typically very small for most devices in production.

370   Micro Electro Mechanical System Design

Figure 10.1c is the *failure frequency distribution* plot, which is a measure of the failure rate of the device at a point in time. This plot can be obtained from the cumulative failure distribution plot (Figure 10.1b). By specifying a time bin size (e.g., 0.1×10^6 h of operation for this example) and noting the number of failures within the time bin as the bin is moved in time across the cumulative distribution plot (Figure 10.1b), the failure frequency distribution plot (Figure 10.1c) is obtained. This method is in essence obtaining a numerical derivative of the cumulative distribution plot to produce the failure frequency distribution plot. This plot shows how the failure rate varies over time.

The distribution plots obtained in Figure 10.1 appear to be somewhat discontinuous and may be indicative of a small sample size. If the sample size is increased from 10 to 100 and the experiments are run again (Figure 10.2), the data trends become more apparent. These data will require computer analysis, which implements the methods discussed for Figure 10.1. The cumulative distribution (Figure 10.1b) appears to be monotonically increasing, and the failure frequency distribution (Figure 10.1c) is approximately constant. These example reliability experiments appear to have a uniform failure frequency distribution. The plots in Figure 10.1 and Figure 10.2 can be put on a probability basis by normalizing the ordinate of the cumulative distribution and frequency distribution plots by the sample size.

These examples have shown that random reliability data need to be analyzed with statistical tools, which will be discussed in the next section. Before delving into statistics, look at the failure rate typically observed for a large class of machinery [1], electronic [2], and MEMS [3] devices. Figure 10.3 is a plot of a failure rate vs. time (i.e., failure frequency distribution) curve called the *bathtub curve*. This curve is composed of three principal regions:

- *Infant mortality*: $0 < t \leq t_{infant}$. The failure rate is initially very high but decreases as the latent defects cause devices to fail.
- *Constant failure rate*: $t_{infant} < t < t_{operation}$. The failure rate for a reliable device is small and constant. Failures in the region are random.
- *Wear-out*: $t_{operation} \leq t \leq t_{wear-out}$. The failure rate is increasing rapidly due to wear-out.

In some cases, a fourth region, called the *depletion region*, occurs beyond the wear-out region. Few operational devices remain in the depletion region, and the failure rate again decreases. The region is of little practical importance because the remaining devices are well beyond the wear-out region.

10.2 ESSENTIAL ASPECTS OF PROBABILITY AND STATISTICS FOR RELIABILITY

The study of reliability is deeply involved with probability and is extremely important to the design of engineered systems. The time at which a specific device

Reliability

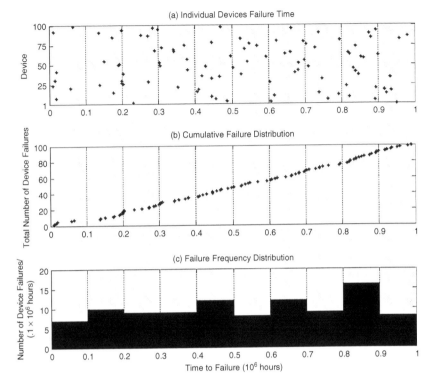

FIGURE 10.2 Device failure and failure distribution plots utilizing a 100-device sample set. (a) Individual device failure vs. time. (b) Cumulative failure distribution. (c) Failure frequency distribution.

will fail is unknown, but it is often possible to determine the probability of failure of that device. In order to be able to quantify the reliability of a device, it is necessary to rely on the framework of probability and statistics. To establish a framework suitable to quantify reliability, a few key concepts of probability theory need to be discussed. It is not possible to provide an in-depth explanation of probability within the scope of this chapter, but several good references [5,6] can provide a more complete background.

An elementary definition of probability involves the relative frequency that particular events will occur. For example, an experiment performed N times had an outcome of event A, N_A times. Therefore, the probability of event A, Pr (A), is defined as the ratio of N_A and N (Equation 10.1). For example, rolling a die has six possible outcomes. The probability that a 2 will be rolled is no different from that for any of the other five numbers. Therefore, the probability of rolling a 2 is Pr (2) = 1/6.

$$\Pr(A) = \frac{N_A}{N} \tag{10.1}$$

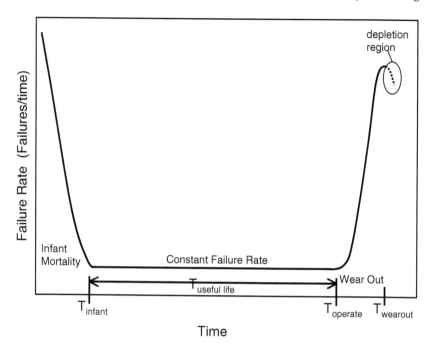

FIGURE 10.3 A failure frequency distribution curve — the "bathtub" curve.

Probability as stated here will have the following axioms:

- Probability is a number between 0 and 1:

$$0 \leq \Pr(A) \leq 1 \tag{10.2}$$

- The probability of a *certain* event, A, is 1:

$$\Pr(A) = 1 \tag{10.3}$$

- The probability of an *impossible* event, A, is 0:

$$\Pr(A) = 0 \tag{10.4}$$

- The probability of a *complete* set of mutually exclusive events is 1. For example, the probability of a complete set of events A, B, C, D, E, F is $\Pr(A) + \Pr(B) + \Pr(C) + \Pr(D) + \Pr(E) + \Pr(F) = 1$.

Probability can be used to describe finite events (e.g., rolling a die) or continuous events (e.g., voltages that vary between 0 and 5 V). A finite event has a countable number of possible outcomes, whereas a continuous event has an infinite number of possible outcomes. In many tests or measurements, the number

Reliability

373

of possible outcomes is not finite. Consider the test of selecting a bolt from a bin with specified failure strength. The actual value is expected to be close to the labeled value, but strength of the bolt will vary by some unknown amount. The variances from the labeled value are due to manufacturing variations, which can assume any value within a specified range. The actual value of the bolt strength is unknown in advance.

Even if the bolt is taken from a bin labeled "70 kpsi," the actual failure strength will vary over a range. The actual probability of selecting a bolt with strength of exactly 70 kpsi is zero. In this example, the bolt strength is a random variable. To study distribution of values of random variables such as bolt strength or the lifetime of a MEMS device, the concept of distributions must be introduced.

A *random variable* is a function whose values are real numbers that depend upon *chance*. In the preceding discussion, the values of a rolled die, voltage, or bolt strength are examples of a random variable. If a random variable, X, can assume any value within a specified range, or possibly an infinite range (e.g., voltage, bolt strength), it is called a continuous random variable. Discrete random variables describe events (e.g., value of a rolled die) that can assume only certain discrete values. Continuous and discrete random variables can be treated with the same concepts of distributions.

Given that X is a random variable and x is any allowed value of the random variable, the *cumulative probability distribution function* (CPDF), F, is the *probability* that the event described by the random variable, X, is less than or equal to a specified value, x:

$$F_X(x) = \Pr(X \le x) \qquad (10.5)$$

A CPDF function, such as the functions in Figure 10.4, must obey the following axioms, which are imposed by the definition of probability on their functional nature:

- The CPDF must range between 0 and 1 because the CPDF is a probability that the random variable, X, is less than a value x:

$$0 \le F_X(x) \le 1 - \infty < x < \infty \qquad (10.6)$$

- It is an impossibility for anything to be less than $-\infty$ and a certainty that all values are less than ∞:

$$F_X(-\infty) = 0 \qquad F_X(\infty) = 1 \qquad (10.7)$$

- By definition of a CPDF (i.e., the probability that X is less than a value x), the CPDF probability increases as x increases. Therefore, $F_X(x)$ is monotonically increasing.

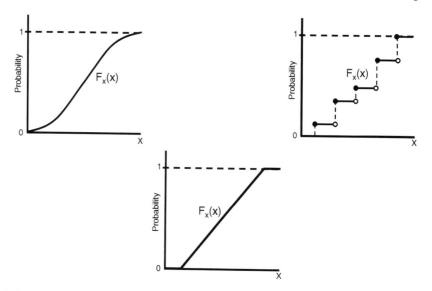

FIGURE 10.4 Example cumulative probability distribution functions.

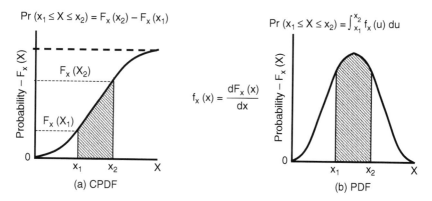

FIGURE 10.5 Using the CPDF and PDF to calculate the probability that a random variable X lies between $x1$ and $x2$.

- The probability that X lies between the values $x1$ and $x2$ can be obtained by subtracting $F_X(x_1)$ from $F_X(x_2)$, which is graphically illustrated in Figure 10.5.

$$\Pr(x_1 < X \le x_2) = F_X(x_2) - F_X(x_1) \tag{10.8}$$

- The probability that the observed random variable X is greater than but not equal to x can be calculated by Equation 10.9 because this is simply the complement of the CPDF. Figure 10.6 illustrates the complement of the CPDF. The complement of a set of values contains all the values

Reliability

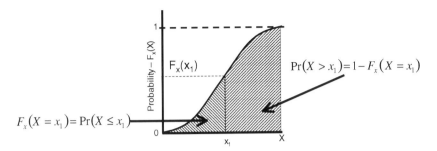

FIGURE 10.6 The CPDF and the complement of the CPDF function.

that are not in the original set, thus yielding the ">" sign in Equation 10.9 because the definition of F_x (Equation 10.5) contains a "≤" sign:

$$\Pr(X > x_1) = 1 - F_X(x) \qquad (10.9)$$

The cumulative probability distribution function, $F(x)$, is a complete distribution of a probability model for a single random variable of interest. Note that the example CPDF functions shown in Figure 10.4 obey the axioms listed previously. The value of the CPDF varies between zero and one and is a monotonically increasing function. However, the rate (i.e., slope) at which the function increases varies.

If the random variable of interest is the number of failures of a device, the failure rate and how it varies over the independent variable x may be of significant interest. For this reason and other useful calculations, it may be preferable to use the derivative of $F_X(x)$, $f_X(x)$ (known as the *probability density function*, PDF), rather than $F_X(x)$. Equation 10.10 is the mathematical expression of the relationship between the CPDF, $F_X(x)$, and the PDF, $f_X(x)$. Equation 10.11 is the converse statement of the relationship between the CPDF, $F_X(x)$, and the PDF, $f_X(x)$.

$$f_X(x) = \frac{dF_X(x)}{dx} \qquad (10.10)$$

$$F_X(x) = \int_{-\infty}^{x} f_X(u) du \qquad (10.11)$$

Figure 10.7 illustrates the relationship between the CPDF and the PDF for a uniformly distributed random variable. On the interval $x_1 < x < x_2$, the CPDF has a constant slope, and the slope is zero for $x < x1$ and $x > x2$. Because the PDF is the derivative of the CPDF, this results in a region on the interval $x_1 < x < x_2$ of the PDF with uniform magnitude $1/(x_1 - x_2)$, which is the slope of the CPDF in the corresponding interval. The PDF function has zero slope and magnitude for

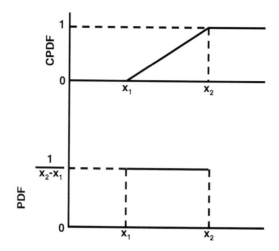

FIGURE 10.7 The CPDF and PDF for a uniform distribution.

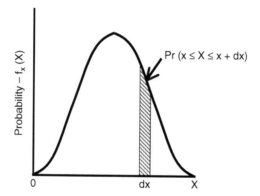

FIGURE 10.8 The physical significance of the probability density function, $fX(x)$, described by the probability element, $fX(x)dx$.

the regions $x < x1$ and $x > x2$. These results are consistent with the relationships expressed in Equation 10.10 and Equation 10.11.

The physical significance of the probability density function can be described by the probability element, $f_X(x)dx$, which is illustrated in Figure 10.8. The probability that the random variable X lies in the range x to $x + dx$ is the area under the PDF between x to $x + dx$ (i.e., the probability element, $f_X(x)dx$). The properties of the PDF as summarized follow:

- The PDF, which is based on probability, is a non-negative function:

$$f_X(x) \geq 0 \quad -\infty < x < \infty \quad non-negative \quad (10.12)$$

Reliability

- The area under the PDF over the entire range of x, $-\infty < x < \infty$ is the probability of the random variable occurrence, which is a certainty, 1:

$$\int_{-\infty}^{\infty} f_X(x)dx = 1 \qquad (10.13)$$

- The area under the PDF over the range $x_1 < x < x_2$ is the probability of occurrence of the random variable:

$$\int_{x_1}^{x_2} f_X(x)dx = \Pr(x_1 < x < x_2) \qquad (10.14)$$

The most significant metrics associated with statistical methods and, in particular, the PDF functions are measures of the "central value" and the "spread" of the distribution. The definitions and significance of the *mean value* and *variance* of a distribution will be discussed.

The mean, μ, and variance, σ^2, of a continuous PDF function are given by Equation 10.15 and Equation 10.16, respectively. For discrete representations, the mean and variance are given by Equation 10.17 and Equation 10.18. The mean, μ, is also known as the *expectation of X* and is denoted by $E[X]$, which is read "the expected value of X." If the random variable, X, is time or number of cycles of a reliability test of a device, the mean, μ, is the average time of failure of the devices. The variance, σ^2, is a measure of the spread or dispersion about the mean. The positive square root of the variance is called the standard deviation, which is denoted by σ. Figure 10.9 shows the relationship of the mean and variance for a Gaussian distribution. Example 10.1 illustrates the calculation of the mean and variance for a given CDPF and PDF.

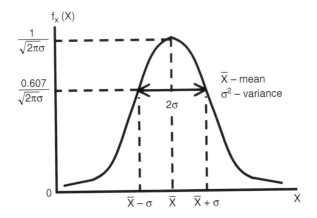

FIGURE 10.9 The mean and variance of a Gaussian distribution.

$$\mu = \int_{-\infty}^{\infty} x f_X(x)\, dx \tag{10.15}$$

$$\sigma^2 = \int_{-\infty}^{\infty} (x-\mu)^2 f(x)\, dx \tag{10.16}$$

$$\mu = \sum_j x_j f(x_j) \tag{10.17}$$

$$\sigma^2 = \sum_j (x_j - \mu)^2 f(x_j) \tag{10.18}$$

Example 10.1

Problem: Figure 10.2 shows discrete data for a reliability test of a switch. Frequently, for reliability test data, a functional fit to the test data is done using the techniques of regression analysis [53]. Because the data in Figure 10.2 are simulated, a uniform distribution model was used. A mathematical model provides enhanced ability for analysis and exploration of the different types of models that may be relevant. The next section will discuss several of the frequently used reliability models. Given the functional formulation for the CPDF and PDF function shown in Figure 10.10 calculate and plot the following: (a) mean; (b) variance; and (c) standard deviation (Figure 10.10).

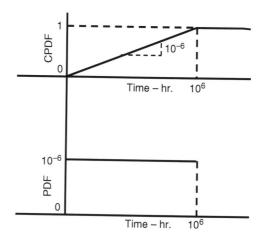

FIGURE 10.10 Functional fit of the CPDF and PDF data of Figure 10.2.

Reliability

Solution: (a) $X(t)$ is the random variable for the failures that is a function of time. Utilizing Equation 10.15, the mean can be calculated as follows:

$$E[X] = \mu = \int_{-\infty}^{\infty} t f_X(t)\, dt$$

$$\mu = \int_{0}^{10^6} t\,(10^{-6})\, dt$$

$$\mu = \frac{10^6}{2}\quad hours$$

(b) Utilizing Equation 10.16, the variance is calculated as follows:

$$\sigma^2 = \int_{-\infty}^{\infty} \left(t - \mu\right)^2 f\left(t\right) dt$$

$$\sigma^2 = \int_{0}^{10^6} \left(t - \mu\right)^2 10^{-6}\, dt$$

$$\sigma^2 = \frac{10^{12}}{12}$$

(c) Utilizing the definition of standard deviation,

$$\sigma = \frac{10^6}{\sqrt{12}} \approx 0.29 x 10^6$$

Another important reliability measure is the *instantaneous failure rate* or *hazard rate*, $h_X(t)$. In other words, the hazard rate is the failure rate of the survivors at time t. The hazard rate is defined as the failure rate, $f_X(t)$ (i.e., PDF of failure), normalized by the probability of devices surviving, $R_X(t)$. The probability of devices surviving, $R_X(t)$, is the complement of the probability of device failures, $F_X(t)$, occurring up to time t. This can be expressed as shown in Equation 10.19. The hazard rate, $h_X(t)$, can be defined in terms of the CPDF and PDF functions previously discussed, as shown in Equation 10.20. The hazard rate is expressed in the FIT units (i.e., failures per billion cycles of operation) previously discussed. The bathtub curve discussed in Section 10.1 is a plot of the hazard rate vs. time or cycles of operation.

$$R_X\left(t\right) = 1 - F_X\left(t\right) \tag{10.19}$$

$$h_X\left(t\right) = \frac{f_X(t)}{R_X(t)} = \frac{f_X(t)}{1 - F_X(t)} \tag{10.20}$$

10.3 RELIABILITY MODELS

Three distribution functions defined and discussed in Section 10.2 are highly relevant to the modeling of reliability:

- Cumulative probability distribution function (CPDF), $F_X(t)$
- Probability density function, $f_X(t)$
- Hazard function or instantaneous failure rate, $h_X(t)$

Reliability can be measured via a well defined experimental program involving a rigorous definition of the method of operation, a definition of failure, and a statistically significant sample set. As in Section 10.1, the data from these reliability experiments can be used empirically to produce the distributions (i.e., CPDF, PDF, hazard function) significant to the study of reliability. These empirical distributions can then be mathematically analyzed [8]. The data from reliability experiments can be approximated by continuous functions [7] whose properties can then be analyzed mathematically to provide insight to device reliability. Three reliability models and their applications to reliability (summarized in Table 10.2) will be discussed:

- Weibull model
- Lognormal model
- Exponential model

10.3.1 WEIBULL MODEL

The Weibull model is widely used in reliability modeling, largely because of its ability to characterize the various regions of the reliability lifetime curve (i.e., bathtub curve) with two parameters, β and λ. The CPDF and PDF and hazard distributions for the Weibull model are given in Equation 10.21 through Equation 10.23, and the mean and variance of the Weibull distribution are shown in Equation 10.24 and Equation 10.25. The two parameters of the Weibull model are:

- λ — Characteristic lifetime parameter ($\lambda > 0$)
- β — shape parameter ($\beta > 0$)

The characteristic lifetime, λ, for the Weibull distribution is the point before which 63.2% of failures occur. This can be seen from Equation 10.22 when t = λ and $F(\lambda) = (1 - 1/e) = 0.632$. Different values of λ change the scale of the time axis without affecting the shape. The shape parameter, β, describes how the failure rate is distributed about the characteristic lifetime:

Reliability

381

TABLE 10.2
Reliability Model Summary

Model	Parameter	Characteristics
Weibull model	Two parameters: λ — characteristic lifetime β — shape parameter	• Parameters have physical meaning in a reliability context • Can model decreasing failure rate typical of infant mortality • Can model an increasing failure rate typical of wear-out • Reduces to the exponential model for $\beta = 1$, which models a constant failure rate; h = constant
Lognormal model	Two parameters: t_{50} — median lifetime σ — standard deviation	• Parameters have physical meaning in a reliability context • Well suited for modeling wearout • Failure rate is neither always increasing nor decreasing, but the failure rate increases on average
Exponential model	One parameter: λ — characteristic lifetime	• Little physical justification for this model • Failure rate, h, is always constant • μ, σ^2, $h(t)$ uniquely defined by λ

- $0 < \beta < 1$: h decreases with time
- $\beta = 1$: reduces to the exponential model; h is constant
- $\beta > 1$: h increases with time

Figure 10.11 is a plot of the CPDF, PDF, and hazard functions for the Weibull model with a characteristic lifetime, $\lambda = 2$, and shape parameters, $\beta = 0.5$, 1.0, and 2.0. All of the CPDF functions have a value of 0.632 at $\lambda = 2$. This is consistent with the definition of characteristic lifetime in the context of the Weibull model. The hazard function varies significantly with the three values of the shape parameter β. For $\beta = 1$, the Weibull reduces to the exponential model, which yields a constant hazard function. For a shape parameter of $\beta = 0.5$ and $\beta = 2.0$, the hazard function is decreasing and increasing, respectively. This ability to model this range of hazard functions allows the Weibull distribution to model the three regions of the bathtub curve (i.e., infant mortality, useful life, and wear-out). The Weibull model describes infant mortality very well but does not model wear-out as well.

Figure 10.12 is a plot of the Weibull distribution for several characteristic lifetimes, $\lambda = 2$, 5, and 8, for a common shape parameter, $\beta = 2.0$. The CPDF function has a value of 0.632 at the various characteristic lifetimes, and the hazard function is increasing for each curve. However, the slope of the hazard function curve varies depending upon λ (Equation 10.23).

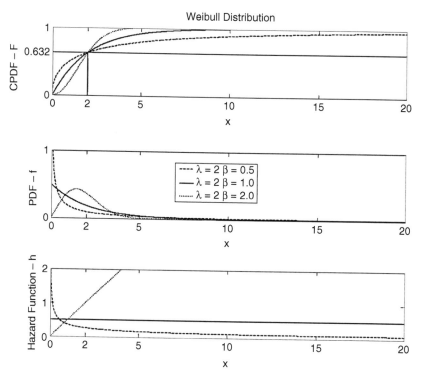

FIGURE 10.11 Weibull distribution with characteristic lifetime of $l = 2$ and a variety of shape parameters ($b = 0.5, 1, 2$).

$$f(t) = \frac{\beta}{\lambda}\left(\frac{t}{\lambda}\right)^{(\beta-1)} \exp\left[-\left(\frac{t}{\lambda}\right)^{\beta}\right] \quad (10.21)$$

$$F(t) = 1 - \exp\left[-\left(\frac{t}{\lambda}\right)^{\beta}\right] \quad (10.22)$$

$$h_X(t) = \frac{\beta}{\lambda}\left(\frac{t}{\lambda}\right)^{\beta-1} \quad (10.23)$$

$$\mu = \lambda \Gamma\left(1 + \frac{1}{\beta}\right) \quad (10.24)$$

Reliability

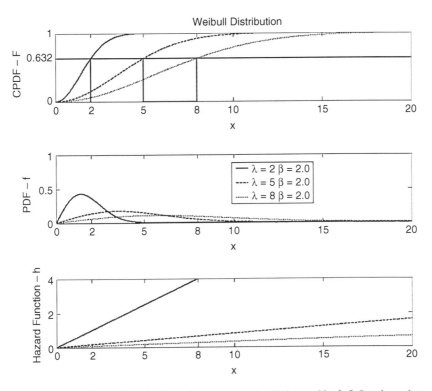

FIGURE 10.12 Weibull distribution with characteristic lifetimes of $l = 2, 5, 8$ and a variety of shape parameters $b = 2.0$.

$$\sigma^2 = \lambda^2 \left\{ \Gamma\left(1 + \frac{2}{\beta}\right) - \left[\Gamma\left(1 + \frac{1}{\beta}\right)\right]^2 \right\} \quad (10.25)$$

The gamma function, $\Gamma(x)$, used in the calculation of the mean, μ, and variance, σ^2, is given by Equation 10.26 and tabulated in many mathematical references [9].

$$\Gamma(x) = \int_0^\infty t^{x-1} e^{-t} dt \quad \text{for} \quad x > 0 \quad (10.26)$$

10.3.2 Lognormal Model

The lognormal model is another two-parameter model widely utilized for reliability applications. The two positive real parameters are t_{50} and σ. The median life is modeled by the t_{50} parameter and the σ parameter is the standard deviation

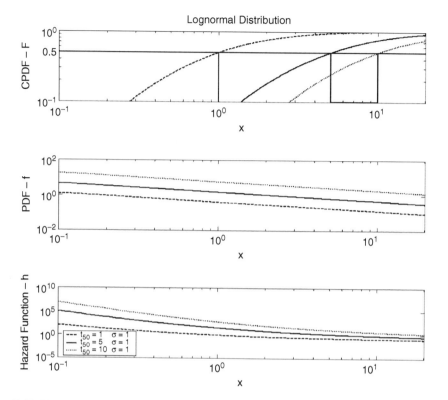

FIGURE 10.13 Lognormal distributions with a variety of median lifetimes of $t50 = 1, 5, 10$ and a standard deviation of $s = 1$.

of the distribution. Equation 10.28 through Equation 10.30 define the CPDF, PDF, and hazard functions for the lognormal model.

The lognormal model arises from a relationship between a random variable, Y, which is Gaussian distributed and defined as a logarithm of another random variable, X (Equation 10.27). This situation physically arises in communication systems in which the attenuation of a signal, Y, in the transmission path is Gaussian distributed and it is logarithmically related to the ratio of the input and output signal powers, X. The situation also occurs in reliability lifetime data in which the logarithm of the time to fail has a Gaussian distribution.

$$Y = \ln X \qquad (10.27)$$

Figure 10.13 and Figure 10.14 are examples of lognormal distributions that illustrate the effect of the model parameters, t_{50} and σ. The lognormal CPDF function has a value of 0.5 at the median life, t_{50}. The failure rate for the lognormal distribution is neither always increasing nor always decreasing. Lognormal dis-

Reliability

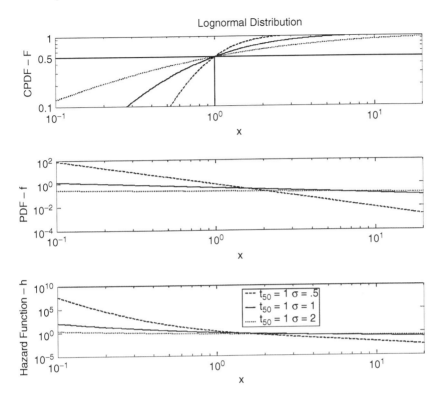

FIGURE 10.14 Lognormal distributions with a variety of standard deviations of $s = 0.5$, 1, 2, and a median lifetime of $t50 = 1$.

tribution is widely used in reliability modeling and is especially well suited for modeling wear-out.

$$f(t) = \frac{1}{\sigma\sqrt{2\pi}} \exp\left[\frac{-1}{2\sigma^2}\left(\ln t - \ln t_{50}\right)^2\right] \quad (10.28)$$

$$F(t) = \frac{1}{2}\left\{1 + erf\left[\frac{1}{\sqrt{2}}\left(\frac{\ln t - \ln t_{50}}{\sigma}\right)\right]\right\} \quad (10.29)$$

$$h(t) = \frac{\frac{1}{\sigma\sqrt{2\pi}} \exp\left[\frac{-1}{2\sigma^2}\left(\ln t - \ln t_{50}\right)^2\right]}{\frac{1}{2}\left\{1 - erf\left[\frac{1}{\sqrt{2}}\left(\frac{\ln t - \ln t_{50}}{\sigma}\right)\right]\right\}} \quad (10.30)$$

The error function, erf(x), used in the calculation of the preceding distributions is given by Equation 10.31 and tabulated in many mathematical references [9].

$$erf(x) = \frac{2}{\sqrt{\pi}} \int_0^x \exp\left(-u^2\right) du \qquad (10.31)$$

10.3.3 EXPONENTIAL MODEL

The exponential model is a very simple, widely used one-parameter, λ, model; however, there is little physical justification. The CPDF, PDF, and hazard functions for the exponential model are defined in Equation 10.32 through Equation 10.34. Equation 10.35 and Equation 10.36 define the mean, μ, and variation, σ^2, of the distribution. Note that the mean, variance, and hazard function for the exponential model are all uniquely defined by the parameter λ.

Because the hazard function for the exponential model is a constant, this model is appropriate for systems involving truly random events such as occur in the useful life portion of the bathtub curve. The model cannot model infant mortality or wear-out. Figure 10.15 is a plot of the CPDF, PDF, and hazard functions for the exponential model with $\lambda = 1.0$, 5.0, and 10.0. The CPDF function has a value of 0.632 at the appropriate characteristic lifetime for each curve plotted. Because this is a one-parameter model, the characteristic lifetime parameter, λ, also defines the hazard function.

$$f(t) = \lambda \exp\left(-\lambda t\right) \qquad (10.32)$$

$$F(t) = 1 - \exp\left(-\lambda t\right) \qquad (10.33)$$

$$h\left(t\right) = \lambda \qquad (10.34)$$

$$\mu = \frac{1}{\lambda} \qquad (10.35)$$

$$\sigma^2 = \frac{1}{\lambda^2} \qquad (10.36)$$

10.4 MEMS FAILURE MECHANISMS

Since World War II, reliability methodology has been increasingly applied to a wide spectrum of devices and systems [10]. The reliability of microelectronic

Reliability

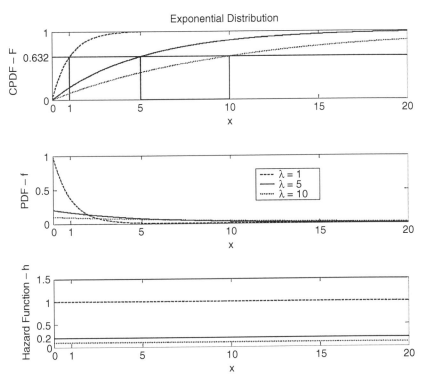

FIGURE 10.15 Exponential distributions with a variety of characteristic lifetimes $l = 1$, 5, 10.

devices has been extensively studied and their failure mechanisms are well understood. Microelectronics share a common set of failure mechanisms, which include hot carriers, oxide breakdown, stress voiding, and electromigration. These failure mechanisms have been physically and statistically studied extensively, thus enabling the prediction of performance in various environments.

MEMS applications cover a very wide spectrum and a large number of device designs have been implemented to address these applications. As a result of the breadth of application, the reliability of MEMS devices is not dominated by the behavior of a "typical" device. The failure mechanisms that may be experienced by MEMS devices are quite broad, especially in comparison to microelectronics. As a result, these failure mechanisms are more varied than those of microelectronics.

Currently, MEMS failure mechanisms are not well characterized or understood; it is not wise to use a macroscale failure mechanism or reliability data directly for a MEMS device because this would require an assumption of similarity that is generally not valid.

The area of understanding the failure mechanisms of MEMS devices in a reliability context is just in its formative stages. However, some published works have documented the first steps of understanding failure mechanisms of MEMS

TABLE 10.3
Common MEMS Failure Mechanisms

Operational failure mechanisms
Wear
Fracture
Fatigue
Charging
Creep
Stiction and adhesion

Degradation mechanisms
Thermal degradation
Optical degradation
Environmental degradation
Stress corrosion cracking

Environmental failure mechanisms
Shock
Vibration
Thermal cycling
Humidity
Radiation
Electrostatic discharge

devices [1,4,11]. MEMS failure mechanisms that have been identified are listed in Table 10.3. The failure mechanisms may be categorized as operational, environmental, and degradation failure mechanisms. *Operational failure mechanisms* are failures due primarily to the device operation. The *environmental failure mechanisms* are failures due primarily to a physical environment that the device may experience. *Degradation failure mechanisms* are subtle failures arising from operational or environmental means that would alter the device performance sufficiently to cause it to fall out of specification.

10.4.1 OPERATIONAL FAILURE MECHANISMS

10.4.1.1 Wear

Wear is the removal of material from a solid surface due to mechanical action. Although the laws of friction are well established, there is no generally accepted theory of wear. A very complex phenomenon involving the mechanics and chemistry of the bodies in contact, wear depends upon a number of variables, such as hardness of materials, contact area, loading, surface speed, etc. (Figure 10.16). Four processes cause wear: *adhesion, abrasion, corrosion,* and *surface fatigue.*

Adhesive wear is caused by one surface pulling material off another as they are sliding. This has been shown to be a primary wear mechanism for polysilicon MEMS devices [12,13]. Figure 10.17 shows wear effects in a rotating MEMS

Reliability

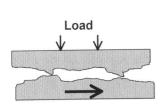
(a) Wear is caused by relative motion of loaded surfaces

(b) Wear debris on a polysilicon surface

FIGURE 10.16 Schematic of wear mechanism and wear debris. (Courtesy of Sandia National Laboratories.)

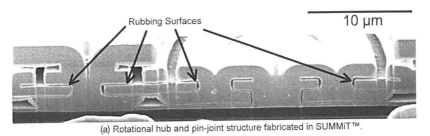
(a) Rotational hub and pin-joint structure fabricated in SUMMiT™.

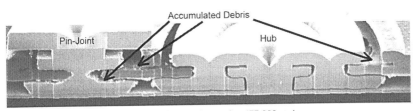
(b) Wear of rubbing surfaces after 477,000 cycles.

FIGURE 10.17 Wear effect in a rotating MEMS device. (Courtesy of Sandia National Laboratories.)

device. Abrasive wear occurs when a hard, rough surface moves relative to a softer surface. Abrasive wear can occur in MEMS [14] when wear particulate is entrained in the motion of moving surfaces. Corrosive wear occurs when surfaces chemically interact and the relative motion removes the reaction products. Corrosive wear may be a wear mechanism involved in biological or fluidic MEMS applications. Surface fatigue wear will occur in rolling applications and will produce surface fatigue cracks in the materials. Friction and wear in MEMS devices is a significant issue because lubrication and bearing systems typical in

390 Micro Electro Mechanical System Design

macrosystems are not feasible in MEMS devices. The self assembled monolayer (SAM) coatings discussed in the packaging chapter (Chapter 9) are methods utilized in MEMS devices to mitigate friction and wear.

10.4.1.2 Fracture

Fracture is a material failure mechanism that results in the structure breaking into separate pieces. The two types of fracture are ductile and brittle. A ductile material (such as many metals) will deform plastically and permanently before ultimate failure when the material fractures. Brittle materials will not plastically deform before failure. Silicon is an example of a brittle material.

The force levels of all but a few MEMS actuators will not be able to fracture MEMS materials directly unaided. Therefore, the fracture failure mechanism will be initiated by an environmental force (e.g., shock or vibration), possibly in combination with a chemical–material interaction to weaken the material. For example, stress corrosion cracking (SCC) is a mechanism observed in polysilicon and single-crystal silicon [15] (Figure 10.18). Silicon and polysilicon have a thin layer of native silicon dioxide covering the surface. If the material is sufficiently stressed to crack the layer of silicon dioxide, it will expose new polysilicon or silicon, which will oxidize to form more silicon dioxide. As this process continues, the material will eventually weaken sufficiently enough for fracture to occur.

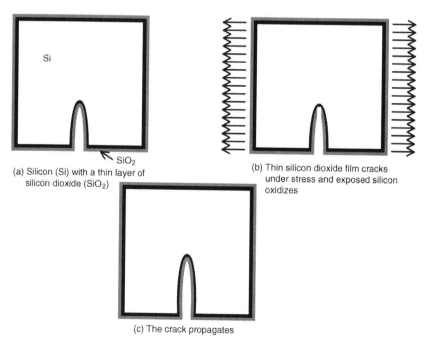

FIGURE 10.18 Stress corrosion cracking (SCC) mechanism in silicon.

Reliability   **391**

10.4.1.3 Fatigue

Fatigue is a failure mechanism caused by the cyclic load of a structure. Vibrating structures such as resonators or a vibratory gyroscope are examples of structures that may be susceptible to fatigue failure. Fatigue material failure occurs by the formation and growth of surface cracks in the material; these grow as the material is cyclically loaded. Fatigue can also affect the internal material damping and electrical resistance of the material, which can also affect device performance. The research on fatigue of MEMS materials to date has shown that fatigue has typically not been observed with silicon materials [19], and the fatigue issues observed with aluminum may be overcome with the appropriate alloy and annealing [20].

10.4.1.4 Charging

Charging is an issue for MEMS devices that contain dielectric layers. Charging effects include sensor drift over time and changes in actuation voltage. Charging may be due to ionizing radiation or the high field strengths required for electrostatic actuation of MEMS devices [16,17]. Charge build-up can be mitigated by utilizing low electric fields or trap-free dielectrics. Another approach is to make the dielectric material leaky so that charge flows away immediately [18].

10.4.1.5 Creep

Creep is a plastic strain of a material that occurs over time (i.e., a slow movement of atoms under mechanical stress). This may be an important issue for MEMS that use metal layers [20]. Creep on the macroscopic scale is generally influenced by operating temperature. A general rule for macroscopic design is that if the operating temperature is less than 0.3 of the melting temperature of the material and the stress is moderate, creep is not an issue [11]. Direct utilization of macroscopic rules of thumb for microscale phenomena and materials is not generally recommended, but the use of higher melting point materials or lower operating temperature will generally result in greater creep resistance.

10.4.1.6 Stiction and Adhesion

Stiction is one of the most important failure mechanisms in MEMS, particularly for surface micromachine fabricated devices. Stiction refers to surfaces coming into contact due to surface forces such as capillary, van der Waals, and electrostatic forces. Due to the scaling effects of MEMS, surface forces become dominant at the microscale level. Some of the initial modeling of capillary forces that occur during the release and drying process of surface micromachining was done by Mastrangelo and Hsu [21]. It has subsequently become very apparent that stiction is a major issue [22,24]. Significant efforts have been undertaken [25] to understand stiction and how its effects may be mitigated. Research has also been undertaken to develop surface coating, which mitigates stiction effects [26,27].

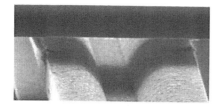 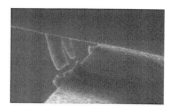

FIGURE 10.19 Anodic oxidation of polysilicon under high electric fields. (R.A. Plass et al., *Proc. SPIE*, 4980, 81–86, Jan. 2003; courtesy of R. Plass, Sandia National Laboratories.)

MEMS device design can also influence the effect of stiction by the use of dimples ("bumpers") to minimize broad area surface contact; increased structure stiffness to prevent collapse; and the minimization of electrostatic surface areas that can interact to cause stiction.

10.4.2 Degradation Mechanisms

A number of material degradation mechanisms can cause device performance to degrade slowly to the point of inoperability. These mechanisms include thermal, optical, and other environmental degradation mechanisms. The stress corrosion cracking [15] discussed earlier is a type of environmental degradation that can possibly occur. Others include degradation of optical coatings or surface due to interaction with the environment. Thermal degradation can occur to the MEMS device or to the MEMS packaging materials. Anodic oxidation of silicon [23] (Figure 10.19) under a high electric field is another long-term material degradation mechanism. Other similar mechanisms occur on MEMS optical coatings.

10.4.3 Environmental Failure Mechanisms

10.4.3.1 Shock and Vibration

Vibration is a deterministic or stochastic continuous force or displacement excitation of a device that may be due to the environment in which it must function. Shock is a single event or pulse applied to a device that may be due to the environment in which it must function. Typically, vibration and shock are measured in acceleration (g). Shock and vibration environments are generally specified [28] by the power spectral density (PSD) of the vibration environment or the shock response spectra (SRS) of the shock environments.

Reliability

Vibration and shock environments could be an issue to MEMS devices by material damage or stiction due to the MEMS surfaces being damaged or forced together. A number of shock and vibration studies for MEMS devices have been performed [29–32]; these show minimal effects on the microscale MEMS device. It has been reported [33] that MEMS inertial sensors have survived shock environments as high as 120,000 g. The type of damage observed has been packaging damage and the movement of particulate.

10.4.3.2 Thermal Cycling

Depending upon the application, thermal environments may be a significant concern for MEMS devices. Space applications may involve temperature environments in the range of –100 to 150°C. Thermal cycling may affect a device through thermal strains between materials that have significantly different coefficients of thermal expansion. For example, metals and silicon may have significantly different coefficients of thermal expansion. The thermal environment may also cause damage of materials used in the MEMS device or its packaging.

10.4.3.3 Humidity

Humidity has been shown to have significant effect on MEMS. Condensation on MEMS surfaces can cause the development of capillary forces that will lead to increased adhesion and stiction effects. Humidity in a cold temperature can cause the formation of ice, which is detrimental to a MEMS device. Wear in MEMS devices has also been shown to be a function of humidity [25,34]. Humidity in MEMS packaging can be controlled through the use of chemical getter technology.

10.4.3.4 Radiation

Radiation can interact with MEMS devices in two ways that will affect device performance. Ionizing radiation can introduce the charging effects discussed previously, which will affect sensor drift, or actuation voltages. Higher levels of radiation can actually cause material damage that can alter material properties. Space application of MEMS devices will encounter radiation environments. Limited work on the radiation effects on MEMS devices [35–37] has been done to date, but indications are that radiation environments will significantly affect MEMS device performance. This is a nontrivial issue that has not been fully addressed.

10.4.3.5 Electrostatic Discharge (ESD)

Electrostatic discharge (ESD) damage, shown in Figure 10.20, will occur when a microelectronic or MEMS device is improperly handled. Large voltages can build up on personnel who are not properly grounded; this can result in significant damage to MEMS devices. Proper handling procedures for packaging and testing personnel are necessary to prevent device damage.

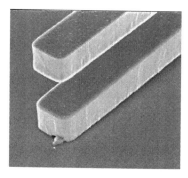

FIGURE 10.20 Electrostatic discharge damage of a MEMS comb finger. (Courtesy of Sandia National Laboratories.)

10.5 MEASUREMENT TECHNIQUES FOR MEMS OPERATIONAL, RELIABILITY, AND FAILURE ANALYSIS TESTING

Like any other device at any scale, a MEMS device requires quantitative experimental data on various aspects of device operation. These data provide necessary information for design or process modification, reliability, and diagnosis of device failure. Providing quantitative data on device operation has a unique set of challenges at the microscale. Macroworld devices can be manipulated, disassembled, and reassembled, and separate discrete instrumentation can be attached to acquire a wide array of data. Due to their scale, MEMS devices are relatively inaccessible via many macroworld test methods. The test methods utilized for MEMS utilize optical noncontact test methods to a large degree. However, a few simple methods for amplitude measurement of dynamic motion (i.e., stroboscopy, blur envelope) or examination of inaccessible portions of a device (i.e., lift-off) are simple, common-sense applications of macroworld methods. An overview of some of the MEMS measurement techniques and their strengths and weaknesses is presented next.

10.5.1 OPTICAL MICROSCOPY

An optical microscope is an easy, cost-effective method for initial examination of a MEMS device. This method can be used for quick examination of defects, debris, textures, stains, fractures, or abnormal displacements of a device. Use of high magnification objectives (e.g., 50×) is good for detection of small deflections (i.e., ~1 μm) of a device. Video taping of device operation is useful for tracking device performance over a long test or off-line detailed image analysis. Figure 10.21 shows a probe station with an optical microscope. The probes are mounted on a three-dimensional micrometer stage and are useful for manual manipulation and applying signals to the MEMS device.

Reliability

FIGURE 10.21 Probe station with an optical microscope.

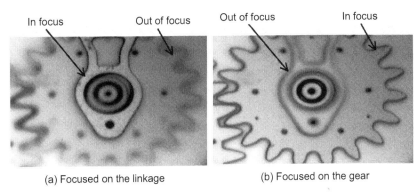

FIGURE 10.22 An optical image illustrating that limited depth of focus will only image structures within the depth of focus clearly. (Courtesy of Sandia National Laboratories.)

A limitation of optical microscopy is the depth of focus, particularly at high magnification. Figure 10.22 shows an out-of-focus image of a MEMS gear. The optical microscope image gives no quantifiable information on the movement perpendicular to the plane of focus.

Typical optical microscopy utilizes *bright field illumination*, which illuminates the specimen parallel with the optical axis of the microscope; thus, objects in the plane of the specimen are bright and inclined objects are dark. Optical microscopy enhancements have been developed that provide enhanced image contrast. These enhancements include *dark field illumination* and *oblique illumination*. Dark field illumination is achieved by insertion of a round patch to block the central rays of the image; therefore, the rays reflected at an angle to the microscope axis are collected to form the image. Oblique illumination is achieved by illuminating the specimen at an angle to the microscope axis that will accentuate any edges on the specimen surface.

Nomarski differential interference contrast is an optical microscopy method that will provide image contrast for surface gradients (i.e., edges and lines) on the sample. Nomarski differential interference contrast is implemented by splitting light into two rays with a Wollaston prism. One of the rays illuminates the sample and both rays are recombined where they can interfere to provide the desired image contrast.

10.5.2 Scanning Electron Microscopy

The scanning electron microscope (SEM) is an extremely useful tool for obtaining information on a MEMS device. The SEM will provide greater magnification and depth of focus than an optical microscope. Also, the MEMS device can be tilted and rotated to obtain unique viewing angles. An SEM can resolve features down to the nanometer (e.g., ~3 nm) scale. An SEM image of a MEMS part will require the part to be placed in a vacuum chamber and possibly flash coated with a thin conductive film to increase image quality (note: the thin conductive film coating may cause the device to become inoperable). The SEM can also be used to assess the electrical continuity of a MEMS device using voltage contrast or resistive contrast imaging [38]. The SEM can image very fine wear debris on a MEMS device and, in conjunction with energy dispersive x-ray spectroscopy (EDX) [39] or an electron energy loss spectroscopy system (EELS) [40], the material composition can be determined.

10.5.3 Focused Ion Beam

Focused ion beam (FIB) systems are extremely valuable tools for experimental evaluation of MEMS devices. FIB systems use a focused beam of ions such as Ga$^+$ at 25 to 50 keV for precise material removal (i.e., sputtering); material deposition (i.e., ion-assisted chemical vapor deposition); and imaging (i.e., detection of secondary electrons or ions generated by the beam exposure). Figure 10.23 shows a cross-section of a SUMMiT™ hub structure, which shows the internal

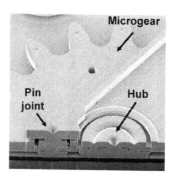

FIGURE 10.23 Focused ion beam cross-section of a SUMMiT hub and pin-joint structure. (Courtesy of Sandia National Laboratories.)

Reliability

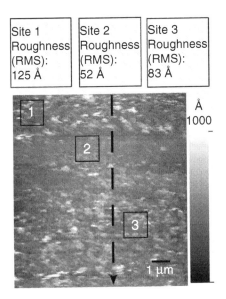

FIGURE 10.24 Atomic force microscope image of surface wear. (D.M. Tanner et al., MEMS reliability: infrastructure, test structures, experiments, and failure modes, Sandia National Laboratories Report, SAND2000-0091, January 2000; courtesy of Sandia National Laboratories.)

layers that would not be otherwise visible. The FIB can be used to free, remove, or connect portions of a device to enable its experimental evaluation [41].

10.5.4 Atomic Force Microscope

The atomic force microscope provides a very detailed topographic image of a sample surface (Figure 10.24). The AFM data on surface topography can provide important information for MEMS surfaces for optical applications as well as diagnosis of wear marks due to device operation.

10.5.5 Lift-Off

Lift-off involves removal of MEMS device elements with a nonconductive laboratory adhesive tape and microsectioning of the structure with a focused ion beam (FIB) to allow examination of the underside of MEMS surfaces (Figure 10.25). This technique can provide information on such things as wear debris, wear marks, and damage to areas that would otherwise be inaccessible.

10.5.6 Stroboscopy

Due to the high frequency of operation enabled by the size scaling of MEMS, the dynamic motion of many MEMS devices is hard to capture and analyze. Stroboscopy is a method from traditional optical measurement [43] that can be

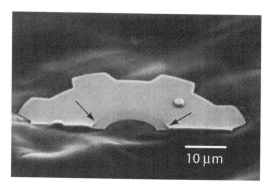

FIGURE 10.25 Lift-off specimen for analysis. (Courtesy of Sandia National Laboratories.)

adapted for MEMS devices. This method utilizes a flash from a stroboscopic light source to freeze the device motion in time. Coordinating the stroboscope flash with the device excitation allows the device motion to be frozen in time or swept through the device operation phase to observe the device dynamic motion. Stroboscopic imaging of a dynamic MEMS system is only applicable to periodic excitation and motion. The maximum detectable frequency is limited by the pulse time of the stroboscopic light source. Stroboscopy can be combined with other optical techniques such as interferometry [44] to obtain out-of-plane dynamic motion as well as full three-dimensional motion analysis [45].

10.5.7 BLUR ENVELOPE

A blur envelope is a simple way to determine the dynamic amplitude of a MEMS device utilizing an optical microscope. The MEMS device is excited to produce cyclic motion and the blurred image is used to determine the resulting dynamic amplitude (Figure 10.26). This technique can be a quick though inaccurate method for determining the amplitude response as the excitation is swept through a range of frequencies.

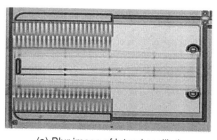
(a) Blur image of lateral oscillation

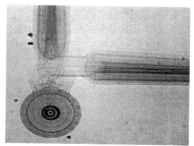
(b) Blur image of rotary motion

FIGURE 10.26 Blur image for determining the device amplitude during operation. (Courtesy of Sandia National Laboratories.)

Reliability

10.5.8 Video Imaging

Video imaging of optical microscope images via data captured by a CCD (charge coupled device) camera may be useful [42] for examination of tests over long periods of time for comparison, documentation, and off-line image analysis.

10.5.9 Interferometry

Interferometry [43] precisely measures distance by comparison of the optical path length between two beams of monochromatic light. The comparison is made by recombining the light beams. If they are in phase, a bright fringe appears; if the beams are out of phase, a dark fringe is observed. Figure 10.27 is a schematic of a Michelson interferometer, and Figure 10.28 is an interferometric image of a MEMS surface. Interferometry will provide a wide field of view topographic image of a MEMS device. Interferometry can be combined with stroboscopy [44] to obtain out-of-plane dynamic motion.

An alternative technique is electronic speckle pattern interferometry (ESPI) [46], which measures surface position relative to a reference position instead of using a flat plane as the reference (used in Micelson interferometry).

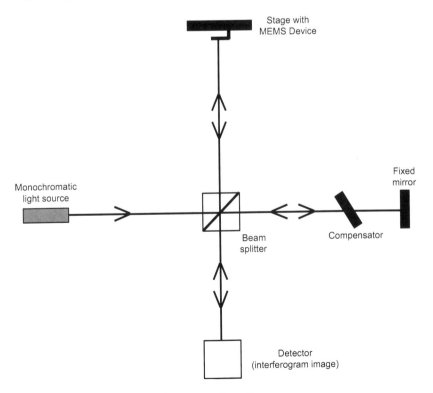

FIGURE 10.27 Michelson interferometer schematic.

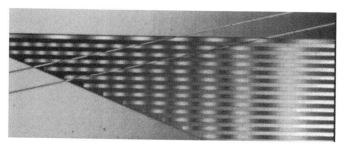

FIGURE 10.28 Interferometric image.

10.5.10 LASER DOPPLER VELOCIMETER (LDV)

A laser Doppler vibrometer (LDV) [47] or velocimeter is based on the detection of the Doppler shift of coherent laser light scattered from a small area of a test object. For a MEMS device, some additional optical elements are necessary to produce a sufficiently small laser beam. The object reflects the laser beam and the Doppler frequency shift is used to measure the component of velocity along the axis of the laser beam. The Doppler frequency shift is caused by the motion of the surface off which the measurement beam is reflected. Because the laser light is very high frequency, a direct demodulator of the light is not possible; however, an interferometer similar to that in Figure 10.27 can be used to mix the reflected laser light and a reference beam. A photodetector can be used to measure the intensity of mixed light whose beat frequency is equal to the difference in frequency between the reference and measurement beams.

10.6 MEMS RELIABILITY AND DESIGN

There is a wide spectrum of MEMS applications and designs to address those applications. An aid in the classification of the various MEMS device with respect to their reliability implications is the *MEMS Device Taxonomy* [48] (Figure 10.29) developed by Sandia National Laboratories. The taxonomy consists of four classes of devices:

- Class 1 devices have no moving parts (e.g., accelerometers, pressure sensors, ink-jet print heads, strain gauges, etc.); parts may flex during the course of their operation, but there is no excited displacement or vibrations inherent to their operation.
- Class 2 devices have moving parts but with no rubbing or impacting surfaces (e.g., vibratory gyroscopes, resonators, mechanical filters, electrostatic comb drives).
- Class 3 devices have moving parts with impacting surfaces (e.g., relays, switches, fluidic valves).
- Class 4 devices have moving parts that contain rubbing and/or impacting surfaces (e.g., gears, slides, rotary hubs).

Reliability

Class I	Class II	Class III	Class IV
No moving parts	Moving parts; no rubbing or impacting surfaces	Moving parts; impacting surfaces	Moving parts; impacting and rubbing surfaces

Ink jet nozzles Strain gauge DNA sequencers	Accelerometer Gyroscope Pressure sensor Resonators Compliant mechanisms	Optical switches DMD™ Relays Valves Pumps	Gear Mechanism Mechanisms Shutters Scanners

FIGURE 10.29 MEMS device taxonomy. (Courtesy of Sandia National Laboratories.)

A further method to correlate MEMS reliability issues with a particular MEMS device is the *product–reliability issue matrix* [48] shown in Figure 10.30. This matrix correlates the reliability issues for a particular device. It is advantageous for a MEMS designer to consider these issues as the design is conceptually developed so that reliability problems can be dealt with in the early stages of design. The devices in each category of the taxonomy have different reliability issues, with devices from category 4 having the most significant reliability issues. To avoid some of the many issues inherent in category 4 MEMS devices, the designer should give consideration to alternative devices that avoid rubbing and

Product	Class	Operational Failure Mechanisms						Degradation Mechanisms				Environmental Failure Mechanisms					
		wear	fracture	fatigue	charging	creep	stiction and adhesion	thermal degradation	optical degradation	environmental degradation	stress corrosion cracking	shock	vibration	thermal cycling	humidity	radiation	electrostatic discharge

FIGURE 10.30 MEMS product-reliability matrix. (MEMS reliability short course, Sandia National Laboratories, Albuquerque, NM.)

402   Micro Electro Mechanical System Design

impacting surfaces. The use of a compliant mechanism [49] is a design alternative and philosophy of utilizing devices that flex in order to perform a function vs. surfaces rubbing, impacting, or sliding.

Reliability-based MEMS design can be accomplished by a disciplined approach to design and consideration of the reliability aspects of the design. The reliability of the design can be aided significantly by considering the following items early in the design process:

- The *MEMS device taxonomy* and the *product–reliability issue matrix* should be considered in the conceptual design of the MEMS device.
- Design *simplicity* should be a prime consideration in the device design. This will generally lead to a reduction in the number of failure modes of the device (i.e., with fewer parts, less can go wrong).
- *Standard components* (i.e., Chapter 4) will reduce time, cost, and the number of reliability issues. Because a standard component will have been analyzed, fabricated, and tested, the design is starting from a known point.
- *Packaging* should be given due consideration in the early stages of design. Packaging can address some of the environmental failure modes of the device.
- *Failure prevention during design.* This is accomplished by performing thorough *failure modes, effects, and criticality analysis* (FMECA) and a *fault-tree analysis* (FTA) of the MEMS device.
- *Design with the expected manufacture variability in mind.* This involves consideration of the expected range of fabrication tolerances and operational environments. Use a probabilistic design approach [51] when the design variables are considered as random variables and the spectrum of design evaluated.

The main purpose of the FMECA and FTA is to identify and eliminate failure modes early in the design cycle when they can be most economically dealt with. The procedure for FMECA is documented in detail and can be found in MIL-STD-1629A (1980) [50]. The FMECA is a *bottom-up* procedure that enables each failure mode of a device to be traced to the effect on the system. The FMECA takes a pessimistic point of view and assumes design weaknesses exist. The FMECA worksheet [50] will record the following information for the analysis:

- Function: definition of the function that the device must perform
- Failure mode: definition of how the device fails to perform the required function
- Failure mechanism: the physical process that causes the failure mode
- Failure cause: how the failure mechanism is activated

Reliability

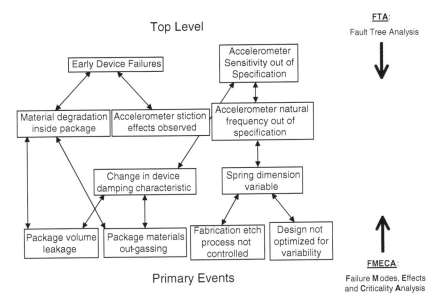

FIGURE 10.31 Schematic of the FTA and FMECA approaches for an example MEMS accelerometer that illustrates the methods.

- Identification of effects of higher level systems: whether the failure model is localized or causes higher level damage
- Criticality rating: a measure of the severity, probability of failure occurrence, and detectability used to assign a priority for the subsequent actions

Fault-tree analysis (FTA) is a widely used technique for system safety and reliability analysis. The analysis is a *top-down* approach that proceeds from a top level or system event to basic device or component failure causes called primary events. The fault tree is a graphical model that portrays the combination of events leading to the top event.

Figure 10.31 is a short example contrasting the application of the FMECA and FTA approaches to a MEMS accelerometer. The *top level* events relate to the accelerometer device performance or lifetime. The *primary* events are the root causes of the occurrence of the top level events, possibly through some intermediate events.

10.7 MEMS RELIABILITY CASE STUDIES

10.7.1 DMD Reliability

The digital micromirror device (DMD) developed by Texas Instruments (TI) [53] over a number of years has made steady progress in performance and reliability

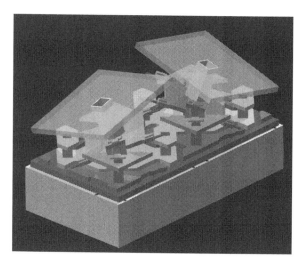

FIGURE 10.32 Illustration of two landed DMD mirrors. (Courtesy of Texas Instruments.)

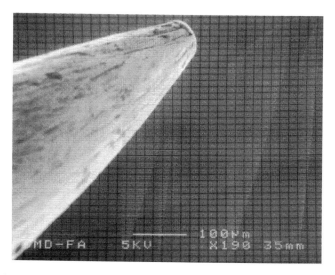

FIGURE 10.33 SEM of a DMD array of mirrors with a pin shown in the foreground for scale. (Courtesy of Texas Instruments.)

to become one of the prime MEMS commercial success stories. The DMD is a bistable mirror (Figure 10.32) used in large arrays (Figure 10.33) as the basis of the TI digital light processing (DLP™) technology utilized in optical projection displays. In order to achieve this commercial success, device performance and reliability are paramount.

From the invention of the DMD in 1987 through production development in 1992 to the design enhancement in 2000 when the mirror size was reduced from a 17- to 14-μm pitch, reliability methods and testing were necessary to achieve

Reliability
405

the performance and reliability results enjoyed by the DMD today. The test and modeling results [52] conclude that the DMD is reliable and robust, with the following qualities:

- DMD mean time between failure (MTBF) > 650,000 h
- DMD lifetime > 100,000 h
- Hinge lifetime > 3×10^{12} mirror cycles (equivalent to >120,000 operating h)
- Environmentally robust

This section will highlight the methods followed to achieve the impressive results attained.

Texas Instruments chose to utilize a FEMA approach for reliability of the DMD. (Many times the acronyms FEMA and FEMCA are used interchangeably; the "C" denotes a criticality analysis of the failure modes identified.) A comprehensive FEMA is a very sizable task, but the results that TI has shown justify its use. The company considered process techniques, design constraints, packaging concerns, test issues, and other failure mode contributors. For each failure mode identified, failure mechanisms and risk to DMD lifetime and failure were assessed. TI rigidly adhered to performing the FEMA process on all new DMD designs, such as the mirror size reduction in 2000.

To perform the testing that was required and identified from the FEMA analysis, TI developed a DMD test capability. At the time during which the DMD was under development, commercially available test equipment was not available. However, a DMD test system was developed that included an X/Y/theta stage, CCD camera, optics, and a computer for interpreting the vision data. The DMD test system is computer controlled to provide flexibility in the test performed and data acquired.

Two examples of tests developed for the DMD are the *bias/adhesion mirror mapping* (BAMM) sweep and a solution space characterization technique [54]. The BAMM test is a parametric test utilizing a parameter referred to as the DMD landing voltage. A typical BAMM curve is a plot of the number of landed mirrors (i.e., mirrors are tipped as shown in Figure 10.32) vs. mirror bias voltage. For example, for a mirror bias voltage up to 15 V, no mirrors are landed. Increasing the bias voltage above 15 V causes an increasing number of mirrors to land to 17 V, where all mirrors are landed. The landing voltage is a function of numerous process and design parameters and is currently used as a metric of the device performance measured on every DMD lot.

The *solution space characterization technique* [52] is a method of graphical correlation of multiple parameters that are varied over a significant operating range to show the operating space of the DMD. Performing the solution space characterization technique before and after exposing the DMD to an environment or operational test will provide an indication of how the DMD solution space may have varied.

With the DMD test station and a suite of parametric tests such as the BAMM and solution space characterization technique developed, a number of possible failure mechanisms were investigated:

- *Hinge fatigue.* The DMD is routinely tested by high-temperature rapid (i.e., much faster than normal operation) mirror cycling tests. The hinge is the flexural element flexed during mirror operation (Figure 10.34). DMDs have demonstrated 3×10^{12} mirror cycles with no hinge fatigue failures. The conclusion has been that hinge fatigue is not a failure mode of concern.
- *Hinge memory.* Hinge memory occurs when the DMD is operated at high temperature and high duty cycles (i.e., the amount of time the mirror is directed to land on one side vs. the other). This mechanism behaves like metal creep, but surface effects [52] may also contribute. It has been also noted that hinge memory is not permanent, but reversible in nature. The reversal of the hinge memory is accomplished by reversing the mirror duty cycle. Hinge memory is the only known life-limiting failure mode [52] exhibited by the DMD.
- *Metal creep* [11,55]. This is the underlying mechanism associated with the hinge memory failure mode of the DMD. The solution to these issues involved the development of other Al compounds that were compatible with the Al etch processes already characterized and in place. The new Al compound also needed to have fewer material slip systems than the existing Al and a higher melting point, which is generally related to metal creep. This work resulted in a patent [56] for the Al compounds (e.g., Al_3Ti, AlTi, AlN) with etch in the Al etch process.
- *Stiction.* The BAMM testing method was also utilized in the stiction studies for the DMD. For stiction characterization, the mirror release voltage was the important metric. The stiction forces on the landed

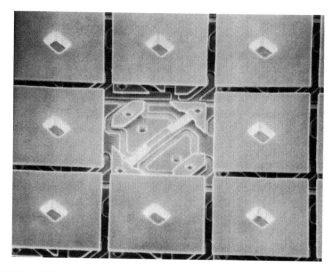

FIGURE 10.34 SEM picture of the DMD with the hinges exposed in the center pixel. (Courtesy of Texas Instruments.)

Reliability 407

mirror would produce variability in the mirror release voltage; thus, the BAMM release curves are an indirect measure of mirror-to-surface adhesion or stiction. Stiction was shown to be not as predictable as hinge memory. A design modification of the spring tips on the DMD design [53] producing positive force to push the mirror off the substrate during operation virtually eliminated stiction failures.

- *Environmental testing.* Subjected to a full range of shock, vibration, thermal, ESD, and optical environment tests, the DMD proved to be quite robust. It proved to be virtually impervious to shock and vibration due to the mirror size and the high natural frequencies (i.e., >100 kHz). The only observable damage was the movement of a few particles.

10.7.2 SANDIA MICROENGINE

Sandia National Laboratories has been involved in the development of MEMS technology for over 15 years. During that time, they have developed a five-level surface micromachined fabrication process, SUMMiT™ (**S**andia **u**ltraplanar, **m**ultilevel **MEMS** **t**echnology) [57] and pursued the development of an array of mechanical MEMS devices [59–62]. This section provides an overview of the initial reliability tests of the Sandia microengine and briefly discusses some of the ongoing work with other MEMS devices and the impact of reliability data on MEMS device design.

The Sandia microengine (Figure 10.35) was the primary actuator in some of the initial Sandia MEMS designs and thus was an appropriate focus for the initial reliability studies [3]. The microengine consists of two orthogonally oriented reciprocating electrostatic comb drive actuators and a linkage connecting the comb drive actuators to a rotating pinion gear, as shown in Figure 10.35.

The ability to acquire reliability data on a statistically significant number of devices is essential to a reliability program. The creation of the SHiMMeR (**S**andia **h**igh-volume **m**easurement of **m**icromachine **r**eliability) system (Figure 10.36) has allowed acquision of reliability data from a large number of packaged parts [58,63]. The SHiMMeR system consists of an X–Y gantry table with a travel area of 500 ×x 540 mm; a video zoom microscope equipped with motorized computer control; and an electrical stimulus system capable of providing arbitrary waveforms of up to 200 V at 10 kHz to 64 24-pin DIP sockets. The optical subsystem and device packages are contained within a humidity-controlled Plexiglas enclosure. The humidity range can be controlled between 2 to 90% RH. The entire system is mounted on a vibration isolation table.

The initial reliability test was performed on 41 microengines and was defined as follows. The engines were operated at 36,000 rpm (chosen to be below the resonance of the comb drives) for a defined number of revolutions and then functionality was observed at 60 rpm. The transition was accomplished by decelerating in one revolution, momentarily stopping for 1 sec, then accelerating in one revolution to the observation speed. An assumption in this definition of operation is that the deceleration, acceleration, and brief stop of the microengine

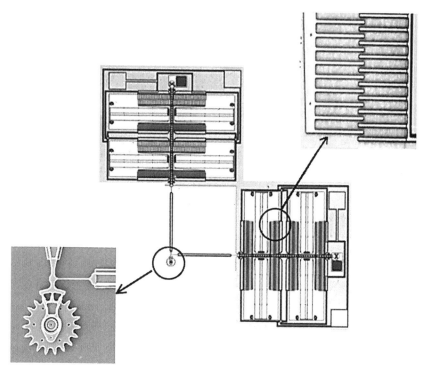

FIGURE 10.35 Sandia microengine. (Courtesy of Sandia National Laboratories.)

FIGURE 10.36 SHiMMeR system with multiple packaged part, computer control, data acquisition, vibration isolation, humidity enclosure, and video microscope. (Courtesy of Sandia National Laboratories, Radiation and Reliability Physics Department.)

do not influence the ultimate failure and lifetime of the microengine. The failure criteria were defined as the inability of the microengine to complete a revolution during the observation period

During the observation period, a pass–fail decision was made; comments on the motion of functioning engines were made; and video was recorded for post test analysis. The test lasted for 28 days with a total of 32 separate stress periods.

Reliability

The initial duration used of high-speed operation was deemed as too short and the high-speed periods were subsequently increased.

The progression to failure was similar in most of the microengines observed [2]. The pinion gear went from rotating smoothly to sticking momentarily but completing the revolution. These behaviors ultimately led to the pinion gear oscillating or freezing in one position. All 41 microengines that started the test eventually failed. The long-term reliability test resulted in over a billion revolutions for the longest running engines with the longest running one operating to 7 billion revolutions. The data showed a decreasing failure rate with no sign of wear-out (i.e., increasing failure rate) evident.

Figure 10.37 shows two plots of the instantaneous failure rate (i.e., hazard function) vs. accumulated cycles curve. Figure 10.37A uses a linear scale, which shows a decreasing failure rate consistent with the first portion of the bathtub curve. Figure 10.37B, which is plotted on a log–log scale, shows the decreasing

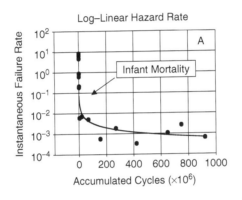

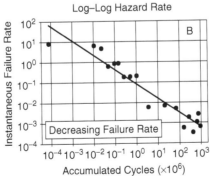

FIGURE 10.37 Instantaneous failure rate curves plotted on different accumulated cycles axes. Curve A shows the infant mortality region and curve B shows a more detailed representation of the decreasing failure rate throughout the life test. (M.S. Rodgers and J.J. Sniegowski, *Tech. Dig. Solid-State Sensor Actuator Workshop*, 144–149, June 1998, Hilton Head Island, SC, 1998; courtesy of Sandia National Laboratories, Radiation and Reliability Physics Department.)

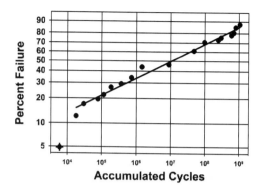

FIGURE 10.38 Microengine cumulative failure distribution data fit to a Weibull distribution. (M.S. Rodgers and J.J. Sniegowski, *Tech. Dig. Solid-State Sensor Actuator Workshop*, 144–149, June 1998, Hilton Head Island, SC, 1998; courtesy of Sandia National Laboratories, Radiation and Reliability Physics Department.)

tendency in failure rate throughout the test. A straight line can fit through the data of Figure 10.37B, indicating that a Weibull distribution that can model decreasing failure rates would be an appropriate model for these data.

Figure 10.37A clearly shows the decreasing failure rate indicative of infant mortality in which a "weeding out" process occurs. This phenomenon will enable a burn-in schedule to be developed in which defective parts can be eliminated from the population. Long-term life tests would need to be performed to assure that wear-out failures do not occur unacceptably soon.

Figure 10.38 shows the fit of the microengine CDF failure data to the Weibull distribution model. The resulting fit shows that this is an appropriate model for the data. The Weibull fit utilized a characteristic life, $\lambda = 66$ million cycles, and a shape parameter, $\beta = 0.22$. The characteristic life parameter, λ, is defined as the point at which 63.2% of the part will have failed. The shape parameter, β, is an indication of dispersion, with lower values indicating greater dispersion in the lifetime. "Typical values of β for production-ready electronic and mechanical products fall in the range of 0.5 to 5" [3,58]. The study summarized here [3,58] is from a very early design of the microengine, which may be indicative of the spread in the data. A value of the shape parameter $\beta < 1$ corresponds to a decreasing failure rate.

Figure 10.39 shows the fit of the microengine CDF failure data to the lognormal distribution model. The resulting fit shows that the lognormal as well as the Weibull distribution are reasonable models for the data. The lognormal fit using the two parameters of the lognormal model results in the median lifetime parameter, $t_{50} = 7.8$ million cycles and the standard distribution parameter, $\sigma = 5.2$. Once again, a spread is indicated in the data by the high value of the standard distribution parameter. Also, the lognormal shows an immature design as a result of the lognormal standard deviation parameter value; this is similar to the result with the Weibull model. "Typical semiconductor products show lognormal stan-

Reliability

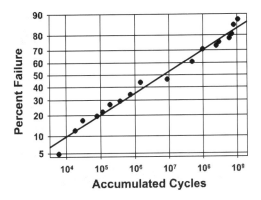

FIGURE 10.39 Microengine cumulative failure distribution data fit to a lognormal distribution. (M.S. Rodgers and J.J. Sniegowski, *Tech. Dig. Solid-State Sensor Actuator Workshop*, 144–149, June 1998, Hilton Head Island, SC, 1998; courtesy of Sandia National Laboratories, Radiation and Reliability Physics Department.)

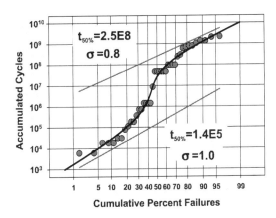

FIGURE 10.40 Microengine cumulative failure distribution data plotted on semilog axes with bimodal analysis. The two populations leading to the bimodal distribution were due to two flexure types in the test population [57]. (M.S. Rodgers and J.J. Sniegowski, *Tech. Dig. Solid-State Sensor Actuator Workshop*, 144–149, June 1998, Hilton Head Island, SC, 1998; courtesy of Sandia National Laboratories, Radiation and Reliability Physics Department.)

dard deviations in the range of 0.1 to 1.0" [3,58]. Increasing and decreasing failure rates cannot be easily discriminated on a lognormal CDF distribution plot such as shown in Figure 10.39. However, varying failure rate over the lifetime of a device as conceptually modeled by the bathtub curve are to be expected.

Upon further examination of the data and replotting (note the semilog scale) the data, as shown in Figure 10.40, a bimodality of the data can be observed. The bimodality of the data indicates the presence of two populations within the data. In fact, two variations of microengines were utilized in the test. The differ-

412   Micro Electro Mechanical System Design

ence was in the thickness of flexures (i.e., 1 μm wide by 25 μm long vs. 2 μm wide by 50 μm long) used in the connecting linkage, which correlated with the population of failures in the bimodal distribution. The lognormal parameters utilized to fit the two modes are $t_{50} = 2.5 \times 10^8$ cycles, $\sigma = 0.8$ for the thin flexure population and $t_{50} = 1.4 \times 10^5$ cycles, $\sigma = 1.0$ for the thick flexure population.

Upon completion of the reliability studies, failure analysis [3,41] of the failed devices was performed. Two failure modes were shown to occur. The primary failure mode was *lateral clamping* of the comb drive, which occurs where the moving combs move transverse to the intended direction of motion and short with the stator combs. The secondary failure mode was the rotating gear sticking to the hub or substrate. This failure was characterized by the presence of motion in the comb drives, but no motion in the gear.

10.8 SUMMARY

This chapter has presented an overview of MEMS reliability and issues that should be considered by the MEMS designer. A brief review of probability theory and distributions relevant to the analysis of reliability data was also presented. The three distributions used to analyze and correlate reliability data are the cumulative probability distribution function (CPDF); probability density function (PDF); and instantaneous failure rate (hazard function). The bathtub curve is a plot of the hazard function vs. time and has been shown to be highly relevant to mechanical and electrical components. Three types of reliability models (i.e., Weibull, lognormal, exponential) that can be used to model reliability data were also discussed. This chapter also presented a review of MEMS failure mechanisms and measurement methods that can be used to assess the device.

The most important concept that the MEMS design engineer should take away from this chapter is that *reliability must be designed into the device*. The most extensive reliability test program will not improve reliability without a device redesign to incorporate any information that is learned. The most efficient (i.e., time, cost) approach is to develop a reliability-based design from the beginning. The steps to achieving a *reliability-based design* include the following:

- MEMS device taxonomy
- Design simplicity
- Utilize standard components when possible
- Packaging consideration in the design phase
- Failure prevention during design via FMECA–FTA analysis
- Design with expected manufacturing variability in mind

QUESTIONS

1. Define product reliability.
2. What are the items involved in the definition of a reliability experiment? Why is their definition important?

Reliability

3. Choose a device (not necessarily a MEMS device) and define a meaningful reliability experiment that could be performed. What three things need to be defined to specify the reliability experiment?
4. What are three types of distributions frequently used in analyzing reliability data? How are these distributions related?
5. What is the bathtub curve? Explain the various regions of the bathtub curve.
6. Why are probability and statistics relevant and important to the study of reliability?
7. Define probability. Given two fair die, what is the probability of rolling a 12? What is the most probable number to be rolled?
8. What are reliability models and why are they used?
9. Discuss the Weibull model and its advantages for use in modeling reliability data.
10. What are FMECA and FTA and what are they trying to accomplish? Explain similarities and differences in the approach to FMECA and FTA.
11. What is the reliability experiment as defined in the Sandia microengine case study? Is the reliability experiment completely and unambiguously defined?
12. What is the difference between a failure mode and a failure mechanism?
13. Why is it important to find the failure mode encountered in a reliability experiment?
14. Why was reliability modeling of the data in the Sandia microengine case study important and what did it reveal?
15. What are six things that should be considered to achieve a reliable design?
16. Review recent articles in the MEMS literature (e.g., *Journal of MEMS, Sensors and Actuators*) and choose a device of interest. Using the taxonomy of MEMS devices, what class is your device? Does the chosen device have any shortcomings or issues for concern when compared to the steps for developing a reliability-based design (Section 10.6)? Perform FMECA and FTA analysis on the device.

REFERENCES

1. S.S. Rao, *Reliability Based Design*, McGraw-Hill, New York, 1992.
2. M.G. Pecht, F.R. Nash, Predicting the reliability of electronic equipment, *Proc. IEEE*, 82(7), July 1994.
3. D.M. Tanner, N.F. Smith, L.W. Irwin, W.P. Eaton, K.S. Helgesen, J. J. Clement, W.M. Miller, J.A. Walraven, K.A. Peterson, P. Tangyunyong, M.T. Dugger, S.L. Miller, MEMS reliability: infrastructure, test structures, experiments, and failure modes, Sandia National Laboratories Report, SAND2000-0091, January 2000.
4. B. Stark, MEMS reliability assurance guidelines for space applications, *JPL Publication 99-1*, Jet Propulsion Laboratory, Pasadena, CA, January 1999.

Micro Electro Mechanical System Design

5. A.W. Drake, *Fundamentals of Applied Probability Theory*, McGraw-Hill Inc., New York, 1967.

6. P. Beckmann, *Elements of Applied Probability Theory*, Harcourt Brace and World Inc., New York, 1968.

7. N. Draper, H. Smith, *Applied Regression Analysis*, 2nd ed., John Wiley & Sons, New York, 1981.

8. P.A. Tobias, D.C. Trinidade, *Applied Reliability*, Van Nostrand Reinhold Co., New York, 1986.

9. M. Abramowitz, I.A. Stegun, *Handbook of Mathematical Functions with Formulas, Graphs, and Mathematical Table*, National Bureau of Standards Applied Mathematics Series 55, June 1964, 10th printing, December 1972.

10. M.G. Pecht, F.R. Nash, Predicting the reliability of electronic equipment, *Proc. IEEE*, 82(7), July 1994.

11. W.M. van Spengen, MEMS reliability from a failure mechanisms perspective, *Microelectronics Reliability*, 43, 1049–1060, 2003.

12. S.L. Miller, G. LaVigne, M.S. Rodgers, J.J. Sniegowski, J.P. Waters, and P.J. McWhorter, Routes to failure in rotating MEMS devices experiencing sliding friction, *Proc. SPIE*, 3324, 24–30, 1997.

13. S.L. Miller, M.S. Rodgers, G. LaVigne, J.J. Sniegowski, P. Clews, D.M. Tanner, and K.A. Peterson, Failure modes in surface micromachined microelectromechanical actuation systems, *Microelectron. Reliability*, 39, 1229–1237, 1999.

14. D.M. Tanner, M.T. Dugger, Wear mechanisms in a reliability methodology, *Proc. SPIE*, 4980, 22–40, January 2003.

15. W.W. Van Arsdell, S.B. Brown, Subcritical crack growth in silicon MEMS, *J. MEMS*, 8(3), 319, 1999.

16. J. Wibbeler, G. Pfeifer, M. Hietschold, Parasitic charging of dielectric surfaces in capacitive microelectromechanical systems (MEMS), *Sens Act A*, 71(1–2), 74–80, Nov. 1998.

17. G.M. Rebeitz, J.B. Muldavin, RF MEMS switches and switch circuits, *IEEE Microw. Mag.*, 59, 2000.

18. J.C. Ehmke, C.L. Goldsmith, Z.J. Yao, S.M. Eshelman, Method and apparatus for switching high-frequency signals, U.S. Patent 6,391,675, issued 21 May 1999.

19. S.B. Brown, W. Van Arsdell, C. Muhlstein, Materials reliability in MEMs devices, *Transducers '97*, 591–593, 1997.

20. M.R. Douglass, Lifetime estimates and unique failure mechanisms of the digitial micromirror device, *Proc. IEEE Int. Reliability Phys. Symp.*, 9–16, 1998.

21. C.H. Mastrangelo, C.H. Hsu, Mechanical stability and adhesion of microstructures under capillary forces — part I: basic theory, *JMEMS*, 2(1), 33–43, March 1993.

22. R. Maboudian, R.T. Howe, Critical review: adhesion in surface micromechanical structures, *J. Vac. Sci. Tech. B.*, 15(1), 1–20, January 1997

23. R.A. Plass, J.A. Walraven, D.M. Tanner, F.W. Sexton, Anodic oxidation-induced delamination of the SUMMiT™ poly 0 to silicon nitride interface, *Proc. SPIE*, 4980, 81–86, January 2003.

24. M. P. de Boer, J.A. Knapp, T.M. Mayer, T.A. Michalske, The role of interfacial properties on MEMS performance and reliability, *Proc. SPIE*, 3825, 2–15, 1999.

25. M.P. de Boer, J.A. Knapp, J.M. Redmond, T.M. Mayer, J.J. Sniegowski, T.A. Michalske, Fundamental mechanisms of micromachine reliability, Sandia National Laboratories, Sandia Report SAND99-3100, January 2000.

Reliability

26. P.F. Man, B.P. Gogoi, C.H. Mastrangelo, Elimination of post release adhesion in microstructures using conformal fluorocarbon coatings, *JMEMS*, 6(1), 25–34, March 1997.
27. U. Srinivasen, M.R. Houston, R.T. Howe, R. Maboudian, Alkyltrichlorosilane-based self-assembled monolayer films for stiction reduction in silicon micromachines, *JMEMS*, 7(2), 252–260, June 1998.
28. C.M. Harris, C.E. Crede, Eds, *Shock and Vibration Handbook*, 2nd ed., McGraw–Hill Book Company, New York, 1976.
29. D.M. Tanner, J.A. Walraven, K.S. Helgesen, L.W. Irwin, D.L. Gregory, J.R. Stake, N.F. Smith, MEMS reliability in a vibration environment, *Proc. IRPS*, 2000, 139–145.
30. D.M. Tanner, J.A. Walraven, K. Helgesen, L.W. Irwin, F. Brown, N.F. Smith, N. Masters, MEMS reliability in shock environments, *Proc. IRPS*, 129–138, 2000.
31. V.T. Srikar, S.D. Senturia, The reliability of microelectromechanical systems (MEMS) in shock environments, *J. MEMS*, 11(3), 206–214, 2002.
32. T.G. Brown, B.S. Davis, Dynamic high-G loading of MEMS sensors: ground and flight testing, *Proc. SPIE*, 3512, 228–235, 1998.
33. A. Lawrence, *Modern Inertial Technology, Navigation, Guidance, and Control*, Springer-Verlag, Heidelberg, 1992.
34. D.M. Tanner, J.A. Walraven, L.W. Irwin, M.T. Dugger, N.F. Smith, W.M. Miller, S.L. Miller, The effect of humidity on the reliability of a surface micromachined microengine, *Proc. IEEE Int. Rel. Phys. Symp.*, 189–197, 1999.
35. L.P. Schanwald; J.R. Schwank; J.J. Sniegowski, M.R. Shaneyfelt, D.S. Walsh, N.F. Smith, K.A. Peterson, P.S. Winokur, J.H. Smith, B.L. Doyle, High-fluence x-ray and charged particle effects in microelectromechanical systems (MEMS) comb-drive actuators, 1998 IEEE Nuclear and Space Radiation Effects Conference (NSREC), July 20, 1998.
36. C.I. Lee, A.H. Johnson, W.C. Tang, C.E. Barnes, Total dose effects on microelectromechanical systems (MEMS): accelerometers, *IEEE Trans Nucl Sci*, 43(6), 3127, 1996.
37. A.R. Knudson, S. Buchner, P. McDonald, W.J. Stapor, A.B. Campbell, K.S. Grabowski, et al., The effects of radiation on MEMS accelerometers, *IEEE Trans. Nucl. Sci.*, 43(6), 3122, 1996.
38. E.I. Cole, J.M. Soden, Scanning electron microscopy techniques for IC failure analysis, in *Microelectronic Failure Analysis Desk Reference*, 3rd ed., 1999.
39. D.B. Williams, J.L. Goldstein, D.E. Newbury, Eds., *X-Ray Spectrometry in Electron Beam Instruments*, Plenum Press, New York, 1995.
40. R.F. Egerton, *Electron Energy-Loss Spectroscopy in the Electron Microscopy*, 2nd ed., Plenum Press, New York, 1996.
41. K.A. Peterson, P. Tangyunyong, A.A. Pimentel, Failure analysis of surface-micromachined microengines, *Proc. SPIE*, 3512, 190–200, 1998.
42. D.M. Freeman, C.Q. Davis, Using video microscopy to characterize micromechanics of biological and manmade micromachines, 1996 Solid State Sensor and Actuator Workshop, Hilton Head, SC, 161–167, June 3–6, 1996.
43. G.L. Cloud, *Optical Methods of Engineering Analysis*, Cambridge University Press, Cambridge, U.K., 1995.
44. M. Hart, R. Conant, K. Lau, R. Muller, Stroboscopic interferometer system for dynamic MEMS characterization, *JMEMS*, 9, 409–418, December 2000.

Micro Electro Mechanical System Design

45. C. Rembe, R. Muller, Measurement system for full three-dimensional motion characterization of MEMS, *JMEMS*, 11(5), 479–488, October 2002.

46. P.Q. Zhang, Q.M. Wang, X.P. Wu, T.C. Huang, Experimental model analysis of miniature objects by optical measurement technique, *Int. J. Analytical Exp. Anal.*, 7(4), 243–253, 1992.

47. K.L. Turner, P.G. Hartwell, N.C. MacDonald, Multidimensional MEMS motion characterization using laser vibrometry, *Transducers '99*, paper 4D1.2, 1999.

48. MEMS Reliability Short Course, Sandia National Laboratories, Albuquerque, NM.

49. L.L. Howell, *Compliant Mechanisms*, John Wiley & Sons, Inc., New York, 2001.

50. MIL-STD-1629A: U.S. Department of Defense, Military standard procedures for performing a failure mode, effects and criticality analysis, November 24, 1980.

51. T.A. Cruse, *Reliability Based Mechanical Design*, Chapter 2, Marcel Dekker, New York, 1997.

52. M.R. Douglass, DMD reliability, a MEMS success story, *Proc. SPIE*, 4980, 1–11, 2003.

53. P. van Kessel, L. Hornbeck, R. Meier, M. Douglass, A MEMS-based projection display, *Proc. IEEE*, 86(8), 1687–1704, August 1998.

54. H. Chu, A. Gonzalez, T. Oudal, R. Aldridge, D. Dudasko, P. Barker, DMD superstructure characterizations, *TI Tech. J.*, 15(3), July–September 1998.

55. A.B. Sontheimer, Digital micromirror device (DMD) hinge memory lifetime reliability modeling, *IRPS 2002*, 118–121.

56. J.H. Tregilgas, Micromechanical device having an improved beam, U.S. Patent 5,552,924, issued 3 September 1996.

57. M.S. Rodgers, J.J. Sniegowski, Five-level polysilicon surface micromachine technology: application to complex mechanical systems, *Tech. Dig. Solid-State Sensor Actuator Workshop*, 144–149, June 1998, Hilton Head Island, SC, 1998.

58. D.M. Tanner, N.F. Smith, D.J. Bowman, W.P. Eaton, K.A. Peterson, First reliability test of a surface micromachined microengine using SHiMMeR, *Proc. SPIE*, 3224, 14–23, 1997.

59. M.A. Polosky, E.J. Garcia, J.J. Allen, Surface micromachined counter-meshing gears discrimination device, *Proc. SPIE*, 3328, 365–373, San Diego, CA, March 1998.

60. J.J. Allen, H.K. Schriner, Micromachine wedge stepping motor, *Int. Mechanical Eng. Congr. Exposition, DSC-*, 66, 317–322, November, 1998.

61. M.S. Rodgers, S. Kota, J. Hetrick, Z. Li, B.D. Jensen, T.W. Krygowski, S.L. Miller, S.M. Barnes, M.S. Burg, A new class of high force, low-voltage, compliant actuation systems, presented at Solid-State Sensor Actuator Workshop, Hilton Head Island, SC, June 4–8, 2000.

62. S.M. Barnes, S.L. Miller, M.S. Rodgers, F. Bitsie, Torsional ratcheting actuating system, *Proc. 3rd Int. Conf. Modeling Simulation Microsyst.*, San Diego, CA, March 27–29, 2000, 273–276.

63. N.F. Smith, W.P. Eaton, D.M. Tanner, J.J. Allen, Development of characterization tools for reliability testing of microelectromechanical system actuators, *Proc. SPIE*, 3880, 156–164.

Appendix A

Glossary

AFM atomic force microscope
AM amplitude modulation
CAD computer aided design
CIF Cal Tech intermediate format; an ASCII graphical data format employed for mask plotting files
CMOS complementary metal oxide semiconductor
CMP chemical mechanical polishing
CPW coplanar waveguide
C-V capacitance voltage characteristic
CVD chemical vapor deposition
DARPA Defense Advanced Research Projects Agency
dB decibel
DCS dichlorosilane
DRC design rule checking
DRIE deep reactive ion etching
EDP ethylene diamine pyrocatecol; silicon etchant, highly toxic
epi epitaxial
EM electromagnetic
FIB focused ion beam
FET field effect transistor
FM frequency modulation
GaAs gallium arsenide
GDSII graphical data stream file format employed for mask plotting files; originally developed and trademarked by CALMA/GE and implies use on a CALMA graphics system http://www.cadence.com
HARM high aspect ratio micromachining
HF hydrofluoric acid; a silicon dioxide etchant
IC integrated circuit
IL insertion loss
IMEMS integrated microelectromechanical systems
KOH potassium hydroxide; an isotropic etchant that attacks silicon at different rates in different directions
LIGA Lithographie Galvanic Abformung (German acronym)

417

418   Micro Electro Mechanical System Design

LPCVD low-pressure chemical vapor deposition

MEMS microelectromechanical system

MOEMS micro-optoelectromechanical systems

MST microsystems technology

MIC microwave integrated circuit

MIG MEMS Industry Group (MIG); a trade association representing the North American MEMS and microstructure industries with the purpose of enabling the exchange of nonproprietary information and increasing commercial development. http://www.memsindustrygroup.org/

MUMP multiuser MEMS process http://www.memsrus.com/

NEXUS Network of Excellence in Multifunctional Microsystems; a nonprofit association to provide access to MEMS/MST information http://www.nexus-emsto.com/

NMOS n-type metal oxide semiconductor

OPC optical proximity correction

PECVD plasma-enhanced chemical vapor deposition

PIC photonic integrated circuit

PM phase modulation

PMMA polymethyl methacrylate; photo resist commonly used in LIGA technology

PSG phosphosilicate glass

PZT lead zirconate titanate

RET resolution enhancement technology

RF radio frequency

RIE reactive ion etching

RTA rapid thermal annealing

SEM scanning electron microscope

SLIGA sacrificial LIGA

SMA shape memory alloy

STM scanning tunneling microscope

SOI silicon on Insulator

SUMMiT Sandia ultraplanar, multilevel MEMS technology http://www.mems.sandia.gov

TEOS tetraethoxysilane

TMAH tetramethylammonium hydroxide; an anisotropic wet etchant

TEM transmission electron microscopy

VPE vapor phase etching

Appendix B
Prefixes

Prefix	Symbol	Factor
tera	T	10^{12} = 1 000 000 000 000
giga	G	10^{9} = 1 000 000 000
mega	M	10^{6} = 1 000 000
kilo	k	10^{3} = 1 000
hecto	h	10^{2} = 100
deka	da	10^{1} = 10
deci	d	10^{-1} = 0.1
centi	c	10^{-2} = 0.01
milli	m	10^{-3} = 0.001
micro	m	10^{-6} = 0.000 001
nano	n	10^{-9} = 0.000 000 001
pico	p	10^{-12} = 0.000 000 000 001
femto	f	10^{-15} = 0.000 000 000 000 001
atto	a	10^{-18} = 0.000 000 000 000 000 001

Appendix C

Micro–MKS Conversions

Parameter	MKS	Multiply by	μMKS
Length	M	10^6	μM
Force	N	10^6	μN
Time	s	1	s
Mass	kg	1	kg
Pressure and stress	$Pa = N/M^2$	10^{-6}	$μN/μM^2$
Density	kg/M^3	10^{-18}	$kg/μM^3$
Current	A	10^{12}	pA
Voltage	V	1	V
Charge	C	10^{12}	pC
Resistivity	Ohm-M	10^{-6}	TOhm-μM
Permittivity	F/M	10^6	pF/μM
Energy	J	10^{12}	pJ
Capacitance	F	10^{12}	pF
Electric field	V/M	10^{-6}	V/ μM
Inductance	H	10^{12}	TH
Permeability	H/M	10^{-18}	TH/μM
Power	W	10^{12}	pW
Thermal conductivity	W/(M °K)	10^6	pW/(μM °K)
Specific heat	J/(kg °K)	10^{12}	pJ/(kg °K)

Appendix D
Physical Constants

Physical constant	Symbol	Value
Angstrom	Å	$Å = 0.1$ nm $= 10,000$ µm
Electronic charge	q	$q = 1.602 \times 10^{-19}$ C
Electron volt	eV	1 eV $= 1.60\ 2 \times 10^{-19}$ J
Permittivity of free space	ε_0	$\varepsilon_0 = 8.854 \times 10^{-14}$ F/cm
Permeability of free space	μ_0	$\mu_0 = 1.2566 \times 10^{-6}$ H/M
Thermal voltage (300 °K)	kT/q	0.0259 V
Boltzmann's constant	k_B	$k = 1.38 \times 10^{-23}$ J/K
Planck's constant	h	$h = 6.626 \times 10^{-34}$ J s
Standard atmospheric pressure		1.013×10^5 Pa

Appendix E

Material Properties

Material	Melting point (°C)	ρ (kg/μM^3)	E (μN/μM^2)	ν	α_T (μ-strain °C)	R (Ω-μm)	ε_r	Thermal conductivity (W/M-°C)	Specific heat (J/(kg-°C)
Polysilicon	1414	2.33×10^{-15}	1.60×10^5	0.23	2.5	23	11.7	70.0	700.0
Silicon dioxide	1713	2.27×10^{-15}	6.90×10^4	0.17	0.5	10^{11}–10^{14}	3.9	1.1	
Silicon nitride	1900	3.17×10^{-15}	2.70×10^5	0.24	4	10^{11}	16		
Silicon carbide	2830	3.20×10^{-15}	4.00×10^5	0.18		10^4	9.7	30	
Diamond (amorphous)		2.9×10^{-15}	8.0×10^5	0.12	2.0	10^8		100.0	600.0
Steel	1425	8.0×10^{-15}	2.07×10^5	0.29	5.1	10^{10}		34.6	500
Au	1064	19.3×10^{-15}	8.00×10^4	0.42	10.7	0.1		297.7	144
Al	660	2.70×10^{-15}	7.00×10^4	0.33	7.2	0.3		155.8	1070
Ti	1668	4.85×10^{-15}	1.10×10^5	0.3	2.9			7.4	605
W	3422	19.3×10^{-15}	4.10×10^5	0.28	1.4			164.4	154
Cu	1084	8.90×10^{-15}	3.10×10^5	0.36	4.9			392.9	428
Ni	1455	8.90×10^{-15}	2.00×10^5		4.0	10^{18}		91.7	512

REFERENCES

J.F. Shackelford, W. Alexander, *CRC Materials Science and Engineering Handbook*, CRC Press, Boca Raton, FL, 2001.

D.R. Linde, Ed., *Handbook of Chemistry and Physics*, CRC Press, Boca Raton, FL, 2003.

R.R. Tummala and E.J. Rymaszewski, Eds., *Microelectronics Packaging Handbook*, Van Nostrand Reinhold, New York, 1989.

American Institute of Physics, *American Institute of Physics Handbook*, McGraw-Hill, New York, 1972.

MEMS clearinghouse: http://www.memsnet.org/material/.

Appendix F
Stiffness Coefficients of Frequently Used MEMS Flexures

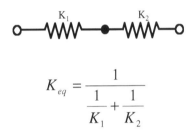

FIGURE F.1 Series spring combination.

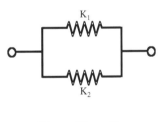

$$K_{eq} = K_1 + K_2$$

FIGURE F.2 Parallel spring combination.

428   Micro Electro Mechanical System Design

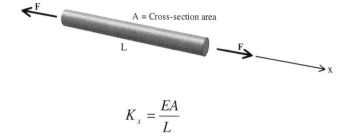

$$K_x = \frac{EA}{L}$$

FIGURE F.3 Axially loaded rod.

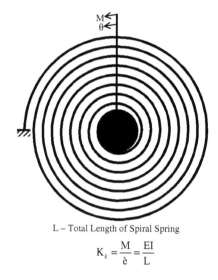

L – Total Length of Spiral Spring

$$K_\theta = \frac{M}{\theta} = \frac{EI}{L}$$

FIGURE F.4 Spiral spring stiffness (moment/angular rotation of the central hub). See Notes 1,2,3.

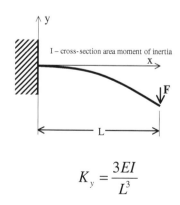

$$K_y = \frac{3EI}{L^3}$$

FIGURE F.5 Fixed-free beam with concentrated force. See Notes 1,2,3.

Stiffness Coefficients of Frequently Used MEMS Flexures

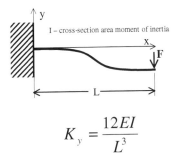

$$K_y = \frac{12EI}{L^3}$$

FIGURE F.6 Fixed-guided (fixed rotation-free deflection) beam. See Notes 1,2,3.

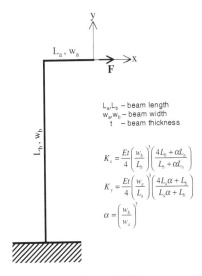

L_a, L_b – beam length
w_a, w_b – beam width
t – beam thickness

$$K_x = \frac{Et}{4}\left(\frac{w_b}{L_b}\right)^3\left(\frac{4L_b + \alpha L_a}{L_b + \alpha L_a}\right)$$

$$K_y = \frac{Et}{4}\left(\frac{w_a}{L_a}\right)^3\left(\frac{4L_a\alpha + L_b}{L_a\alpha + L_b}\right)$$

$$\alpha = \left(\frac{w_b}{w_a}\right)^3$$

FIGURE F.7 Crab leg flexure [1]. See Notes 1,2,3,4.

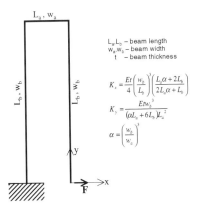

L_a, L_b – beam length
w_a, w_b – beam width
t – beam thickness

$$K_x = \frac{Et}{4}\left(\frac{w_b}{L_b}\right)^3\left(\frac{L_a\alpha + 2L_b}{2L_a\alpha + L_b}\right)$$

$$K_y = \frac{Etw_b^3}{(\alpha L_a + 6L_b)L_a^2}$$

$$\alpha = \left(\frac{w_b}{w_a}\right)^3$$

FIGURE F.8 Folded flexure [1,2]. See Notes 1, 2, 3, 4.

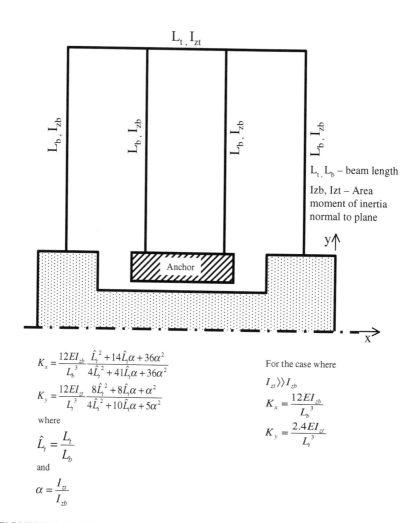

FIGURE F.9 Double folded flexure [1,2]. See Notes 1,2,3,4.

Stiffness Coefficients of Frequently Used MEMS Flexures

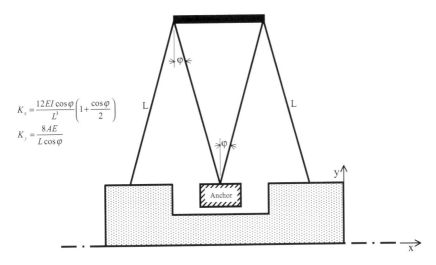

FIGURE F.10 Double V beam flexure [3]. See Notes 1,2,3.

NOTES

1. Youngs Modulus, E, should be obtained from the specific fabrication process utilized. Appendix E and its references can provide approximate values.
2. The area moment of inertia, I, is calculated for the bending axis perpendicular to the plane of the page.
3. Tables G.2 and G.3 provide methods to calculate I in terms of beam dimensions.
4. These equations were developed [1,2] with Castigliano's Theorem which included bending and torsion strain energy terms only. Due to the neglected axial strain energy terms, the K_y stiffness may be significantly underestimated. This will occur when bending is not the dominant contributor to deflection in some truss members.

REFERENCES

1. G.K. Fedder, Simulation of microelectromechanical systems, Ph.D. thesis, Dept. of Electrical Engineering and Computer Sciences, University of California at Berkeley, Berkeley, CA, 1994.
2. W.C. Tang, Electrostatic comb drive for resonant sensor and actuator applications, Ph.D. thesis, Dept. of Electrical Engineering and Computer Sciences, University of California at Berkeley, Berkeley, CA, 1990.
3. L. Saggere, S. Kota, and S.B. Crary, *Int. Mechanical Eng. Congr. Exhibition*, ASME, DSC-55, 1994.

$$K_x = \frac{12EI\cos\varphi}{L^3}\left(1+\frac{\cos\varphi}{2}\right)$$

$$K_y = \frac{8AE}{L\cos\varphi}$$

Appendix G

Common MEMS Cross-Section Properties

TABLE G.1
Area Polar Moment of Inertia and Torsional Constant

Cross section	Area Polar Moment of Inertia	Torsion Constant - J
r — circle, d	$\dfrac{\pi d^2}{32} = \dfrac{\pi r^2}{8}$	$\dfrac{\pi d^2}{32} = \dfrac{\pi r^2}{8}$
square, side a	$0.1667a^4$	$0.1406a^4$
rectangle, $2a$ high, $2b$ wide	$\dfrac{4}{3}(ba^3 + ab^3)$	$ab^3\left(\dfrac{16}{3} - 3.36\dfrac{a}{b}\left(1 - \dfrac{b^4}{12a^4}\right)\right)$

TABLE G.2
Area Properties for a Rectangular Cross-Section

Cross-Section	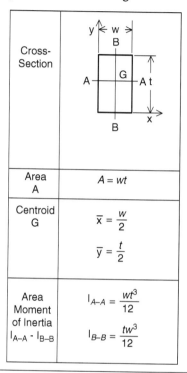
Area A	$A = wt$
Centroid G	$\bar{x} = \dfrac{w}{2}$ $\bar{y} = \dfrac{t}{2}$
Area Moment of Inertia I_{A-A} - I_{B-B}	$I_{A-A} = \dfrac{wt^3}{12}$ $I_{B-B} = \dfrac{tw^3}{12}$

Common MEMS Cross-Section Properties

TABLE G.3
Area Properties for Common MEMS Cross-Sections

Cross Section	
Area A	$A = \dfrac{t(w_1 + w_2)}{2}$
Centroid G	$\bar{x} = w_1/2$ $\bar{y} = \dfrac{t}{3}\dfrac{2w_2 + w_1}{w_2 + w_1}$
Area Moment of Inertia $I_{A-A} - I_{B-B}$	$I_{A-A} = \dfrac{t^3(w_1^2 + 4w_1 w_2 + w_2^2)}{36\,(w_1 + w_2)}$ $I_{B-B} = \dfrac{tw_2^3}{12} + \quad\ldots$ $\dfrac{t(w_2 - w_1)\,(33\,w_2^4 + 3w_1^4 + 48\,w_2^3 w_1 - 12\,w_2 w_1^3 + 8\,t^2 w_2^2)}{144\,(w_2 + w_1)^2}$

Appendix H

Appendix H shows the functions and files to perform Lagrange's equation and the SUGAR simulations discussed in Chapter 7. Appendix H.1 and H.2 are the MATLAB™ functions that perform the computations for Lagrange's equations (Equation 7.13 and Equation 7.14). LagEqn.m (Appendix H.1) is the main function, which calls maxderiv.m (Appendix H.2) as necessary. These functions require MATLAB and the Symbolic Math Toolbox. Appendix sections H.3 through H.7 are the MATLAB files used for Example 7.1 through Example 7.5. Sections H.8 through H.10 are files used for the SUGAR simulation in Example 7.6.

H.1 LAGEQN.M

```
function [eqns]=LagEqn(T,U,D,W,G,Gcoord,Tvar);
% [eqns]=LagEqn(T,U,D,W,G,Gcoord,Tvar)
% Lagrange's equation will be used to find the equations
    of motion
% Equation #.11
%
% The equations of motion will be found using symbolic
% manipulation given the following input data:
% T-Kinetic Energy
%    (e.g., T=1/2*m*Dx^2; or undefined if T=0;)
% U-Potential Energy
%    (e.g., U=1/2*k*x^2; or undefined if U=0;)
% D-Raleigh Dissipation Function
%    (e.g., D=1/2*c*Dx^2; or undefined if D=0)
% W-Virtual Work Vector
%     (e.g., W=sym([f*x;f*y]); or W=[0;...] ncordx1;)
% G-Constraint Eqns
%     (e.g., G=r-r0; or undefined if G is not relevant;)
% Gcoord-vector of Generalized Coordinates
%    (e.g., Gcoord=sym([x; y]); required)
% Tvar-vector of time dependent variable symbols
```

437

```
%    (e.g., Tvar=sym('[m;k]'); or undefined if Tvar is
   not relevant)
%
% Output Data:
% eqns- a vector of equations, where eqns=0
%
% Notes:
% 1. A capital D preceding a generalized coordinate
   means derivative
%    (e.g., Dx - first derivative of the generalized
   coordinate x)
%    (e.g., D2x - second derivative of the generalized
   coordinate x)
%
% 2. LAM# (e.g., LAM1, LAM2) is a reserved variable
   name associated with
% the Lagrange Multipliers for the constraint eqns.
%
% 2. IN ALL CASES, The following symbols must be
   defined before a call
% to this function.
%    syms T U D W G Gcoord Tvar
%
% 3. If any of the symbols are not used in a particular
   problem, DO NOT set
% the symbol to a value. For example if T=0 -> Do not
   set T to a value.
%
% ©J.J. Allen 2004

[nG,nc]=size(G);            %nG= # constraint equations
[nGcoord,nc]=size(Gcoord);   %nGcoord = # generalized
                                        coordinates

if nc~=1
    error('LagEqn: Gcoord should be a symbolic column
    vector (nGcoord,1)');
end

[nTvar,nc]=size(Tvar);     %# time dependent variables
[nVW,nc]=size(W);
if nVW~=nGcoord & nVW~=0
    error('LagEqn: W vector should be a symbolic
    vector(nGcoord,1');
end
```

Appendix H

```
eqns=sym(zeros(nGcoord,1));          %initialize equations

%fully expand Functionals
T=expand(T);
U=expand(U);
D=expand(D);
G=expand(G);

for ne=1:nGcoord
    GC=Gcoord(ne); %symbol for generalized coordinate
    eval(['syms D' char(GC)])
    eval(['Dx=D' char(GC) ';'])

    x=Gcoord(ne);

    dTdx=diff(T,x);
    dTdDx=diff(T,Dx);
    dUdx=diff(U,x);
    dFdDx=diff(D,Dx);
    dWdx=diff(W(ne),x);
    dGdx=diff(G,x);
    Dmax=maxderiv(dTdDx);
    dTdDxdt='0';

%differentiate generalized coordinates,
%and all higher derivatives wrt time
    for ig=1:nGcoord
        GC=Gcoord(ig);
        %order 0
        eval(['syms D' char(GC)])
        eval(['dTdDxdt=dTdDxdt + expand(diff(dTdDx,GC)
        * D' …
        char(GC) ');' ])

        %order1
        eval(['syms D2' char(GC)])
        eval(['dTdDxdt=dTdDxdt + '...
        'expand(diff(dTdDx,D' char(GC) ') * D2'
        char(GC) ');' ])

        %derivative wrt to order 2 or greater
        for id=2:Dmax
            eval(['dTdDxdt=dTdDxdt +
            expand(diff(dTdDx,D' …
```

```
                      int2str(id) char(GC) ') * D'
                      int2str(id+1) char(GC) ');' ])
           end
    end

%differentiate time dependent variables,
%and all higher derivatives wrt time
    for it=1:nTvar
        Tv=Tvar(it);
        %order 0
        eval(['syms D' char(Tv)])
        eval(['dTdDxdt=dTdDxdt + expand(diff(dTdDx,Tv)
        * D' ...
            char(Tv) ');' ])

        %order1
        eval(['syms D2' char(Tv)])
        eval(['dTdDxdt=dTdDxdt + expand(diff(dTdDx,D'
        ... char(Tv) ') * D2' char(Tv) ');' ])

        %derivative wrt to order 2 or greater
        for id=2:Dmax
            eval(['dTdDxdt=dTdDxdt +
            expand(diff(dTdDx,D'...
                int2str(id) char(Tv) ') * D'
                int2str(id+1) char(Tv) ');' ])
        end
    end

%Constraint force terms - Cforce
    syms Cforce
    Cforce=0;
    for ic=1:nG
        eval(['syms LAM' int2str(ic)])
        eval(['Cforce=Cforce+ LAM' int2str(ic)
        '*dGdx(ic);' ])
    end

    %Form Lagrange's equation
    eqns(ne)=dTdDxdt-dTdx +dUdx+dFdDx-dWdx-Cforce;
end
```

Appendix H

441

H.2   MAXDERIV.M

```
function [Dmax]=maxderiv(expr);
%[Dmax]=maxderiv(expr)
%This function will find the maximum derivative for
    any variable
% in the symbolic expression, expr.

expr=char(expr);

indx=find(expr=='D');

num=length(indx);
if num==0
    Dmax=0;
else
    Dmax=1;
    for i=1:num
        deg=str2num(expr(indx(i)+1));
        if max(size(deg))~=0
            Dmax=max([Dmax; deg]);
        end
    end
end

return
```

H.3 XCEL1.M

```
clear all
diary off
delete xcel1.dia
diary xcel1.dia
echo on
clc

syms T U D W G Gcoord Tvar %declare symbolic functionals
syms M C K F %declare symbolic constants in problem
syms x Dx %declare symbolic generalized coordinates &
    derivatives

T=1/2*M*Dx^2;
U=1/2*K*x^2;
D=1/2*C*Dx^2;
W=[F*x];
Gcoord=[x];

[eqns]=LagEqn(T,U,D,W,G,Gcoord,Tvar);

%display results
disp('Generalized Coordinates')
pretty(Gcoord)
disp('equations of motion')
pretty(eqns)

echo off
diary off
```

Appendix H

443

H.3   CKT1.M

```
clear all
diary off
delete ckt1.dia
diary ckt1.dia
echo on
clc

syms T U D W G Gcoord Tvar %declare symbolic functionals

syms R C L V %declare symbolic constants in problem
syms Q DQ %declare symbolic generalized coordinates &
    derivatives

T=1/2*L*DQ^2;
U=1/2*1/C*Q^2;
D=1/2*R*DQ^2;
W=[V*Q];
Gcoord=[Q];

[eqns]=LagEqn(T,U,D,W,G,Gcoord,Tvar);

%display results
disp('Generalized Coordinates')
pretty(Gcoord)
disp('equations of motion')
pretty(eqns)

echo off
diary off
```

444   Micro Electro Mechanical System Design

H.4 M2OSCL.M

```
diary m2oscl.dia
echo on
clc

syms T U D W G Gcoord Tvar %declare symbolic functionals

syms Mx My Kx Ky Cy g e0 A Er El %declare symbolic
    constants in problem
syms x Dx y Dy %declare symbolic generalized coordinates
    & derivatives

T=1/2*Mx*Dx^2 + 1/2*My*Dy^2;
U=1/2*(4*Kx)*x^2+1/2*(4*Ky)*(x-y)^2...
 +1/2*e0*A*El^2/(g+x)+1/2*e0*A*Er^2/(g-x);
D=1/2*Cy*Dy^2;
W=[0; 0];
Gcoord=[x; y];

[eqns]=LagEqn(T,U,D,W,G,Gcoord,Tvar);

%display results
disp('Generalized Coordinates')
pretty(Gcoord)
disp('equations of motion')
pretty(eqns)

echo off
diary off
```

Appendix H

H.5 RODWALL.M

```
clear all
diary off
delete RodWall.dia
diary RodWall.dia
echo on
clc

syms T U D W G Gcoord Tvar %declare symbolic functionals
syms LAM1 %declare Lagrange multipliers if there are
    constraints

syms Lx Ly L M I K F %declare symbolic constants in
    problem
syms x Dx y Dy %declare symbolic generalized coordinates

U=1/2*K*y^2;
W=[F*x; 0];
G=[(Lx-x)^2+(Ly+y)^2-L^2];
Gcoord=[x;y];

[eqns]=LagEqn(T,U,F,W,G,Gcoord,Tvar);

%display results
disp('Generalized Coordinates')
pretty(Gcoord)
disp('equations of motion')
pretty(eqns)
disp('constraint equations')
pretty(G)

echo off
diary off
```

H.6 PARALLELRLC.M

```
clc
diary off
delete parallelRLC.dia
diary parallelRLC.dia

%Parallel RLC circuit
syms T U D W G Gcoord Tvar %declare symbolic functionals
syms R C L I            %declare symbolic constants
    in problem
syms lam Dlam            %declare symbolic generalized
    coordinates & derivatives

T=1/2*C*Dlam^2;
U=1/(2*L)*lam^2;
D=1/(2*R)*Dlam^2;
W=[I*lam];
Gcoord=[lam];

[eqns]=LagEqn(T,U,D,W,G,Gcoord,Tvar);

%display results
disp('Generalized Coordinates')
pretty(Gcoord)
disp('equations of motion')
pretty(eqns)

echo off
diary off
```

Appendix H

H.7 SOLENOID.M

```
clear all
diary off
delete solenoid.dia
diary solenoid.dia
echo on
clc

syms T U D W G Gcoord Tvar %declare symbolic functionals

syms M K R C L E L0 x0 %declare symbolic constants
    in problem
syms Q DQ x Dx              %declare symbolic generalized
    coordinates & derivatives

L=L0/(1+(x/x0)^2)

T=1/2*L*DQ^2+1/2*M*Dx^2;
U=1/2*K*x^2;
D=1/2*R*DQ^2+1/2*C*Dx^2;
W=[E*Q; 0];
Gcoord=[Q; x];

[eqns]=LagEqn(T,U,D,W,G,Gcoord,Tvar);

%display results
disp('Generalized Coordinates')
pretty(Gcoord)
disp('equations of motion')
pretty(eqns)

echo off
diary off
```

H.8 SUMMIT.M

```
process poly = [
    Poisson = 0.23          %Poisson's Ratio = 0.3
    thermcond = 2.33            %Thermal conductivity Si
                                = 2.33e-6/C
    viscosity = 1.78e-5         %Viscosity (of air) =
                                1,78e-5
    fluid = 2e-6                %Between the device and
                                the substrate.
    density = 2300          %Material density = 2300
                                kg/m^3
    Youngsmodulus = 160e9 %Young's modulus = 1.60e11
                                N/m^2
    permittivity = 8.854e-12%permittivity F/m
]

process p1: poly = [
    h = 1e-6 %Layer height of Summit poly1 = 1e-6 m
]

process p2: poly = [
    h = 1.5e-6 %Layer height of Summit poly2 = 1.5e-6 m
]

process p12: poly = [
    h = 2.5e-6 %Layer height of Summit poly2 = 2.5e-6 m
]

process p3: poly = [
    h = 2.25e-6 %Layer height of Summit poly3 = 2.25e-
    6 m
]

process p4: poly = [
    h = 2.25e-6 %Layer height of Summit poly4 = 2.25e-
    6 m
]
```

Appendix H

H.9 LEV_BEND.NET

```
uses summit.net
uses stdlib.net

param Lelec=0
param Lcenter=0

gap3de p12 [a b aa bb] [l=Lelec w1=2.5u w2=2.5u h=10u
     gap=2u
            R1=10 R2=10 ox=pi/2]
beam3de p12 [b c] [l=Lcenter w=10u R=10 ]
beam3de p12 [c d] [l=Lcenter w=10u R=10 ]
gap3de p12 [d e dd ee] [l=Lelec w1=2.5u w2=2.5u h=10u
     gap=2u
            R1=10 R2=10 ox=pi/2]

anchor p12 [a] [l=10u w=10u h=2.5u R=10 ]
anchor p12 [e] [l=10u w=10u h=2.5u R=10 ]

anchor p12 [aa] [l=10u w=10u R=10 ]
anchor p12 [bb] [l=10u w=10u R=10 ]
anchor p12 [dd] [l=10u w=10u R=10 ]
anchor p12 [ee] [l=10u w=10u R=10 ]

eground * [aa] []
eground * [bb] []
eground * [dd] []
eground * [ee] []

param Vactuate=0

Vsrc * [a gnd] [V=Vactuate sv=0.1 sph=0]
eground * [gnd] []
```

H.10 LEV_BEND.M

```
clear all
clc

Le=100;
Lc=300;
param.Lelec=Le*1e-6;
param.Lcenter=Lc*1e-6;

Vvec=[];
Zvec=[];
V=0;
for V=0:5:200

    param.Vactuate=V;

    net=cho_load('Lev_bend.net',param);

    [dq,conv]=cho_dc(net);

    if conv==0
        disp('did not converge -> break out of the
    loop')
        break
    end
    cx=dq(lookup_coord(net,'c','x'));
    cy=dq(lookup_coord(net,'c','y'));
    cz=dq(lookup_coord(net,'c','z'));

    Vvec=[Vvec; V];
    Zvec=[Zvec; cz/1e-6];

    disp(['V = ' num2str(V) ' Z = ' num2str(cz/1e-6)])

end

cho_display(net,dq)
```

Appendix H   **451**

H.11 RRITZ_FFBEAM.M

```
%Raliegh Ritz solution of a fixed-fixed beam with a
    distributed load
%Lagranges equations are used to obtain the governing
    equations.
%Using 10 terms in the solution.
clear all
clc
disp('Fixed Fixed Euler Beam a distributed load')
syms T U D W G Gcoord Tvar
syms EI L Y phi a0 a1 a2 a3 a4 a5 a6 a7 a8 a9 x w

phi=[ cos(2*pi*x/L)-1 cos(2*2*pi*x/L)-1
    cos(2*3*pi*x/L)-1...
    cos(2*4*pi*x/L)-1 cos(2*5*pi*x/L)-1
    cos(2*6*pi*x/L)-1...
    cos(2*7*pi*x/L)-1 cos(2*8*pi*x/L)-1
    cos(2*9*pi*x/L)-1...
    cos(2*10*pi*x/L)-1];
D2phi=diff(phi,'x',2);

a= [ a0; a1; a2; a3; a4; a5; a6; a7; a8; a9];

Y=phi * a;

%Strain Energy, U
U= EI/2*int((D2phi*a)^2,0,L);

%non-potential Energy, W
% w - distributed load
W=[int(w*phi(1),0,L)*a0; int(w*phi(2),0,L)*a1;...
    int(w*phi(3),0,L)*a2; int(w*phi(4),0,L)*a3;...
    int(w*phi(5),0,L)*a4; int(w*phi(6),0,L)*a5;...
    int(w*phi(7),0,L)*a6; int(w*phi(8),0,L)*a7;...
    int(w*phi(9),0,L)*a8; int(w*phi(10),0,L)*a9];

Gcoord=[a0; a1; a2; a3; a4; a5; a6; a7; a8; a9];

eqns=LagEqn(T,U,D,W,G,Gcoord,Tvar);
```

Index

A

Abrasive wear, 389
Absolute viscosity, 222
Accelerometers, 7, 314–319
  amplitude and phase distortion, 316
  closed-loop, 318–319
  damping effects, 193, 316
  layout, 157
  open-loop, 317
Acoustic wave devices, 329
Activation energy, 24
Actuators, 273–289
  electromagnetic field effects and scaling, 130–134
  electrostatic actuation, 273–284, *See also* Capacitors
    compliant displacement mechanisms, 282–283
    design tradeoffs, 280
    electrostatic spring softening, 277
    inchworm approach, 283
    interdigitated comb capacitor, 278–280
    parallel plate capacitor, 273–278, 280
    pull-in instability, 276, 283–284
    scratch drive, 283
  Lorentz force, 288
  oscillator model, 250–252
  thermal actuation, 285–288
Adhesive wear, 388
Airbag sensor application, 314
Alcohol breath analyzer, 330
Aluminum, 341, 425
Amorphous carbon, 108–109, 425
Amorphous materials, 19, 194
Amorphous silicon piezoresistivity, 304
Ampere, 229
Amplification factor, 219
Amplitude distortion, 316
Amplitude modulation, 292–294
Analog Devices, Inc., 6
Anchor area, 182
Anchoring layers, 159–164
Anisotropic stress strain relation, 199
Annealing, 18, 51–53, 194
Anodic bonding, 51

Antimony (Sb), 26
Archimedes spiral, 101
Area polar moment of inertia, 433
Area properties, common cross-sections, 433–435
Area to volume ratio, 116
Arsenic (As), 26, 59, 299
Art-to-part, 189
Assembly, 345–348
Atmospheric pressure CVD, 36
Atomic force microscope, 102, 140, 397
Atomic scale modeling, 188
Atomic structure, 18
AutoCAD, 157
Automated design rule checking, 182–183
Automotive MEMS applications, 9, 314
Axial loading, 211, 428
Axial rod structural mechanics, 201–203

B

Back end of the line (BEOL) processing, 340–350, *See also* Postfabrication processing
Ball grid array (BGA) packages, 351
Band limited signal, 309
Bathtub curve, 370, 372
Beam bending, 205–208, 265–267, 284
Beam stiffness coefficients, 205, 428–429
Bending
  beam, 205–208
  flat plate, 208–210
  leveraged beam, 265–267, 284
Bent beam thermal actuator, 286
Bias/adhesion mirror mapping (BAMM), 405
Bimorph actuator, 286
Biological applications
  compatible coatings, 342
  concentration scale effects, 135–136
  sensitive coatings, 342
  sensors, 328
Biomedical MEMS applications, 9–10
Biot number (Bi), 124
Blur envelope, 398
Bonding processes, 50–51, 353
Bond pad spacing, 355

453

454

Micro Electro Mechanical System Design

Bonds, interatomic, 17–18, 194
Boron (B) doping, 26, 39, 55, 59, 299
Boron nitride (BN), 32
Boundary element (BEM) methods, 262–265
Breakdown voltage, 145
Breath analyzers, 330
Bright field illumination, 395
Brittle fracture, 390
Brownian noise, 144–145, 313
Bulk acoustic wave, 329
Bulk micromachining, 4, 6, 65, 66, 68–79, *See also* Etching processes
   comparison of MEMS and conventional technologies, 69
   masks, *See* Masks
   PennSOIL silicon-on-insulator process, 76–79
   plasma etching, 72–74
   SCREAM technology, 74–76
   wet etching, 70–72
Buried silicon dioxide (BOX), 29
Burst noise, 313

C

C4, 352
Calorimetric chemical sensors, 329
Cal Tech Intermediate Format (CIF), 155
Capacitance (C), 234–235, 273, 275
   parasitic, 290–291
Capacitors, 234–235
   accelerometers, 315
   actuator applications, *See also* Actuators
      interdigitated comb capacitor, 278–280
      parallel plate capacitator, 273–278, 280
   differential, 290
   electrostatic sensing applications, 290–298, *See also under* Sensors
   pull-in instability, 276, 283–284
   switched capacitor techniques, 296–298
Cap and post hub, 167
Castigliano's theorem, 258–259
Charge integrator, 294–295
Charging, 391
Chemical concentration, scale effects, 135–136
Chemically compatible coatings, 342
Chemically sensitive coatings, 342
Chemical mechanical polishing (CMP), 53, 86–87
Chemical sensors, 328–330
Chemical vapor deposition (CVD), 35–37, 85, 88

Chip scale packaging, 352
Chlorine-based plasma etching, 73
Chromium (Cr) coatings, 342
Closed-loop accelerometer, 318–319
Closed polygon approximation, 156–157
CMOS transistors, 130
Coatings, 340, 342–344
   effects, 343–344
   types, 342
Coefficient of thermal expansion, 98, 200, 285
Cold working, 18–19, 194
Column structural mechanics, 211–213
Combustible gas sensor, 331–332
Common-mode rejection, 293
Compliant mechanisms, 282–283, 402
Component standardization, 183–184, 402
Computational issues of scale, 137–139
Conduction, 121
Conductivity, thermal, 18
Constant failure rate, 370
Constraint equations, 247–248
Continuum mechanics assumptions, 142–144, 220–221
Convection, 121
Copper (Cu), 425
Coriolis acceleration, 321–322
Corrosive wear, 389
Couette damping, 226–227
Coulomb, 229
Coulomb damping, 217
Coulomb's law, 229
Covalent bonds, 17
Crab leg flexure, 429
Creep, 391, 406
Critical damping, 219
Cross-section properties, 433–435
Crystal growth
   Czochralski growth process, 25–27
   dopants, 26–27
   float zone process, 27
   oxygen precipitates, 27
   post-crystallization processing, 27–28
Crystalline materials and structures, 18, 19–21, 194
   imperfections, 23–24
   Miller indices, 21–23
Cubic crystal system, 19
Cumulative failure distribution plot, 368
Cumulative probability distribution function (CPDF), 369, 373–380
Current, 229
Current source, 238
Czochralski growth process, 25–27

Index

455

D

Damascene process, 50, 107
Damping, 193, 215–228
  accelerometer design and, 316
  continuum assumptions and, 220–221
  Couette model, 226–227
  critical, 219
  hysteresis loop, 217
  Lagrange's equations, 246–247
  mechanisms, 220–222
  oscillatory mechanical systems and,
    217–219
  Raleigh dissipation function, 246
  resonance, 218–219
  slide film damping, 224, 226–227
  squeeze film model, 224–226
  Stokes model, 226, 227
  viscous, 222–224, 246–247
Damping ratio (ζ), 218–219
Darkfield illumination, 395
Degradation mechanisms, 392
Degrees of freedom (DOF), 187, 244
Demodulation, 295–296, 323
Depletion region, 370
Depth of focus, 86, 180–181, 395
Design analysis model, 265
Design, reliability considerations and, 400–403,
    See also Reliability
Design realization tools, 155–189
  anchoring layers, 159–164
  beam and substrate connection, 170
  closed polygon approximation, 156–157
  design rules, *See* Design rules
  discrete hinges, 170–175
  geometric editing, 157
  masks, *See* Masks
  MEMS analysis, 186–188
  MEMS visualization tools, 184–186, *See*
    also Measurement and testing;
    Visualization tools
  modeling, *See* Modeling and simulation
  rotational hubs, 164–170
  standard components, 183–184, 402
Design rules
  area of anchor, 182
  checking, 182–183
  etch compatibility, 179
  etch pattern uniformity, 178
  etch release holes, 181
  litho depth of focus, 180–181
  mask alignment registration errors, 178
  MEMS packaging layout, 355–356
  patterning limits, 176

  stiction, 181
  stringers and floaters, 179–180
Design synthesis modeling, 187, 243–244, *See*
    also Modeling and simulation
Design uncertainty, 267–268
Detailed design model, 187–188
Diamond and diamond-like carbon, 108–109,
    425
Diaphragm, 326
Die assembly, 345
Die attach, 340, 352–353
Dielectric charging effects, 391
Differential capacitors, 290
Diffusion, 24
Diffusion-based doping, 56
Digital mirror device (DMD), 341
  packaging, 355, 357–359
  reliability methods and testing, 403–407
Dimples, 167, 181
Direct-write E-beam lithography, 43
DMD, *See* Digital mirror device
DNA sequencing chip, 9
Dopants and doping, 26–27, 54–61
  annealing and activation, 51–52
  diffusion, 55–60
  implantation, 55, 60–61
  silicon piezoresistivity and, 299
Double folded flexure, 430
Double V beam flexure, 431
Drying, 340, 341
Ductile fracture, 390
Dynamic absorber, 252

E

Edge dislocation, 24
Edge effects, 178
Elastic limit, 195
Electrical and magnetic fields, 130–134,
    229–233
Electrical charge, 229
Electrical energy functions, 245
Electrical energy sources, 238
Electrical-fluidic packaging, 355, 359–361
Electrical potential, 230
Electrical resistance, *See* Resistance
Electrical shorts, coatings and, 343
Electrical system dynamics, 228–241
  capacitor, *See* Capacitors
  circuit interconnection, 238–241
  electrical and magnetic fields, 229–233
  energy sources, 238
  Faraday's law, 232–233

456 Micro Electro Mechanical System Design

inductors, 235–236
Kirchoff's law, 238–240
Lagrange's equations and governing
    equations, 249–257
    parallel circuit, 254–255
    resonator, 250–252
    simple RLC circuit, 249–250
    solenoid, 255–257
Maxwell's equations, 228, 231
quantities and units, 228, 229
resistors, 236–237
right-hand rule, 233
Electrical system scaling, 129–134
Electrochemical sensors, 329
Electromechanics, 193–241
    damping, 193, 215–228, *See also* Damping
    electrical circuit elements, 228–241, *See
        also* Electrical system dynamics
    structural mechanics, 193–215, *See also*
        Structural mechanics
Electron energy loss spectroscopy (EELS), 396
Electronic nose, 329–330
Electronic speckle pattern interferometry
    (ESPI), 399
Electron tunneling, 145, 306–308
Electrostatic actuation, 273–284, *See also under*
    Actuators
Electrostatic bonding, 51
Electrostatic capacitance sensing, 290–298, *See
    also under* Sensors
Electrostatic discharge, 393–394
Electrostatic spring softening, 277
Electrothermal-compliant (ETC) microdevices,
    76–77
EMDIP, 355, 359–361
Encapsulation, 348–350
Energy dispersive x-ray spectroscopy (EDX),
    396
Energy sources, 238
ENIAC, 1
Envelope detector, 295
Environmental degradation, 392
Environmental failure mechanisms, 392–393
Environmental testing, 407
EPON SU-8, 109
Epoxy die attach, 352–353
Etching processes, 38–43, 83–84
    electrochemical stop, 71
    end-point detection, 73
    ion milling, 43
    material compatibility, 179
    pattern uniformity, 178
    PennSOIL technology, 76–79

plasma, 39–43, 72–74
process dimensional control parameters,
    96–97
process modeling tools, 185
reactive ion etching, 42, 73–74, 88
release holes, 157–158, 181–182
SCREAM, 74–76
wet, 38–39, 70–72, 88
Euler-Bernouli beam, 205–208
Euler column theory, 212
Eutectic die attach, 352–353
Evaporation-based deposition, 32–34
Exponential model, 386

F

Fabrication processes, 17–63
    annealing, 18, 51–53, 194
    assembly, 345–348
    back end of the line, 340–350, *See also*
        Postfabrication processing
    chemical mechanical polishing, 53
    chemical vapor deposition, 35–37, 85, 88
    comparison of MEMS and conventional
        technologies, 69
    Czochralski growth process, 25–27
    design rules, 176–183, *See also* Design rules
    doping, *See* Dopants and doping
    etching, 38–43, *See also* Etching processes
    evaluation and selection of, 68
    evaporation, 32–34
    historical development, 4–5
    ion milling, 43
    LIGA, *See* LIGA
    materials, 17–24, *See also* Material
        properties; *specific materials*
    patterning, 43–50
        damascene process, 50
        lift-off process, 48–49
        lithography, 43–48, *See also*
            Lithography
    physical vapor deposition, 30–35
    post-crystallization processing, 27–28
    relative tolerance, 267–268
    scaling issues, 139–140
    silicon-on-insulator, 28–30
    sputtering, 34–35
    substrates, 25–30
    surface micromachining, *See* Surface
        micromachining
    technology development, 6
    wafer bonding, 50–51

Index

Fabrication technologies, *See* Bulk micromachining; LIGA; Surface micromachining

Failure
cumulative distribution plot, 368
defining, 367–368
degradation mechanisms, 392
design and prevention, 402
environmental failure mechanisms, 392–393
frequency distribution plot, 370
operational mechanisms, 388–392
charging, 391
creep, 391
fatigue, 391
fracture, 390
stiction and adhesion, 391–392
wear, 388–390
stress at (ultimate strength), 195

Failure in time (FIT), 369
Failure modes, effects, and criticality analysis (FMECA), 402, 405
Faraday's law, 232–233
Fatigue, 391
Fault-tree analysis, 402, 403
Feature size, 176
Feynman, Richard, 2–3
Finite difference (FDM) methods, 262–265
Finite element analysis (FEM), 138, 187, 262–265
SUGAR, 263–267
Flat plate bending, 208–210
Flexure stiffness coefficients, 429–431
Flicker noise, 311–312
Flip-chip bonding, 345, 351–352
Floaters, 180
Float zone process, 27
Fluidic damping, 217
Fluidic printed wiring board, 360–361
Fluid mechanics, scale effects, 124–129
Fluoride etching, 42
Flux linkage (λ), 235–236
Focus depth, 86, 180–181, 395
Focused ion beam (FIB), 396–397
Folded flexures, 429–430
Fracture, 390
Freeze drying, 341
Frequency response function (FRF), 310
Frick's law diffusion, 56–59
Friction forces, 245–246
coatings and, 342
Fusion bonding, 51

G

Gallium arsenide (GaAs), 21, 25, 27–28
Gas flow field, 125–126
Gas production (outgassing)
coatings and, 344
epoxy and, 353
Gas sensors, 330–332
GDSII, 155–158
GeneChip, 9
Geometric editing, 157
Geometric emulation, 185–186
Geometric scaling, 115–116
Germanium
crystal structure, 21
Si-Ge polycrystalline alloys, 108
Getters, 344, 393
Glasses, 194
Glossary, 417–418
Gold (Au), 425
Au-Sn solder, 353
coatings, 342
Grain boundary, 18
Gram molecular weight (MW), 135
Granularity, scaling issues, 142–144
Grashof number (Gr), 124
Gravity forces and scale, 145
Gyroscopes, 7, 314, 319–324
demodulation, 323
precession, 319
rotational rate sensors, 320–324
Sagnac effect, 320–321
vibratory, 321–324

H

Hagen-Poiseulle flow, 226
Hazard rate, 379, 380
Heat transfer
thermal stresses and strains, 200–201
thermal system scaling, 121–124
High-density plasma etching (HDP), 42–43, 73
Hinges, 170–175
fatigue, 406
memory, 406
Historical perspectives
MEMS technology development, 3–6
microelectronics technology development, 1–3
Hooke's law, 196–198
Humidity, 393
Hydrodynamic system scaling, 124–129

458 Micro Electro Mechanical System Design

Hydrofluoric (HF) acid, 39, 84, 87, 341
Hysteresis loop, 217

I

IMaP, 106
IMEMS, 6, 94–95
Inchworm approach, 283
Indium phosphide, 21
Inductance, 130
Inductors, 235–236
Inertial sensors, *See* Accelerometers;
  Gyroscopes
Infant mortality, 370, 410
Instantaneous failure rate, 379, 380
Integrated circuits (IC)
  fabrication technologies and scale,
    139–140
  historical perspectives, 1–2
  package design, 350–352
Integrated MEMS (IMEMS) technology, 6,
    94–95
Integrator, 293, 294–295
Interconnection, 340, 353
  package selection and design, 350–352
Interdigitated comb capacitor, 278–280
Interferometric fiber-optic gyro (IFOG), 321
Interferometry, 399
Interstitial defect, 23
Ionic bonds, 17
Ion implantation, 60
Ionization gauge, 325–326
Ion milling, 43
Irreversible reaction, 328
Isotropic stress-strain relation, 198

J

Johnson noise, 144–145, 313

K

Kinematic viscosity, 222
Kinetically controlled CVD, 36
Kinetic energy, Lagrange's equations, 244–245,
    249
Kinetic theory of gases, 33
Kirchhoff's law, 238–240
Knudsen number (Kn), 125–126

L

Lagrange multipliers, 247
Lagrange's equations, 244–257
  constrained linkage motion, 252–254
  constraint equations, 247–248
  generalized coordinates, 244
  MATLAB functions and files for, 437–451
  multibody resonator, 250–252
  parallel RLC circuit, 254–255
  Raleigh-Ritz method, 260–262, 451
  simple mechanical system, 248–250
  simple RLC circuit, 249–250
  solenoid, 255–257
Laminar flow, 125
Laser Doppler velocimeter (LDV), 400
Lateral electrostatic resonator, 102
Lateral thermal actuator, 286
Latin hypercube, 270
Layout design rules, *See* Design rules
Layout design tools, 155–175, *See also* Design
    realization tools
Legendre transformation, 235
Leveraged beam bending model, 265–267, 284
Lift-off process, 48–49
  measurement techniques, 397
LIGA, 6, 65–66, 79–83
  comparison of MEMS and conventional
    technologies, 69
  masks, *See* Masks
  microdrive application, 80–83
  resist material, 79
  technology development, 5–6
Line defects, 23, 24
Line width and space design rule, 176
Linkage motion model, 252–254
Lithographie Galvanoformung Abformung, *See*
    LIGA
Lithography, 43–48
  aligner, 47–48
  depth of focus, 180–181
  LIGA, *See* LIGA
  masks, 45–46, *See also* Masks
  optical system design, 44–45
  photoresist, 43, 46–47
  resist material, 109
  SUMMiT surface micromachining process,
    88
Lodestone, 232
Lognormal model, 383–386
Lorentz force actuation, 288
Lorentz force law, 232
Low-clearance hub, 167
Low-pressure CVD, 36

Index **459**

Lumped parameter models, Lagrange's
   equations applications, 248–257

M

Machine accuracy, 138
Macromodels, 187
Magnetic field density, 231
Magnetic fields, 6–8, 129, 130–134, 229–233
   inductors and, 235–236
Magnetic flux density, 132, 231
Magnetron sputtering, 35
Manhattan geometries, 182–183
Masks, 45–46, 155
   alignment registration error, 178
   design tools, 157–158
   etch compatibility, 179
   SUMMiT definitions, 160
Mass and volume reduction, 116–117
Mass transport controlled CVD, 36
Material properties, 17–24, 95–106, 425, *See
      also specific materials*
   alternative MEMS materials, 106–109
      diamond and diamond-like carbon,
         108–109
      polycrystalline silicon-germanium, 108
      SiC, 106–108
      SU-8, 108–109
   atomic and crystal properties (table), 20
   crystal lattices, 19–21
   doping, *See* Dopants and doping
   electrical resistance, 103–105
   etch compatibility, 179
   failure mechanisms, 388–394, *See also*
      Failure
   interatomic bonds, 17–18
   measurement for process control, 105–106
   mechanical properties, 194–199
   residual stresses, 51, 98–101
   scaling issues, 141–144
   stiffness, *See* Stiffness coefficient
   strength, 102–103
   structure, 18–19
   Young's modulus, *See* Young's modulus
MATLAB applications, 249, 257, 261
   functions and files for, 437–451
   SUGAR, 263–267, 437, 448–451
Maxwell's equations, 228, 231
Mean value, 377
Measurement and testing
   blur envelope, 398
   focused ion beam, 396–397
   interferometry, 399

laser Doppler velocimeter, 400
lift-off, 397
material mechanical properties, 105–106
microscopy, *See* Microscopic techniques
optical microscopy, 394–395
scaling issues, 138–139
stroboscopy, 397–398
video imaging, 399
visualization, *See* Visualization tools
Mechanical energy functions, 245
Mechanical properties, 194–199, *See also*
      Material properties; *specific
      properties*
   scaling issues, 117–121
   stiffness, *See* Stiffness coefficient
MEMCAD, 185, 188
MEMS design tools, *See* Design realization
      tools
MEMS Device Taxonomy, 400, 402
MEMS web sites, 8
Metallic bonds, 18
Metal-oxide semiconductor field effect
      transistor (MOS-FET), 25
Michelson interferometer, 399
Microdrive applications, 80–83
Microelectromechanical systems (MEMS),
      fabrication process, *See*
      Fabrication processes
Microelectromechanical systems (MEMS),
      integrated systems (IMEMS), *See*
      Integrated MEMS (IMEMS)
      technology
Microelectromechanical systems (MEMS)
      technology
   applications, 7–11
   challenges, 12
   comparison with microelectronics, 12
   fabrication technologies, *See* Bulk
      micromachining; LIGA; Surface
      micromachining
   historical perspectives, 1–6
   information sources, 6–8
   technology characterization, 95–106, *See
      also* Technology characterization
Microelectronic technology, historical
      perspective, 1–3
Micro-optical-gyro (MOG), 321
Microphones, 328
Microsatellites, 10–11
Microscopic techniques
   atomic force, 102, 140, 397
   optical microscopy, 394–395
   scanning tunneling, 140, 306
   SEM, 396

460   Micro Electro Mechanical System Design

Miller indices, 21–23, 194
Miniaturization scaling issues, *See* Scaling
    issues
Modal coordinate transformation, 217
Modeling and simulation, 243–271
  atomic scale, 188
  continuous system mechanics, 257–262
  design analysis, 243, 265
  design synthesis, 187, 243–244
  detailed design, 187–188
  geometric emulation, 185–186
  Lagrange's equations, *See* Lagrange's
    equations
  lumped parameter governing equations,
    248–257
    multibody resonator, 250–252
    parallel RLC circuit, 254–255
    simple mechanical system, 248–250
    simple RLC circuit, 249–250
    solenoid, 255–257
  MEMS analysis, 186–188
  numerical modeling, 262–267
  phenomenological, 187
  process modeling tools, 184–185
  reliability, 380–386
  scaling issues, 137–139
  uncertainty, 267–270
Mole, 135
Molybdenum-polysilicon (Mo/Si) coating, 342
Monte Carlo method, 270
Moore's law, 4
MOS-FET, 25
Motor application, 80–83
M-TEST, 105–106
Multichip module (MCM), 345, 352

N

Nanoindenter, 102
Nanoscale manipulation, 139–140
Natural frequency, 267
Navier-Stokes equations, 124–125
Nested anchor, 164
Nickel (Ni), 425
n-junctions, 54
Noise, 308–314
  1/f noise, 312
  Brownian, 144–145, 313
  burst, 313
  flicker, 311–312
  frequency response function, 310
  power spectral density function, 309
  shot, 311

  signal to noise ratio, 310
  sources, 311–314
  summary of models and mechanisms, 312
  thermal, 313
  white, 309
Nomarski differential interference contrast, 396
Nonpotential forces, 245–247
n-type semiconductors, 26, 55, 299

O

Oblique illumination, 395
Ohm (unit), 237
Ohm's law, 237
Oil condition sensor, 7
Open-loop sensors, 317
Operational amplifier circuit, 293
Optical chemical sensors, 329
Optical coatings, 342
Optical degradation, 392
Optical MEMS applications
  digital mirror device packaging, 355,
    357–359
  retinal prosthesis, 10
  rotational rate sensor, 320–321
  scaling effects, 134–135
Optical microscopy, 394–395
Orthorhombic lattice, 21
Orthotropic stress-strain relations, 198–199
Oscillatory mechanical systems
  damping and, 217–219
  Lagrange's equations for resonator,
    250–252
Oxygen, 27, 30
OYSTER, 185

P

Packaging, 339–362
  assembly, 345–348
  back end of the line, 340–350, *See also*
    Postfabrication processing
  commercial microelectronic technology,
    355–357
  cost, 348
  die attach, 340, 352–353
  encapsulation, 348–350
  functions, 339
  MEMS layout rules, 355–356
  multichip module, 345, 352
  R&D prototype, 355–357
  reliability design, 402

Index

461

Sandia's EMDIP, 355, 359–361
sealing, 340, 353
selection and design, 340, 350–352
TI's digital mirror device (DMD), 355, 357–359
wire bond, 340, 353
Parallel plate capacitor, 273–278, 280
Parallel spring configuration, 427
Parasitic capacitances, 290–291
Partial derivative rule for uncertainty approximation, 269
Part-to-art, 189
Paschen's effect, 145
Paschen's law, 133
Patterning, 43–50, *See also* Fabrication processes
Patterning limit design rules, 176
PennSOIL technology, 76–79
Permanent magnet, 232
Permittivity, 229
Phase distortion, 316
Phenomenological models, 187
Phosphorus (P) doping, 26, 54, 59, 299
Photolithography, *See* Lithography
Photoresist, 43, 46–47, 86
LIGA, 79
SU-8, 109
Physical constants, 423
Physical vapor deposition (PVD), 30–35
Piezoresistivity
accelerometers, 315
doping, 54
microphones, 328
polycrystalline and amorphous silicon, 304
signal detection applications, 304–306
single-crystal silicon, 299–304
Pin joints, 167
Pirani gauge, 325
p-junctions, 54
Planetary gear system, 82
Plasma-enhanced CVD, 36
Plasma etching, 39–43, 72–74
Plastic strain, 195
Plate theory, 208–210
Plate-to-plate hinge, 172
p-n junction, 71
Point defects, 23
Poisson's ratio, 197, 198
Polishing, chemical mechanical, 53
Polycrystalline materials, 18, 194
granularity of microscale/nanoscale properties, 144
Polycrystalline silicon
Mo/Si polylayer coating, 342

piezoresistivity, 304
properties, 425
Si-Ge alloys, 108
stress corrosion cracking, 390
SUMMiT process, 88–91, 341
surface micromachining, 84
Polyimide, 79
Polymerization, 42, 73
Polymethylmethacrylate (PMMA), 79–80
Popcorn noise, 313
Postfabrication processing, 340–350
coating, 340, 342–344
drying, 340, 341
release, 340, 341
Potassium hydroxide (KOH), 70
Potential, 230
Potential energy, Lagrange's equations, 244–245, 249
Power spectral density function (PSD), 309, 392
Precession, 319
Prefixes, 419
Pressure sensors, 7, 314, 324–328
diaphragm, 326
microphones, 328
vacuum sensor, 324–326
Probability and statistics for reliability, 370–379
Probability density function (PDF), 375–380
Process dimensional control parameters, 96–97
Process modeling tools, 184–185
Process monitoring, technology characterization and, 95–96
Product-reliability issue matrix, 401, 402
p-type semiconductors, 26, 55, 299
Pull-in instability, 276, 283–284
Pure random sampling, 270

Q

Quadrature error, 324
Quality factor, 217

R

Radiation effects, 393
Radiation heat transfer, 121
Raleigh dissipation function, 246, 249
Raleigh-Ritz method, 260–262, 451
Random sampling, 270
Random variable, 373
Rapid thermal annealing, 52–53
Rate limiting step, 38
Reactive ion etching (RIE), 42, 73–74, 88

462 Micro Electro Mechanical System Design

Reactive sputtering, 35
Rectangular cross-section area properties, 434
Registration errors, 97, 178
Relative tolerance, 267–268
Release, 340, 341
Reliability, 367–413
  coating stability, 344
  design and, 400–403
  exponential model, 386
  failure distribution plots, 368–370
  failure mechanisms, 386–393, *See also*
    Failure
  FMECA, 402, 405
  instantaneous failure rate, 379
  lognormal model, 383–386
  measurement and testing techniques,
    394–400
  mechanical property scale effects, 119–121
  MEMS Device Taxonomy, 400, 402
  nondeterministic factors, 368
  probability and statistics, 370–379
  product-reliability issue matrix, 401, 402
  sample size and, 368
  Sandia microengine system, 407–412
  solution space characterization technique,
    405
  Texas Instruments' digital mirror device
    (DMD), 403–407
  theory and terminology, 367–370
  Weibull model, 380–383, 410
Research and development (R&D) packaging,
  355–357
Residual stresses, 51, 98–101
  coatings and, 344
Resist, *See* Photoresist
Resistance, 103–105, 130, 237
  piezoresistivity, 298–306
Resistivity, 103, 130
Resistors, 236–237
  switched capacitor techniques, 296–298
Resolution errors, 97
Resonance, 218–219, 316
Resonator, Lagrange's equations, 250–252
Retinal prosthesis, 10
Reversible reaction, 328
Reynolds number (Re), 125
RF applications, 345
Right-hand rule, 233
Ring diffusion, 24
RLC circuit models
  parallel circuit, 254–255
  simple circuit, 249–250
Robotic applications, 10–11
Robotic visual servoing, 347

Rotational hub design realization, 164–170
Rotational rate sensors, 7, 320–324
Round-off error, 138
Ruggedness, scale effects and, 119–121

S

Sacrificial surface micromachining (SSM), 6,
    65, 66, 83–94, *See also* Surface
    micromachining
  comparison of MEMS and conventional
    technologies, 69
Sagnac effect, 320–321
Sample size, 368
Sandia
  electrical-fluidic packaging, 355, 359–361
  MEMS Device Taxonomy, 400, 402
  microengine reliability testing, 407–412
  ultraplanar multilevel MEMS technology,
    See SUMMiT
Scaling issues, 3, 115–152
  breakdown voltage, 145
  Brownian noise, 144–145
  chemical and biological system
    concentration, 135–136
  computation, 137–139
  continuum assumptions and, 142–144
  electrical system, 129–134
  electron tunneling, 145
  fabrication technologies, 139–140
  fluid mechanics, 124–129
  force-generating phenomena, 146
  geometric scaling, 115–116
  mass and volume relations, 116–117
  material properties, 141–144
  mechanical properties, 117–121
  optical systems, 134–135
  summary (table), 147–148
  thermal system, 121–124
  units and terminology, 4, 6, 138–139
Scanning electron microscopy (SEM), 396
Scanning tunneling microscope (STM), 140,
    306
Scratch drive actuator, 283
SCREAM, 74–76
Sealing, 353
Second-order oscillator differential equations,
    249
Segregation coefficient, 26
Self-assembled monolayer (SAM) coatings,
    342, 349, 390
Self-shadowing, 343

Index

Sensors, 290–333
  accelerometers, 314-319, *See* also
      Accelerometers
  biological, 328
  capacitor applications, 290–298
      amplitude modulation, 292–294
      demodulation, 295–296
      parasitic capacitances, 290–291
      switched capacitor techniques, 296–298
  chemical, 328–330
  closed-loop, 318–319
  demodulation, 323
  electronic nose, 329–330
  electron tunneling, 306–308
  gas
      combustible gas, 331–332
      Taguchi, 330–331
  gyroscopes, 314, 319–324
  integrated MEMS and electronics, 94–95,
      See also Integrated MEMS
      (IMEMS) technology
  microphones, 328
  noise, 308–314
  open-loop, 317
  physical sensors, 314–328
  piezoresistive, 298–306
  pressure, 7, 314, 324-328
  rotational rate, 7, 320–324
Serial microassembly, 347
Series spring configuration, 427
Shadow mask, 343
Shear modulus, 197
Shear stress and strain, 197
  scale effects, 120–121
Sheet resistance, 103–105
SHiMMeR, 407
Shock, 119, 392–393
Shock response spectra (SRS), 392
Shot noise, 311
Shuffle motor, 283
Signal to noise ratio (SNR), 310
Silane (SiH_4), 35–37
Silica crucible, 27
Silicon, 2, 25, 26, *See also* Polycrystalline
      silicon
  crystal structure, 21
  Czochralski growth process, 25–27
  doping, *See* Dopants and doping
  etching, *See* Etching processes
  fusion bonding, 51
  MEMS fabrication, *See* Fabrication
      processes
  oxygen precipitates, 27
  piezoresistivity, 298–304

post-crystallization processing, 27–28
  stress corrosion cracking, 390
  wafer bonding, 50–51
      anodic (electrostatic), 51
      fusion, 51
Silicon carbide (SiC), 106–108, 425
Silicon dioxide, 2, 19, 71
  buried (BOX), 29
  etching, 70
  properties, 425
  stringer artifacts, 164
  SUMMiT process, 88–91, 341
  surface micromachining, 84
  wet etch process, 38–39
Silicon-germanium polycrystalline alloys, 108
Silicon nitride, 70, 71, 88, 425
Silicon-on-insulator (SOI), 25, 28–30
  PennSOIL bulk micromachining process,
      76–79
SIMOX process, 30
Simulation, *See* Modeling and simulation
Single-crystal reactive etching and metallization
      (SCREAM), 74–76
Size scale issues, *See* Scaling issues
Slide film damping model, 224, 226–227
Solder balls, 345, 351
Solenoid, 255–257, 447
Solution space characterization technique, 405
Space applications of MEMS, 10, 393
SPICE, 263
Spiral spring stiffness, 428
Spring configurations, stiffness coefficients,
      427, 428
Spring-mass-damper, 217
Sputtering, 34–35
Squeeze film damping model, 224–226
Squeeze number (σ), 225
Staggered anchor, 164
Standard components, 183–184, 402
Staple and pin hinge, 172
Stiction, 86, 181
  coatings and, 342
  drying and, 341
  operational failure mechanism, 391–392
  reliability methods and testing, 406
Stiffness coefficient, 203, 204, 208, 213–215,
      427–431
Stiffness scaling, 117–121
Stokes damping, 226, 227
Stoney equation, 98
Strain
  axial rod model, 202
  definition, 194
  shear, 197

464 Micro Electro Mechanical System Design

thermal, 200–201
three-dimensional characterization,
198–199
Strain gauges, 298
Strength of materials, 102–103, 195
Stress
definition, 194
at failure (ultimate strength), 195
piezoresistivity and, 300
reliability and scale effects, 120–121
residual stresses, 98–101, 344
shear, 197
thermal, 200–201
three-dimensional characterization,
198–199
Stress corrosion cracking, 390, 392
Stress-strain curve, 195–196
Stringer artifacts, 86, 164, 179–180
Stroboscopy, 397–398
Structural damping, 216
Structural mechanics, 193–215
axial loading, 211
axial rod, 201–203
beam bending, 205–208, 265–267
columns, 211–213
flat plate bending, 208–210
Hooke's law, 196–198
material models, 194–199
stiffness coefficients, 213–215, *See also*
Stiffness coefficient
stress-strain definitions and relationships,
194–199
thermal strains, 200–201
torsion rod, 203–204
SU-8, 109
Sublimation, 32, 341
Substitutional defect, 23
Substrates, 25–30, *See also* Crystal growth;
Silicon
SUGAR, 263–267, 437, 448–451
SUMMiT, 86, 88–94, 157, 158–175, 407
anchoring layers, 159–164
automated design rule checking, 182–183
beam and substrate connection, 170
design rules, 176
discrete hinges, 170
hub structure SEM, 396
mask and layer definitions, 160
MATLAB files, 447
release process, 341
rotational hubs, 164–170
standard components, 183–184
Supercritical drying, 341
Surface acoustic wave, 329

Surface fatigue, 389
Surface micromachining, 6, 65, 66, 83–94
anchoring layers, 159–164
artifacts, 86
beam and substrate connection, 170
chemical mechanical polishing, 53
design tools, *See* Design realization tools
dimples, 181
discrete hinges, 170–175
etch compatibility, 179
etch release holes, 157–158
integrated MEMS and electronics, 94
masks, *See* Masks
release, 340, 341
rotational hubs, 164–170
Sandia reliability program, 407–412
stiction, 87, *See also* Stiction
stringer artifacts, 86, 164, 179–180
SUMMiT, 87, 88–94
technology development, 4–5
Surface mount device (SMD), 351
Surface tension, 87, 126–129, 145
Switched capacitor techniques, 296–298
Synchronous demodulation, 296
Synchronous motor, 82
System degree of freedom, 244

T

Taguchi gas sensor, 330–331
Technology characterization, 95–106
process dimensional control parameters,
96–97
process material properties, *See* Material
properties
Texas Instruments
digital mirror device (DMD), 355, 357–359,
403–407
IMEMS technology, 6
Thermal actuation, 285–288
Thermal conductivity, 18
absolute pressure gauge, 325
Thermal cycling, 393
Thermal degradation, 392
Thermal expansion, 200, 285
actuation application, 285–288
coefficient mismatch and residual stress, 98
Thermal noise, 144–145, 313
Thermal strains and stresses, 200–201
Thermal system scaling, 121–124
Thermocompression bond, 353
Thermosonic bonding, 353
Thin-film material properties, 95–96

Index

Thin-film shell encapsulation, 349
Through-hole device (THD), 351
Timoshenko beam theory, 208
Tin dioxide, 330
Titanium (Ti)
    coatings, 342
    properties, 425
Torsion constant (J), 433
Torsion rod structural mechanics, 203–204
Transduction, 273, 274, 315, *See also* Actuators;
             Sensors
Transistor technology development, 1–2, 4
Trap sites, 311–312
Triode sputtering, 35
Truncation error, 138
Tungsten (W), 425
Tuning fork gyro (TFG), 324
Tunneling tip methods, 306–308
Turbulent flow, 125

U

Ultimate strength, 195
Ultrasonic bond, 353
Uncertainty, 267–270
Unibond process, 30
Unit cell, 19
Units and measures, 138–139
    electrical quantities, 228, 229
    micro-MKS conversions, 421
    physical constants, 423
    prefixes, 419

V

Vacancy defect, 23
Vacuum pickup, 355
Vacuum sensor, 324–326
Van der Pauw method, 104–105
Van der Waals forces, 18
Variance, 377
Vibration absorber, 252
Vibration environments, 392–393
Vibratory gyroscope, 321–324
Video imaging, 399
Virtual work function, 249

Viscosity, 222
Viscosity coefficient, 224
Viscous damping, 222–224, 246–247
Visualization tools, 184–186, *See also*
             Measurement and testing;
             Modeling and simulation
    geometric emulation, 185–186
    interferometry, 399
    microscopy, *See* Microscopic techniques
    part-to-art, 189
    process modeling tools, 184–185
    video imaging, 399
Voltage buffer, 293–294
Voltage source, 238
Volume reduction, 116–117

W

Wafer bonding, 50–51
Wear, 388–390
Wear-out, 370, 385
Weber number (We), 126
Web sites, 7
Weibull model, 380–383, 410
Wet chemical etching, 38–39, 70–72, 88
Wheatstone bridge circuits, 305
Wheel speed sensor, 7
White noise, 309
Wire bond, 340, 353, 356

X

X-ray lithography, 43, 79

Y

Yield stress, 195
Young's modulus, 101–102, 196
    orthotropic material, 198
    scale effects, 117

Z

Zincblende, 21